Lecture Notes in Artificial Intelligence 9343

Subseries of Lecture Notes in Computer Science

More information about this series at http://www.springer.com/series/1244

Eyke Hüllermeier · Mirjam Minor (Eds.)

Case-Based Reasoning
Research
and Development

23rd International Conference, ICCBR 2015
Frankfurt am Main, Germany, September 28–30, 2015
Proceedings

 Springer

Editors
Eyke Hüllermeier
Institut für Informatik
Universität Paderborn
Paderborn
Germany

Mirjam Minor
Institut für Informatik
Goethe-Universität Frankfurt
Frankfurt/Main
Germany

ISSN 0302-9743   ISSN 1611-3349 (electronic)
Lecture Notes in Artificial Intelligence
ISBN 978-3-319-24585-0   ISBN 978-3-319-24586-7 (eBook)
DOI 10.1007/978-3-319-24586-7

Library of Congress Control Number: 2015949320

LNCS Sublibrary: SL7 – Artificial Intelligence

Printed on acid-free paper

Springer International Publishing AG Switzerland is part of Springer Science+Business Media (www.springer.com)

Preface

This volume comprises the papers presented at ICCBR 2015, the 23rd International Conference on Case-Based Reasoning (http://www.iccbr.org/iccbr15), which took place at the Forschungskolleg Humanwissenschaften of Goethe University Frankfurt in Bad Homburg, Germany, during September 28–30, 2015. There were 37 submissions from 15 countries spanning North and South America, Europe, and Asia. Each one was reviewed by three Program Committee members. The committee accepted 16 papers for oral presentation and 10 papers for poster presentation at the conference.

The International Conference on Case-Based Reasoning is the pre-eminent international meeting on case-based reasoning (CBR). Previous ICCBR conferences have been held in Sesimbra, Portugal (1995), Providence, USA (1997), Seeon Monastery, Germany (1999), Vancouver, Canada (2001), Trondheim, Norway (2003), Chicago, USA (2005), Belfast, UK (2007), Seattle, USA (2009), Alessandria, Italy (2010), London, UK (2011), Lyon, France (2012), Saratoga Springs, USA (2013) and most recently in Cork, Ireland (2014).

The first day of ICCBR featured topical workshops on current aspects of CBR including case-based agents, e-CBR: building cyberinfrastructure for the CBR community, experience and creativity, CBR in the health sciences, and computer cooking. The Doctoral Consortium involved presentations by 12 graduate students in collaboration with their respective senior CBR research mentors. The first day also hosted the Computer Cooking Contest, the aim of which is to promote the use of AI technologies such as case-based reasoning, information extraction, information retrieval, and semantic technologies.

The second and third day consisted of scientific paper presentations on theoretical and applied CBR research as well as invited talks from two distinguished scholars: Qiang Yang, the head of the Computer Science and Engineering (CSE) department at Hong Kong University of Science and Technology (HKUST), where he is a New Bright Endowed Chair Professor of Engineering, and Michael M. Richter, Professor of Computer Science at the University of Kaiserslautern, Germany, until 2003 and presently Adjunct Professor at University of Calgary, Canada. Qiang Yang gave a keynote address on advances in transfer learning where target problem domains are complex and have fewer data to work with than a source domain. The keynote talk of Michael M. Richter was devoted to case-based reasoning and stochastic processes. The presentations and posters covered a wide range of CBR topics of interest both to researchers and practitioners including advanced retrieval, plans, processes, scalability, adaptability, maintenance, recommender systems, and robotics.

Many people helped make ICCBR 2015 a success. Eric Kübler and Jenny Quasten, Goethe University Frankfurt am Main, Germany, served as local organizers with Eyke Hüllermeier, University of Paderborn, and Mirjam Minor, Goethe University Frankfurt am Main, Germany, as program co-chairs. We would like to thank Joseph Kendall-Morwick, Capital University, USA, who acted as a workshop chair. Our thanks also go to Nirmalie Wiratunga, The Robert Gordon University, UK, and Sarah Jane Delany, Dublin Institute of Technology, Ireland, for organizing the Doctoral Consortium. We thank Emmanuel Nauer, LORIA, France, and David Wilson, University of North Carolina at Charlotte, USA, who were responsible for the Computer Cooking Competition. We thank Pascal Reuss, Stiftung Universität Hildesheim, Germany, who served as a publicity chair. We want to acknowledge the support of the team of the conference venue Forschungskolleg Humanwissenschaften. We are also very grateful to all our funding providers, which at the time of printing included the National Science Foundation, the International Joint Conference on Artificial Intelligence, the Goethe University Frankfurt am Main, Empolis Information Management GmbH, and LORIA.

We thank the Program Committee and the additional reviewers for their timely and thorough participation in the review process. We appreciate the time and effort put in by the local organizers. Finally, we acknowledge the support of EasyChair in the submission, review, and proceedings creation processes, and we thank Springer for its continued support in publishing the proceedings of ICCBR.

July 2015   Eyke Hüllermeier
                                                               Mirjam Minor

Organization

Program Chairs

Eyke Hüllermeier   University of Paderborn, Germany
Mirjam Minor   Goethe University Frankfurt, Germany

Local Organization

Eric Kübler   Goethe University Frankfurt, Germany
Jenny Quasten   Goethe University Frankfurt, Germany

Workshop Chair

Joseph Kendall-Morwick   Capital University, USA

Doctoral Consortium Chairs

Nirmalie Wiratunga   The Robert Gordon University, UK
Sarah Jane Delany   Dublin Institute of Technology, Ireland

Computer Cooking Chairs

Emmanuel Nauer   Université de Lorraine, France
David Wilson   University of North Carolina at Charlotte, USA

Publicity Chair

Pascal Reuss   Stiftung Universität Hildesheim, Germany

Program Committee

Agnar Aamodt   NTNU, Norway
David W. Aha   Naval Research Laboratory, USA
Klaus-Dieter Althoff   DFKI/University of Hildesheim, Germany
Kerstin Bach   Verdande Technology, USA
Derek Bridge   University College Cork, Ireland
Isabelle Bichindaritz   State University of New York at Oswego, USA
Ralph Bergmann   University of Trier, Germany
William Cheetham   GE Global Research, USA
Alexandra Coman   Northern Ohio University, USA
Amélie Cordier   LIRIS, France

Susan Craw	The Robert Gordon University, UK
Sarah Jane Delany	Dublin Institute of Technology, Ireland
Belen Diaz-Agudo	Universidad Complutense de Madrid, Spain
Michael Floyd	Knexus Research, USA
Mehmet H. Göker	Salesforce, USA
Ashok Goel	Georgia Institute of Technology, USA
Pedro González Calero	Complutense University of Madrid, Spain
Joseph Kendall-Morwick	Capital University, USA
Deepak Khemani	Indian Institute of Technology, India
Luc Lamontagne	Laval University, Canada
David Leake	Indiana University, USA
Jean Lieber	LORIA - Inria Lorraine, France
Ramon Lopez De Mantaras	IIIA - CSIC, Spain
Cindy Marling	Ohio University, USA
Stewart Massie	The Robert Gordon University, UK
Lorraine McGinty	University College Dublin, Ireland
David McSherry	University of Ulster, UK
Alain Mille	LIRIS, France
Stefania Montani	Università Piemonte Orientale, Italy
Emmanuel Nauer	Université de Lorraine, France
Santiago Ontañón	Drexel University, USA
Miltos Petridis	CEM, Brighton University, UK
Enric Plaza	IIIA - CSIC, Spain
Luigi Portinale	Universitá Piemonte Orientale, Italy
Ashwin Ram	PARC, USA
Juan Recio-Garcia	Complutense University of Madrid, Spain
Michael Richter	University of Calgary, Canada
Thomas Roth-Berghofer	University of West London, UK
Jonathan Rubin	PARC, USA
Antonio Sánchez-Ruiz	Complutense University of Madrid, Spain
Barry Smyth	University College Dublin, Ireland
Frode Soermo	Verdande Technology, USA
Ian Watson	University of Auckland, New Zealand
Rosina Weber	Drexel University, USA
David Wilson	University of North Carolina at Charlotte, USA
Nirmalie Wiratunga	The Robert Gordon University, UK

Additional Reviewers

Xavier Ferrer Arran
Pascal Reuss

Contents

Case Base Maintenance
in Preference-Based CBR

Amira Abdel-Aziz[1] and Eyke Hüllermeier[2]($\boxtimes$)

[1] Department of Mathematics and Computer Science,
University of Marburg, Marburg, Germany
amira@mathematik.uni-marburg.de
[2] Department of Computer Science,
University of Paderborn, Paderborn, Germany
eyke@upb.de

Abstract. In preference-based CBR (Pref-CBR), problem solving experience is represented in the form of contextualized preferences, namely, preferences between candidate solutions in the context of a target problem to be solved. Since a potentially large number of such preferences can be collected in the course of each problem solving episode, case base maintenance clearly becomes an issue in Pref-CBR. In this paper, we therefore extend our Pref-CBR framework by another component, namely, a method for dynamic case base maintenance. The main goal of this method is to increase efficiency of case-based problem solving, by reducing the size of the case base, while maintaining performance. To illustrate the effectiveness of our approach, we present a case study in which Pref-CBR is used for the repetitive traveling salesman problem.

1 Introduction

In the recent years, we have been working toward a methodological framework for case-based reasoning on the basis of formal concepts and methods for reasoning with *preferences* [1,2,10]. Deviating from the common representation of experiences in terms of problem/solution tuples (x, y), preference-based CBR (or Pref-CBR for short) proceeds from weaker "chunks of information" $y \succ_x z$, namely, preferences between competing solutions "contextualized" by problems: y is (likely to be) more preferred than z as a solution for x.

Problem solving in Pref-CBR is realized as a search process, in which candidate solutions are iteratively improved. In each step, the current best solution y is compared with another, slightly modified/adapted solution z, and the better one is retained. Since a single comparison is assumed to be costly, the number of adaptation steps is limited. Nevertheless, each step gives rise to a piece of information $y \succ_x z$. Therefore, a single *case* eventually consists of a problem x together with a set of (pairwise) preferences over solutions (instead of merely a single solution, like in conventional CBR).

It is clear that simply storing each encountered problem along with a set of associated preferences is not advisable, especially since a case base of that

E. Hüllermeier and M. Minor (Eds.): ICCBR 2015, LNAI 9343, pp. 1–14, 2015.
DOI: 10.1007/978-3-319-24586-7_1

type may quickly become too large and hamper efficient case retrieval; besides, many of the preferences collected in a problem solving episode will be redundant to some extent. In CBR, this problem has been addressed by methods for *case base maintenance* [11,18]. Such methods seek to maintain the problem solving competence of a case base thanks to *case base editing* strategies, including the removal of misleading (noisy) or redundant cases. Case base maintenance (CBM) proved essential to guarantee the efficiency and performance of CBR systems. According to the aforesaid, it might be even more critical for preference-based than for conventional CBR.

In this paper, we therefore address the problem of case base maintenance in Pref-CBR. To this end, we develop a CBM strategy that extends our Pref-CBR framework so far. Despite being inspired by existing CBM techniques for conventional CBR, our strategy is specifically tailored to our framework and exploits properties of the underlying preference-based representation of problem solving experience.

The remainder of the paper is organized as follows. Related work on case base maintenance is briefly recalled in Sect. 2. By way of background, and to assure a certain level of self-containedness, we also recall the essentials of Pref-CBR in Sect. 3. Our approach to case base maintenance for Pref-CBR is then detailed in Sect. 4. To illustrate the effectiveness of this approach, Sect. 5 presents a case study using the (repetitive) traveling salesman problem as a problem solving domain. The paper ends with some concluding remarks and an outlook on future work in Sect. 6.

2   Related Work

To increase efficiency while maintaining the competence of a case base, several CBR methods implement strategies that focus on choosing which cases to delete from the case base. The simplest strategy is random deletion, which is initiated once a given limit of the size of the case base is exceeded [21]; obviously, this method guarantees a bound on the size but no preservation of the competence of the case base. A more principled approach is utility deletion, where the utility of a case is measured by its performance benefits (e.g., given by Minton's utility metric); cases with negative utility are removed [20]. There are other methods such as footprint deletion and footprint utility deletion, which specify the cases to be deleted based on their competence contributions [19]. The cases are categorized into pivotal, spanning, support and auxiliary; pivotal cases have highest effect on competence, while auxiliary cases have lowest effect [13]. Modifying the idea of coverage (set of target problems a case can solve) and reachability (the set of cases that can provide a solution for a target problem) of a case as introduced in [19], cases are identified by their coverage and reachability values based on rough set theory for categorizing data in [17], and accordingly relevance of each case is extracted.

Other maintenance methods focus more on an increase in efficiency, in terms of memory storage size and computation time of solving problems [8]. This increase in efficiency could in return cause some degradation in performance. One well-known method is based on the condensed nearest neighbor (CNN) rule by [9], where a subset of the case base is selected, which should perform almost as well as the original case base in classifying new cases. CNN was then extended by selective nearest neighbor (SNN); any case in the original case base must be closer to a case in the formed subset belonging to the same class, than to any case in the original case base belonging to a different class [16]. Reduced edited nearest neighbor (RENN) method further extends CNN by removing noisy cases, which have a different class than the majority of their nearest neighbors; it is computationally more expensive than CNN [5] though. Also described in [5], the blame based noise reduction (BBNR) method deletes cases that cause other cases to be misclassified. A case base can also be reduced as explained by [15], where a subset of the case base is formed in which selection of cases is based on some "justifications". These justifications are being output from using a (lazy) machine learning method; this selection procedure resembles the competence selection of cases in [21], but in the former the selection of cases is based on the justifications rather than the competence.

Additionally, adaptation-guided case base maintenance methods base the selection of cases to be retained in the case base on both their value in solving problems and on their value in generating new adaptation rules; these adaptation rules contribute to the knowledge for later problem solving [11]. Complexity-informed maintenance is another method presented in [4]; it provides redundancy reduction and offers a compromise between a smaller case base and greater accuracy. Case complexity enables varying levels of aggressiveness in redundancy and error reduction maintenance algorithms, thus compromising between amount of reduction and correspondingly level of performance. The higher the aggressiveness, the more reduction in case base size and correspondingly the lower the performance level.

The previously listed methods are used to either increase the efficiency of the case-based reasoning system while maintaining its competence, or having a trade-off between an increase in the level of efficiency and a decrease in the level of performance. The case base is maintained when a certain size limit is reached, or by setting periodic time slots for the maintenance to be performed. As pointed out by [14], to tackle performance problems of a CBR system, the goal would be to update the existing case base while maintaining problem solving competence. This is also the goal of our method, which is specifically designed for the Pref-CBR problem solving framework.

3   Preference-Based CBR in a Nutshell

In this section, we briefly recall the basics of Pref-CBR, which is essential for understanding our CBM strategy to be introduced in the next section. For further details of Pref-CBR, we refer to [1, 2, 10].

3.1   Basic Setting and Notation

Let a problem space $\mathbb{X}$ and a solution space $\mathbb{Y}$ be given. We assume $\mathbb{X}$ to be equipped with a similarity measure $S_X : \mathbb{X} \times \mathbb{X} \to \mathbb{R}_+$ or, equivalently, with a (reciprocal) distance measure $\Delta_X : \mathbb{X} \times \mathbb{X} \to \mathbb{R}_+$. Thus, for any pair of problems $x, x' \in \mathbb{X}$, their similarity is denoted by $S_X(x, x')$ and their distance by $\Delta_X(x, x')$. Likewise, we assume the solution space $\mathbb{Y}$ to be equipped with a similarity measure S_Y or, equivalently, with a (reciprocal) distance measure Δ_Y.

In preference-based CBR, problems $x \in \mathbb{X}$ are not associated with single solutions but rather with preferences over solutions, that is, with elements from a class of preference structures over the solution space $\mathbb{Y}$. Here, we assume this class to consist of all linear order relations $\succ$ on $\mathbb{Y}$, and we denote the relation associated with a problem x by $\succ_x$. More precisely, we assume that $\succ_x$ has a specific form, which is defined by an "ideal" solution[1] $y^* \in \mathbb{Y}$ and the distance measure Δ_Y: The closer a solution y to $y^* = y^*(x)$, the more it is preferred; thus, $y \succ_x z$ iff $\Delta_Y(y, y^*) < \Delta_Y(z, y^*)$. In conjunction with the regularity assumption that is commonly made in CBR, namely, that similar problems tend to have similar (ideal) solutions, this property legitimates a preference-based version of this assumption: *Similar problems are likely to induce similar preferences over solutions.*

3.2   Case-Based Inference

The key idea of preference-based CBR is to exploit experience in the form of previously observed preferences, deemed relevant for the problem at hand, in order to support the current problem solving episode; like in standard CBR, the *relevance* of a preference will typically be decided on the basis of problem similarity, i.e., those preferences will be deemed relevant that pertain to similar problems. An important question that needs to be answered in this connection is the following: Given a set of observed preferences on solutions, considered representative for a problem x_0, what is the underlying preference structure $\succ_{x_0}$ or, equivalently, what is the most likely ideal solution y^* for x_0?

We approach this problem from a statistical perspective, considering the true preference model $\succ_{x_0}$ associated with the query x_0 as a random variable with distribution $\mathbf{P}(\cdot \,|\, x_0)$, where $\mathbf{P}(\cdot \,|\, x_0)$ is a distribution $\mathbf{P}_\theta(\cdot)$ parametrized by $\theta = \theta(x_0) \in \Theta$. The problem is then to estimate this distribution or, equivalently, the parameter θ on the basis of the information available. This information consists of a set $\mathcal{D}$ of preferences of the form $y \succ z$ between solutions.

The basic assumption underlying nearest neighbor estimation is that the conditional probability distribution of the output given the input is (approximately) locally constant, that is, $\mathbf{P}(\cdot \,|\, x_0) \approx \mathbf{P}(\cdot \,|\, x)$ for x close to x_0. Thus, if the above preferences are coming from problems x similar to x_0 (namely, from the nearest neighbors of x_0 in the case base), then this assumption justifies considering

---

[1] The solution y^* could be a purely imaginary solution, which may not exist in practice.

$\mathcal{D}$ as a representative sample of $\mathbf{P}_\theta(\cdot)$ and, hence, estimating θ via maximum likelihood (ML) inference by

$$\theta^{ML} = \arg\max_{\theta \in \Theta} \mathbf{P}_\theta(\mathcal{D}). \tag{1}$$

An important prerequisite for putting this approach into practice is a suitable data generating process, i.e., a process generating preferences in a stochastic way. Our data generating process is based on the idea of a discrete choice model as used in choice and decision theory. More specifically, we assume the *logit* model of discrete choice:

$$\mathbf{P}(\boldsymbol{y} \succ \boldsymbol{z} \,|\, \boldsymbol{y}^*) = \frac{1}{1 + \exp\left(-\beta\big(\Delta_Y(\boldsymbol{z}, \boldsymbol{y}^*) - \Delta_Y(\boldsymbol{y}, \boldsymbol{y}^*)\big)\right)} \tag{2}$$

Thus, the probability of observing the (revealed) preference $\boldsymbol{y} \succ \boldsymbol{z}$ depends on the degree of suboptimality of $\boldsymbol{y}$ and $\boldsymbol{z}$, namely, their respective distances to the ideal solution, $\Delta_Y(\boldsymbol{y}, \boldsymbol{y}^*)$ and $\Delta_Y(\boldsymbol{z}, \boldsymbol{y}^*)$: The larger the difference $\Delta_Y(\boldsymbol{z}, \boldsymbol{y}^*) - \Delta_Y(\boldsymbol{y}, \boldsymbol{y}^*)$, i.e., the less optimal $\boldsymbol{z}$ in comparison to $\boldsymbol{y}$, the larger the probability to observe $\boldsymbol{y} \succ \boldsymbol{z}$. The coefficient β can be seen as a measure of precision of the preference feedback. For large β, $\mathbf{P}(\boldsymbol{y} \succ \boldsymbol{z})$ converges to 0 if $\Delta_Y(\boldsymbol{z}, \boldsymbol{y}^*) < \Delta_Y(\boldsymbol{y}, \boldsymbol{y}^*)$ and to 1 if $\Delta_Y(\boldsymbol{z}, \boldsymbol{y}^*) > \Delta_Y(\boldsymbol{y}, \boldsymbol{y}^*)$; this corresponds to a deterministic (error-free) information source. The other extreme case, namely $\beta = 0$, models a completely unreliable information source reporting preferences at random.

The probabilistic model outlined above is specified by two parameters: the ideal solution $\boldsymbol{y}^*$ and the (true) precision parameter $\beta^* \in \mathbb{R}_+$. Depending on the context in which these parameters are sought, the ideal solution might be unrestricted (i.e., any element of $\mathbb{Y}$ is an eligible candidate), or it might be restricted to a certain subset $\mathbb{Y}_0 \subseteq \mathbb{Y}$ of candidates.

Now, to estimate the parameter vector $\theta^* = (\boldsymbol{y}^*, \beta^*) \in \mathbb{Y}_0 \times \mathbb{R}^*$ from a given set $\mathcal{D} = \{\boldsymbol{y}^{(i)} \succ \boldsymbol{z}^{(i)}\}_{i=1}^N$ of observed preferences, we refer to the maximum likelihood estimation principle. Assuming independence of the preferences, the likelihood of $\theta = (\boldsymbol{y}, \beta)$ is given by

$$\ell(\theta) = \ell(\theta \,|\, \mathcal{D}) = \prod_{i=1}^N \mathbf{P}\left(\boldsymbol{y}^{(i)} \succ \boldsymbol{z}^{(i)} \,|\, \theta\right). \tag{3}$$

The ML estimation $\theta^{ML} = (\boldsymbol{y}^{ML}, \beta^{ML})$ of θ^* is given by the maximizer of (3):

$$\theta^{ML} = \left(\boldsymbol{y}^{ML}, \beta^{ML}\right) = \arg\max_{\boldsymbol{y} \in \mathbb{Y}_0,\, \beta \in \mathbb{R}_+} \ell(\boldsymbol{y}, \beta) \tag{4}$$

The problem of finding this estimation in an efficient way is addressed in [10].

3.3 CBR as Preference-Guided Search

Case-based inference as outlined above realizes a "one-shot prediction" of a promising solution for a query problem, given preferences in the context of similar problems encountered in the past. In a case-based problem solving process, this prediction may thus serve as an initial solution, which is then adapted step by step. An adaptation process of that kind can be formalized as a search process, namely, a traversal of a suitable space of candidate solutions [3].

In the spirit of preference-based CBR, we implement case-based problem solving as a search process that is guided by preference information collected in previous problem solving episodes. To this end, we assume the solution space $\mathbb{Y}$ to be equipped with a topology that is defined through a *neighborhood structure*: For each $y \in \mathbb{Y}$, we denote by $\mathcal{N}(y) \subseteq \mathbb{Y}$ the neighborhood of this candidate solution. The neighborhood is thought of as those solutions that can be produced through a single modification of y, e.g., by applying one of the available adaptation operators to y.

Our case base **CB** stores problems x_i together with a set of preferences $\mathcal{P}(x_i)$ that have been observed for these problems. Thus, each $\mathcal{P}(x_i)$ is a set of preferences of the form $y \succ_{x_i} z$, which are collected while searching for a good solution to x_i.

We conceive preference-based CBR as an iterative process in which problems are solved one by one. In each problem solving episode, a good solution for a new query problem is sought, and new experiences in the form of preferences are collected. In what follows, we give a high-level description of a single problem solving episode:

(i) Given a new query problem x_0, the K nearest neighbors $x_1, \ldots, x_K$ of this problem (i.e., those with smallest distance in the sense of Δ_X) are retrieved from the case base **CB**, together with their preference information $\mathcal{P}(x_1), \ldots, \mathcal{P}(x_K)$.

(ii) This information is collected in a single set of preferences $\mathcal{P}$, which is considered representative for the problem x_0 and used to guide the search process.

(iii) The search for a solution starts with an initial candidate $y^\bullet \in \mathbb{Y}$, for example the "one-shot prediction" (4) based on $\mathcal{P}$, and iterates L times. Restricting the number of iterations by an upper bound L accounts for our assumption that an evaluation of a candidate solution is costly.

(iv) In each iteration, a new candidate y^{query} in the neighbourhood of $y^\bullet$ is determined, based on (4) with $\mathbb{Y}_0 = \mathcal{N}(y^\bullet)$, and given as a query to an (external) information source, which we refer to as the "oracle". Thus, the oracle is asked to compare y^{query} with the current best solution $y^\bullet$. The preference reported by the oracle is memorized by adding it to the preference set $\mathcal{P}_0 = \mathcal{P}(x_0)$ associated with x_0, as well as to the set $\mathcal{P}$ of preferences used for guiding the search process. Moreover, the better solution is retained as the current best candidate.

(v) When the search stops, the current best solution $y^\bullet$ is returned (as an approximation of y^*), and the case $(x_0, \mathcal{P}_0)$ is added to the case base.

The preference-based guidance of the search process is realized in (iii) and (iv). Here, our case-based inference method is used to find the most promising candidate among the neighborhood of the current solution $y^\bullet$, based on the preferences collected in the problem solving episode so far. By providing information about which of these candidates will most likely constitute a good solution for x_0, it (hopefully) points the search into the most promising direction.

4 Case Base Maintenance for Pref-CBR

Most methods for case base maintenance make use of two important criteria for case addition or removal, namely, noise and redundancy. A "noisy" case is a case that differs significantly from its (nearby) neighbors and, therefore, violates the regularity assumption underlying CBR. Retrieving such a case and using it to solve a new problem should obviously be avoided, whence it should better not be stored in the case base. A redundant case, on the other side, is very similar to its neighbors and, therefore, does hardly provide additional information, at least if enough other cases have already been stored. Such cases can often be removed to reduce the size of the case base without compromising performance.

In Pref-CBR, a case does not only contain a single solution, like in conventional CBR, but rather a set of preferences. Thus, instead of either retaining or removing a complete case, there is in principle the possibility to retain or remove a *part* of a case, simply by retaining or removing a part of the pairwise preferences. In fact, as will be seen later on, both noise and redundancy can occur on the level of a single case as well as on the level of the case base.

First of all, however, one should clarify what noise and redundancy may actually refer to in the context of Pref-CBR. In fact, it is important to note that a piece of information is not noisy or redundant per se. First, it can only be noisy or redundant when being considered jointly with other information. Moreover, what also needs to be taken into consideration is the way in which the information will be (re-)used: What is the influence of the information on future problem solving episodes?

4.1 Noise and Redundancy in Pref-CBR

To answer this question, recall the key idea and basic inference principle of Pref-CBR: An observed preference $y \succ z$ provides a kind of "directional hint" in the solution space $\mathbb{Y}$:[2] It suggests moving toward those solutions y^* for which the probability $\mathbf{P}(y \succ z \mid y^*)$ in (2), is large, hence making y^* likely as a solution for the problem at hand, and away from those solutions for which the probability of observing this preference is small. Likewise, a whole set of preferences $\mathcal{P}$ suggest moving toward those solutions for which the combined likelihood (3) is large, and away from those solutions for which this likelihood is small. Roughly speaking, the likelihood function combines the individual hints into a single one.

Now, we propose the following distinction between noise and redundancy on the level of a single case and the level of the case base.

---

[2] The notion of "direction" should not be taken literally. In fact, the mathematical structure of $\mathbb{Y}$ will normally not allow for defining a direction in a geometrical sense.

- **Intra-case Redundancy:** Pairwise comparisons collected during a problem solving episode can obviously be redundant to some extent, in particular because the same solutions will be shared among many of these comparisons. Moreover, as we just explained, each comparison $y \succ z$ provides a directional hint in the solution space. Therefore, two preferences can also be redundant in the sense of suggesting similar directions.
- **Intra-case Noise:** According to (2), preference feedback is correct only with a certain probability. Thus, even if unlikely, one may thoroughly observe $y \succ z$ although $\mathbf{P}(y \succ z \mid y^*) < \mathbf{P}(z \succ y \mid y^*)$. According to what we just said, a preference of that kind will guide the search in the wrong direction and, therefore, could be considered as "noise".
- **Inter-case Redundancy:** Instead of looking at a single preference, we now look at the whole set of preferences $\mathcal{P} = \mathcal{P}(x)$ that have been collected for a problem x, because this is the information to be reused later on. Again, as explained above, these preferences provide a "directional hint" in the solution space. Therefore, just like in the case of individual preferences, two sets of preferences $\mathcal{P}$ and $\mathcal{P}'$ can be redundant in the sense of suggesting similar directions in the solution space. Note that this type of redundancy is likely to occur for two problems having similar ideal solutions y^*. Yet, even in that case the preferences are not necessarily redundant, because they might have been collected in different parts of the solution space.
- **Inter-case Noise:** Just as a case may appear redundant in the context of other cases, it can be noisy in the sense that its preferences are inconsistent with those of the others. Here, inconsistency means that the preferences suggest very different directions in the solution space.

4.2 Maintenance Strategies

Our general maintenance strategy is incremental and essentially consists of deciding, for each new case $(x_0, \mathcal{P}_0)$ produced, whether or not that case should be stored—and perhaps which parts thereof. To this end, each of the aforementioned types of noise and redundancy have to be handled in a proper way.

As already explained, the similarity or discrepancy between preferences or sets of preferences depends on the similarity or dissimilarity of the "directional hints" they provide. But how to quantify the latter? The direction suggested to the search process is a local property that depends on the current search state in $\mathbb{Y}$—as such, it is difficult to quantify in a single value. Reasoning on a more global level, the arguably most appropriate way to compare two sets of preferences $\mathcal{P}_1$ and $\mathcal{P}_2$ is to compare the respective globally optimal solutions y_1^{ML} and y_2^{ML}, i.e., the likelihood estimates (3) with $\mathbb{Y}_0 = \mathbb{Y}$. Such a comparison could easily be done using Δ_Y. However, finding the global likelihood maximizer might be very costly—this is why our search procedure is local. Besides, when comparing single preferences as a special case, the likelihood is often unbounded. In the following, we therefore propose approximate strategies that circumvent these difficulties and that are computationally more efficient.

Intra-case Redundancy: Consider a case $(x, \mathcal{P})$, and let $y^\bullet$ denote the solution the problem solving process ended up with—again, recall that $y^\bullet$ will in general differ from $y^*(x)$, either because the latter was not reached or because it may not even exist. Now, consider a single preference $y \succ z$ in $\mathcal{P}$. How redundant is that preference? To answer this question, we should compare the likelihood function (3) with and without the preference, i.e., the functions $\ell(\cdot \mid \mathcal{P})$ and $\ell(\cdot \mid \mathcal{P}')$ with $\mathcal{P}' = \mathcal{P} \setminus \{y \succ z\}$. Of course, comparing the functions globally is very difficult. Moreover, as explained above, we may not be able to compare their respective global maximizers either. What we could do, for example, is checking whether or not the locally restricted optimum in the neighborhood of $y^\bullet$ would change, i.e., whether the local optimum for $\mathcal{P}$ is the same as the optimum for $\mathcal{P}'$. If not, then $y \succ z$ has an important influence and should certainly not be removed.

Intra-case Noise: As explained earlier, we consider a preference $y \succ z$ as noise if $\mathbf{P}(y \succ z \mid y^*) < 1/2$. This property cannot be checked, however, because y^* is not known. Yet, using $y^\bullet$ as a proxy, we could at least check $\mathbf{P}(y \succ z \mid y^\bullet) < 1/2$.

Inter-case Redundancy: Consider two cases $(x_0, \mathcal{P}_0)$ and $(x_1, \mathcal{P}_1)$ with solutions $y_0^\bullet$ and $y_1^\bullet$, respectively. How redundant are these cases or, more specifically, how redundant is the new case $(x_0, \mathcal{P}_0)$ with respect to the previous case $(x_1, \mathcal{P}_1)$? Again, for the reasons explained above, a comparison of the likelihood functions $\ell(\cdot \mid \mathcal{P}_0)$ and $\ell(\cdot \mid \mathcal{P}_1)$ or their maximizers may not be feasible. Instead, we again refer to the actually found solutions $y_0^\bullet$ and $y_1^\bullet$ as surrogates of these maximizers. More specifically, we compare the probability (3) of the preferences $\mathcal{P}_0$ under the associated (ML) parameters $(y_0^\bullet, \beta)$ with the probability under $(y_1^\bullet, \beta)$, i.e., when replacing $y_1^\bullet$ by $y_0^\bullet$. If

$$\frac{\ell(y_1^\bullet, \beta \mid \mathcal{P}_0)}{\ell(y_0^\bullet, \beta \mid \mathcal{P}_0)} \geq t \tag{5}$$

for a threshold $t > 0$, this indicates that the preferences $\mathcal{P}_0$ are not only hinting at $y_0^\bullet$ but also at $y_1^\bullet$ (just like $\mathcal{P}_1$), which in turn can be interpreted as a sign of redundancy. Moreover, since not only the preferences but also the solutions themselves are reused, we additionally require

$$\Delta_Y(y_0^\bullet, y_1^\bullet) \leq v \tag{6}$$

for a second threshold $v \geq 0$. If both conditions are met, $(x_0, \mathcal{P}_0)$ is considered redundant with respect to $(x_1, \mathcal{P}_1)$.

Inter-case Noise: We just gave two conditions which, in conjunction, suggest the similarity (and hence the potential redundancy) of two cases. It is natural, then, to consider the cases as dissimilar if the opposite of at least one of the conditions holds, i.e., if either the ratio in (5) is smaller than some (small) threshold or the distance in (6) is larger than some threshold. If a new case $(\boldsymbol{x}_0, \mathcal{P}_0)$ is dissimilar in this sense to all of its neighbors, we may consider it as being exceptional or at least non-representative.

4.3   Method

As a first step, we realized a "light" version of the above approach to maintenance in Pref-CBR by implementing the strategy for inter-case redundancy. More specifically, our strategy consists of the following steps:

- Given a new problem $\boldsymbol{x}_0 \in \mathbb{X}$ to be solved, Pref-CBR is used to find a solution $\boldsymbol{y}_0^{\bullet} \in \mathbb{Y}$. In addition to the solution itself, Pref-CBR returns a set of preferences $\mathcal{P}_0$ (see [1] for a detailed description of the problem solving process on the level of pseudo-code).
- To decide whether the new case should be stored, the K nearest neighbors of $\boldsymbol{x}_0$ are retrieved from the current case base: $(\boldsymbol{x}_1, \mathcal{P}_1), \ldots, (\boldsymbol{x}_K, \mathcal{P}_K)$.
- The two criteria (5) and (6) are checked for $(\boldsymbol{x}_0, \mathcal{P}_0)$ and each of the cases $(\boldsymbol{x}_i, \mathcal{P}_i)$, $i = 1, \ldots, K$.
- If the criteria are fulfilled for at least one of the K cases, $(\boldsymbol{x}_0, \mathcal{P}_0)$ is considered redundant and not stored; otherwise, it is added to the case base **CB**.

Note that this strategy has three parameters, namely, the number of neighbours K and the thresholds t and v in (5–6).

5   Case Study

We conducted an experimental study with the traveling salesman problem (TSP), i.e., with TSP instances as problems and tours as candidate solutions. Needless to say, our ambition is not to develop new state-of-the-art solvers for this NP-hard optimization problem—obviously, our completely generic problem solving framework cannot compete with specialized TSP solvers. Nevertheless, combinatorial optimization problems such as TSP provide an interesting test bed for Pref-CBR:

- In practice, such problems often need to be solved repeatedly (imagine, for example, a conveyance planning a tour every day), suggesting a reuse of previous solutions [12]; interestingly, the TSP problem has already been tackled by means of CBR by other authors [6,7].
- The solution space $\mathbb{Y}$ is non-trivial but typically equipped with a natural structure, on which reasonable distance measures $\Delta_\mathbb{Y}$ can be defined.
- One of the key assumptions of Pref-CBR, namely, that the optimality of a solution cannot be guaranteed, is often fulfilled—this is due to the hardness of such problems, calling for heuristic approximations.
- Nevertheless, a comparison between two candidate solutions is often possible. In TSP, for example, a preference between two tours can easily be created by computing and comparing their lengths[3].

---

[3] Actually, we could even create more than a *qualitative* preference, because the numerical values of the solutions (lengths of the tours) are known as well. This is indeed additional information we are not exploiting in this application.

Another assumption of Pref-CBR, namely that a comparison is costly (and hence the number of adaptations and queries to the oracle limited), is admittedly not fulfilled in the case of TSP. Yet, one can easily imagine practically relevant generalizations of the problem for which this assumption applies. For example, suppose we replace a precise evaluation criterion such as *length* of a tour by a more "soft" criterion such as *comfort* or *convenience*. Then, to compare two candidates, it may indeed be necessary to practically try both of them (e.g., to walk a hiking tour), which might be time-consuming and involve input of a human expert (playing the role of the "oracle" then). In such cases, comparing two candidates qualitatively may also be simpler than rating them individually.

5.1   Setting

The components of our Pref-CBR setting are specified as follows:

- The problem space $\mathbb{X}$ is the set of all subsets $x \subset \mathcal{X}$ of size $|x| = 30$, where $\mathcal{X} \subset \mathbb{R}^2$ is a randomly created reference set of 75 points on the plane; each point can be thought of as the location of a city.
- The distance $\Delta_X(x, x')$ between two problems is defined in terms of the average squared distance between points in an optimal geometric superposition of the two point sets. This measure is computed using the "Procrustes Rotation" method, which is implemented in the "vegan" library of R[4], subsequent to an optimal assignment of the points that is obtained by solving the linear assignment problem with Euclidean distance as a cost measure.
- Solutions are represented as permutations specifying the order of cities/points in a tour. Thus, $\mathbb{Y}$ is the set of all permutations of $\{1, \ldots, 30\}$. This space is equipped with a local neighbourhood structure by connecting each solution y with 200 "perturbations" of this solution, each of which is obtained by randomly switching the position of a small number (2, 4 or 6) of points.
- To define the distance $\Delta_Y(y, y')$ between two solutions, each solution is first mapped to a feature vector with the following entries: path length, mean distance between each city to its nearest neighbors, standard deviation of distances of each city to its nearest neighbor. Then, the corresponding feature vectors are compared in term of their Euclidean distance.
- The parameters of Pref-CBR were set as follows: number of nearest neighbors $K = 17$, number of adaptation steps $L = 40$.

5.2   Experiment

In our experimental study, we compared Pref-CBR with and without case base maintenance. As additional baselines, we included a two random deletion policies, one that removes each newly observed case with a fixed probability (RCD) of $1/2$, and one that removes individual preferences with the same fixed probability (RPD). We generated a sequence of 200 instances of the TSP problem

---

[4] http://cran.r-project.org/web/packages/vegan/index.html.

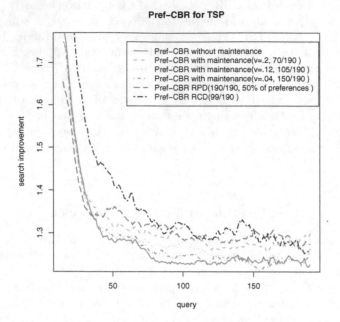

Fig. 1. Performance of Pref-CBR with and without maintenance on TSP data.

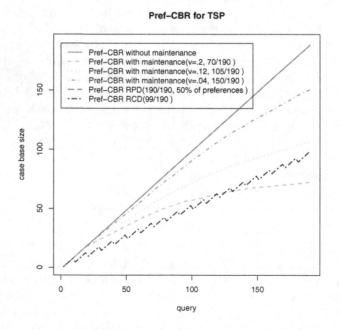

Fig. 2. Case base size for Pref-CBR search with and without maintenance of TSP data.

(using the "tspmeta" library in R), giving rise to the same number of problem solving episodes. Each time a solution has been produced, we measure performance by computing the ratio between the corresponding tour length and the optimal tour length found by the "cheapest_insertion" TSP solver. Since the sequence of performance values thus produced is rather noisy, we average over a larger number of repetitions of this experiment to produce smoother curves.

These curves are shown in Fig. 1, both for Pref-CBR without maintenance and Pref-CBR with maintenance and different values of the parameter v (while the threshold t was fixed to 1). Moreover, Fig. 2 shows the evolution of the size of the case base. As can be seen, the desired effect is indeed achieved: The size of the case base is significantly reduced while performance is maintained (in contrast to the random deletion policies). Moreover, the larger v, the stronger the tendency to delete cases. Thus, this parameter can be used to control the size of the case base.

6   Summary and Outlook

This paper extends our framework of preference-based CBR by a method for dynamic case base maintenance. The main goal of this method is to increase efficiency of case-based problem solving while maintaining performance. The effectiveness of our approach was illustrated in a case study with the traveling salesman problem.

So far, the implementation of our maintenance method includes only a part of the strategies discussed in Sect. 4, namely, a strategy for handling what we called inter-case redundancy. For future work, we therefore plan to extend this method by incorporating additional strategies for handling intra-case redundancy as well as intra- and inter-case noise.

References

1. Abdel-Aziz, A., Cheng, W., Strickert, M., Hüllermeier, E.: Preference-based CBR: a search-based problem solving framework. In: Delany, S.J., Ontañón, S. (eds.) ICCBR 2013. LNCS, vol. 7969, pp. 1–14. Springer, Heidelberg (2013)
2. Abdel-Aziz, A., Strickert, M., Hüllermeier, E.: Learning solution similarity in preference-based CBR. In: Lamontagne, L., Plaza, E. (eds.) ICCBR 2014. LNCS, vol. 8765, pp. 17–31. Springer, Heidelberg (2014)
3. Bergmann, R., Wilke, W.: Towards a new formal model of transformational adaptation in case-based reasoning. In: Prade, H. (ed.) ECAI-98, 13th European Conference on Artificial Intelligence, pp. 53–57 (1998)
4. Craw, S., Massie, S., Wiratunga, N.: Informed case base maintenance: a complexity profiling approach. In: Proceedings AAAI-2007, Twenty-Second National Conference on Artificial Intelligence, 22–26 July 2007, Vancouver, British Columbia, Canada, pp. 1618–1621 (2007)
5. Cummins, L., Bridge, D.: On dataset complexity for case base maintenance. In: Ram, A., Wiratunga, N. (eds.) ICCBR 2011. LNCS, vol. 6880, pp. 47–61. Springer, Heidelberg (2011)

6. Cunningham, P., Smyth, B., Hurley, N.: On the use of CBR in optimisation problems such as the TSP. Technical report TCD-CS-95-19, Trinity College Dublin, Department of Computer Science (1995)
7. Erfani, H.: Integrating case-based reasoning, knowledge-based approach and TSP algorithm for minimum tour finding. J. Appl. Math. Islam. Azad Univ. Lahijan **3**(9), 49–59 (2006)
8. Gates, G.W.: The reduced nearest neighbor rule. IEEE Trans. Inf. Theor. **18**(3), 431–433 (1972)
9. Hart, P.: The condensed nearest neighbor rule. IEEE Trans. Inf. Theor. **14**(3), 515–516 (1968)
10. Hüllermeier, E., Schlegel, P.: Preference-based CBR: first steps toward a methodological framework. In: Ram, A., Wiratunga, N. (eds.) ICCBR 2011. LNCS, vol. 6880, pp. 77–91. Springer, Heidelberg (2011)
11. Jalali, V., Leake, D.: Adaptation-guided case base maintenance. In: Proceedings AAAI, National Conference on Artificial Intelligence (2014)
12. Kraay, D.R., Harker, P.T.: Case-based reasoning for repetitive combinatorial optimization problems, part I: framework. J. Heuristics **2**, 55–85 (1996)
13. Lawanna, A., Daengdej, J.: Hybrid technique and competence-preserving case deletion methods for case maintenance in case-based reasoning. Int. J. Eng. Sci. Technol. **2**(4), 492–497 (2010)
14. Lupiani, E., Juarez, J.M., Palma, J.: Evaluating case-base maintenance algorithms. Knowl. Based Syst. **67**, 180–194 (2014)
15. Ontañón, S., Plaza, E.: Justification-based selection of training examples for case base reduction. In: Boulicaut, J.-F., Esposito, F., Giannotti, F., Pedreschi, D. (eds.) ECML 2004. LNCS (LNAI), vol. 3201, pp. 310–321. Springer, Heidelberg (2004)
16. Salamó, M., Golobardes, E.: Rough sets reduction techniques for case-based reasoning. In: Aha, D.W., Watson, I. (eds.) ICCBR 2001. LNCS (LNAI), vol. 2080, pp. 467–482. Springer, Heidelberg (2001)
17. Salamo, M., Golobardes, E.: Hybrid deletion policies for case base maintenance. In: Proceedings of FLAIRS-2003, pp. 150–154 (2003). Enginyeria Arquitectura, and La Salle
18. Smiti, A., Elouedi, Z.: Overview of maintenance for case based reasoning systems. Int. J. Comput. Appl. **32**(2), 49–56 (2011)
19. Smyth, B., Keane, T.: Remembering to forget. In: Mellish, C.S. (ed.) Proceedings International Joint Conference on Artificial Intelligence, pp. 377–382, Morgan Kaufmann (1995)
20. Smyth, B.: Case-base maintenance. In: del Pobil, A.P., Mira, J., Ali, M. (eds.) Tasks and Methods in Applied Artificial Intelligence. LNCS, vol. 1416, pp. 507–516. Springer, Heidelberg (1998)
21. Zhu, J., Yang, Q.: Remembering to add: competence-preserving case-addition policies for case-base maintenance. In: Proceedings IJCAI-99, 16th International Joint Conference on Artificial Intelligence, pp. 234–239. Morgan Kaufmann Publishers Inc, San Francisco, CA, USA (1999)

Learning to Estimate:
A Case-Based Approach to Task Execution Prediction

Bryan Auslander[1](✉), Michael W. Floyd[1], Thomas Apker[2], Benjamin Johnson[3],
Mark Roberts[3], and David W. Aha[2]

[1] Knexus Research Corporation, Springfield, VA, USA
{bryan.auslander,michael.floyd}@knexusresearch.com
[2] Navy Center for Applied Research in Artificial Intelligence,
Naval Research Laboratory (Code 5514), Washington, DC, USA
{thomas.apker,david.aha}@nrl.navy.mil
[3] NRC Postdoctoral Fellow,
Naval Research Laboratory (Code 5514), Washington, DC, USA
{benjamin.johnson.ctr,mark.roberts.ctr}@nrl.navy.mil

Abstract. A system that controls a team of autonomous vehicles should be able to accurately predict the expected outcomes of various subtasks. For example, this may involve estimating how well a vehicle will perform when searching a designated area. We present CBE, a case-based estimation algorithm, and apply it to the task of predicting the performance of autonomous vehicles using simulators of varying fidelity and past performance. Since there are costs to evaluating the performance in simulators (i.e., higher fidelity simulators are more computationally expensive) and in deployment (i.e., potential human injury and deployment expenses), CBE uses a variant of local linear regression to estimate values that cannot be directly evaluated, and incrementally revises its case base. We empirically evaluate CBE on Humanitarian Assistance/Disaster Relief (HA/DR) scenarios and show it to be more accurate than several baselines and more efficient than using a low fidelity simulator.

1 Introduction

Humanitarian Assistance/Disaster Relief (HA/DR) missions can occur without warning and require a rapid response to minimize damage and preserve human life. Additionally, they often occur in remote areas (e.g., an avalanche site) or dangerous locations (e.g., flooded towns, cities damaged by earthquakes, active wildfires), so it may be difficult for human relief workers to safely assist. Instead, autonomous vehicles can be used in place of, or in collaboration with, humans to allow for quicker and safer deployments.

We present Case-Based Estimator (CBE), a utility component of a larger HA/DR system that assigns autonomous vehicles to search areas in disaster zones. CBE estimates the performance of numerous vehicle-zone pairings and allows a human operator or automated mission manager to make informed decisions about how best to allocate the vehicles. Missions vary in their properties (i.e., type of disaster, location, terrain, type of vehicles, size of relief team). Thus, CBE may lack knowledge about how the autonomous vehicles will perform and must instead rely on simulators with varying fidelity.

© Springer International Publishing Switzerland 2015
E. Hüllermeier and M. Minor (Eds.): ICCBR 2015, LNAI 9343, pp. 15–29, 2015.
DOI: 10.1007/978-3-319-24586-7_2

However, given the real-time nature of HA/DR missions there may not be time to evaluate every vehicle-zone pairing in every simulator. Instead, CBE will need to use information from the lower fidelity, less computationally expensive simulators to predict performance on the higher fidelity simulators and select a subset of vehicle-zone pairs to examine in more detail. This process employs regression to estimate the performance in successively higher fidelity simulators and allows the decision maker (e.g., Operator or automated mission planner) to make informed decisions on which tasks to assign to vehicles. We report an empirical study in which CBE yields more accurate results than lower fidelity simulators and outperforms unfiltered regression approaches.

In this paper we describe CBE and how it uses data from simulators (introduced in Sect. 3) of varying fidelity to predict the performance of autonomous vehicles. Section 2 examines related work in the areas of case-based estimation and agent deployment. Section 3 describes the HA/DR domain. Section 4 briefly summarizes our HA/DR command system. Section 5 focuses on how we use CBR to estimate the performance of autonomous HA/DR vehicles. We evaluate our approach in Sect. 6, followed by a discussion of our results in Sect. 7 and concluding remarks in Sect. 8.

2   Related Work

Our current work focuses on online numeric prediction; we compute a linear regression equation from a subset of the most similar cases' outcomes using an online algorithm. This is an example of locally weighted regression (LWR) (Cleveland and Devlin 1988), and in particular of algorithms that compute local estimates of the regression surface (Atkeson et al. 1997a). These popular algorithms have a long history of use in, for example, robotics control tasks (Atkeson et al. 1997b). Many variants have been examined in the CBR literature, including in the context of case-based reinforcement learning techniques (e.g., Aha and Salzberg 1993; Gabel and Riedmiller 2007; Molineaux et al. 2008). Given a problem p, LWR algorithms identify the set K of p's k-nearest neighbors and compute a linear or nonlinear regression equation from K's (numeric) solution values. These are often similarity-weighted, where the most similar neighbors exert more influence on the derivation of the equation. This equation is then used to predict a solution value for p. Our algorithm, CBE, computes a simple unweighted linear regression model to make predictions, but where the value of k is not fixed (it varies depending on which cases exceed a similarity threshold). We have found it to perform well in our application, and leave the investigation of other LWR methods for future work. There are also similarities to two-stage retrieval models such as MAC/FAC in Forbus et al. (1995). It uses a simple similarity metric to identify a subset of cases to evaluate with a more comprehensive structural analysis. This is similar to CBE, which uses a function based estimate of simulation performance to retrieve promising candidates for further evaluation using more rigorous simulation models.

CBR has previously been studied for robotics applications. For example, Likhachev et al. (2002) use CBR to learn parameter settings for the behavior-based control of a ground robot in environments that change over time. While they focus on motion control for a single robot, we instead focus on the high-level control of robot teams. Ros et al.

(2009) focus on action selection for RoboCup soccer, and use a sophisticated representation and reasoning method. However, this body of research focuses on motion planning for relatively short-term behaviors, whereas we focus on longer duration plans that are monitored by a goal reasoning (GR) module (see Sect. 4).

GR agents that employ CBR techniques have been used for other control tasks, such as formulating the goals for team coordination (Jaidee et al. 2013), predicting the behavior of hostile agents (Borck et al. 2015), and recognizing the plans of an agent's teammates (Gillespie et al. 2015). However, in contrast to these other integrations, our focus is on predicting the outcomes of a plan executed by a set of robots.

In (Auslander et al. 2014) we described a CBR algorithm that sets the parameter values of complex HA/DR plans involving a heterogeneous set of unmanned autonomous vehicles that search multiple Areas of Interest (AOI). We represented cases using a similar (problem, solution, outcome) tuple. Our algorithm found solution parameter settings that performed well by adapting similar cases and using their outcome metrics to vote on parameter settings. When executing plans generated using our case-based algorithm on problems with high uncertainty, it outperformed plans generated using baseline approaches. In this paper, we instead focus on a complementary problem: estimating similar outcomes given a problem and solution parameter settings. These two approaches can potentially be combined in the future to improve parameter setting by estimating the performance of a proposed solution.

Finally, CBR has previously been studied for military applications, including disaster response. For example, Abi-Zeid et al. (1999) studied incident prosecution, including real time support for situation assessment in search and rescue missions. Their ASISA system uses CBR to select hierarchical information-gathering plans for situation assessment. Muñoz-Avila et al.'s (1999) HICAP instead uses conversational CBR to assist operators with refining tasks in support of noncombatant evacuation operations. SiN (Muñoz-Avila et al. 2001) is an extension that integrates a planner to automatically decompose tasks where feasible. However, while these systems use planning modules to support rescue operations, they do not predict the outcomes of a given plan's execution, nor focus on coordinating robot team behaviors.

3   Humanitarian Assistance/Disaster Relief Operations

HA/DR operations (O'Connor 2012) are performed by several countries in response to events such as Hurricane Katrina (August 2005), the Haiti earthquake (January 2010), and Typhoon Haiyan (November 2013). Before any personnel can begin operations, information about the Area of Operations must be acquired (e.g., locations of survivors, infrastructure condition, viable ingress points, and evacuation routes). This information will also need to be continuously updated as the situation develops. Each Area of Operation is composed of one or more Areas of Interest that need to be searched.

Current operations employ remotely controlled drones and human-piloted helicopters to gather this information. We are developing methods for deploying a heterogeneous team of autonomous unmanned vehicles with appropriate sensor platforms to automate much of this process, so as to reduce time and cost. This should enable responders to perform critical tasks more quickly for HA/DR operations. Independently of which system is used, an Operator given a list of missions must be able to prioritize which missions should be planned for and scheduled.

We focus on a method for comparing potential mission outcomes to enable the Operator or mission planner to select which missions to perform. These missions can be automatically generated from our goal reasoning system or provided by operators. This module's task is to provide estimates of the outcome metrics, which can be used to make more informed decisions on what to dispatch. This may yield better plans.

We use three simulations of varying fidelity in the CBE: an inexpensive function-based approach, a quick low fidelity simulator, and a slower high fidelity simulation. The first can estimate a metric without simulation (e.g., by computing the path a vehicle might take and dividing the path length by the vehicle's speed to estimate time required). To ensure efficiency, these estimates do not account for important factors such as wind and fuel levels, but they do provide instant, initial results.

Our low fidelity simulation is MASON (Luke et al. 2005), a discrete-event multi-agent simulator that models all physics behaviors and physicomimetics control (Martinson et al. 2011). MASON models the physical movements of generic agents acting in the environment. However, it lacks specific physical models of its actors and does not account for detailed problem factors such as the effects of wind. MASON's low fidelity allows it to more quickly generate results, but these are likely to be less accurate because it does not model all features.

Open AMASE (Duquette 2009) is the highest fidelity simulation we use, and in this paper we use it as a substitute for a real-world environment. This simulation models small tactical unmanned aircraft systems (STUAS) using a kinematic flight dynamics model that includes environmental effects (e.g., wind) on performance. AMASE also has facilities for modelling the field of view of cameras mounted on the STUAS based on the vehicles' six degree of freedom pose. This allows AMASE to calculate a metric for coverage defined as the area the sensor observed at a specified resolution. The lower fidelity models cannot produce this metric, and instead assume the paths followed produced full coverage. Figure 1 displays an example problem using both real-world data and AMASE's representation.

4   Situated Decision Process

To intelligently act in domains like HA/DR, a team of autonomous agents must continually monitor, evaluate, and dispatch new tasks or goals. To this end, we have designed a system architecture called the *Situated Decision Process* (SDP) (Roberts et al. 2015). In the SDP, a centralized Mission Manager subsystem assigns primitive goals to teams of autonomous agents, based on the input of an Operator and the vehicles' observations during execution. Intelligent, autonomous evaluation and selections of goals or tasks during execution requires rapid, accurate estimation of multiple scenario parameters.

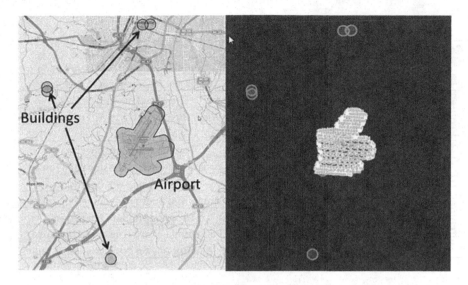

Fig. 1. Left: Representation of a problem set using OpenStreetMap data. Right: Same problem shown in AMASE with tracks for the airport region.

In HA/DR scenarios, the vehicles must quickly react to changes in the perceived environment, as well as to changes to the Operator's inputs. Doing so requires the rapid evaluation of such changes; it requires the ability to predict the effect of performing tasks more quickly than can be simulated with high fidelity and with more accuracy than can be achieved with low fidelity. This led us to consider using CBR to quickly and accurately estimate the parameter settings used by the Mission Manager to intelligently evaluate the utility of the vehicles' goals and tasks.

At the individual vehicle and sub-team level, CBE's estimates can be used in *motivators* for goal selection in a goal reasoning algorithm (Wilson et al. 2013). This would help us implement the situated portion of the SDP by permitting decision making on vehicles without direct access to the Mission Manager. This enables vehicles to choose predictable actions that should provide locally optimal results.

5   Case-Based Performance Estimation

To provide the data necessary for the Mission Manager to make informed decisions about its various vehicle deployment options, we use the CBE to evaluate mission options in HA/DR scenarios. We describe its case representation in Sect. 5.1 and case similarity metric in Sect. 5.2. The CBR algorithm and knowledge acquisition technique are presented in Sect. 5.3.

Table 1. CBE's case representation

Case Component		Attribute Name	Description
Problem (p)	Problem Description	Total Area of AOI	Total area of AOI in m²
		Distance to AOI	Distance from vehicle to AOI in m
		Wind Speed	The speed of wind in m/s
		Wind Direction	Wind angle relative to AOI orientation
	Solution		STUAS Configuration
Outcomes (O_{all})	O_e	Duration	Time to complete operation
		Energy	Amount of energy consumed in Joules
	O_l $O_{l_{estimate}}$	Duration	Time to complete operation
		Energy	Amount of energy consumed in Joules
	O_h $O_{h_{estimate}}$	Duration	Time to complete operation
		Coverage	Percentage of area observed by sensor
		Energy	Amount of energy consumed in Joules

5.1 Case Representation

We represent a case $C = \langle p, O_{all} \rangle$ as a problem p and the set of all outcomes O_{all} when that problem is evaluated using models of different fidelity. In this paper, we use three models of increasing complexity: an evaluation function, a low fidelity simulator (MASON), and a high fidelity simulator (AMASE). Similarly, we are also interested in estimating the performance when using the simulators (e.g., if the simulator is unavailable or computationally expensive). As such, the case contains the outcomes generated by the evaluation function (O_e), the low fidelity simulator (O_l), and the high fidelity simulator (O_h), and estimates of the low and high fidelity simulations $\left(O_{l_{estimate}} \text{ and } O_{h_{estimate}} \right) : C = \langle p, O_e, O_l, O_{l_{estimate}}, O_h, O_{h_{estimate}} \rangle$.

Table 1 provides detail on this representation. A problem p is composed of a problem description and proposed solution. The problem description is further divided into four features that characterize an aerial search task. *Total Area of AOI* is the total size of the area of interest (AOI) (i.e., the area being searched) in square meters. The *Distance to the AOI* is a measure of how far the search vehicle would have to travel to reach the center of the area. This becomes important as the trip time becomes a significant cost of the operation. *Wind Speed* is a measure of the magnitude of the wind in meters per second. Wind is a large source of error between the low and high fidelity simulations and tracking it enables a system to separate cases by the wind magnitude. *Wind Direction* is a measure of the alignment of the wind relative to the search area; it is a value in $[0°, 90°]$.

A solution represents the configuration of the search vehicles that will be assigned to the search area (e.g., vehicle types, the number of vehicles, sensor configurations). Here we focus on problems where a single vehicle of a fixed type is assigned to perform the search. (See (Auslander et al. 2014) for more complex solutions.)

Initially, each case contains only the problem description with unknown values for each outcome. As more information is obtained (i.e., evaluating the problem using the evaluation function or one of the simulators, or estimating the outcomes), it is added to the case. Only promising problems identified from the performance estimates are evaluated at the higher, more computationally expensive fidelities, so not all cases will have values for all outcomes. All outcomes have measurements for the search *duration* (seconds) and search *energy* (joules), while search *coverage* (percent of area observed with sensors) can be measured only by the high fidelity simulator and is therefore only contained in its outcomes. The estimated values may be continually overwritten, if new data becomes available that modifies these values, while the data obtained from simulation is recorded only once. The estimates serve as an inexpensive temporary measurement until the actual simulation is run; they are no longer used after the actual values are known.

5.2   Case Similarity

Case similarity is calculated using a weighted comparison of the problem features in two problems. Given two problems p_1 and p_2, the similarity metric (Eq. 1) calculates a similarity between 0 and 1. Each problem contains n features, and each feature f_i is given a weight w_i (*maxValue(i)* and *minValue(i)* represent the maximum and minimum value the ith feature can take). In CBE, there are four problem features: *Total Area of AOI, Distance to AOI, Wind Speed*, and *Wind Direction. Total Area of AOI* and *Distance to AOI* are assigned weights of 2.0, whereas other features are assigned weights of 1.0 to enable better separation of regions of varying sizes and locations. A weighted approach is used to allow more flexibility in discriminating among cases (e.g., emphasizing the geometric properties of the domain).

$$sim\left(p_1, p_2\right) = \frac{1}{\sum_{i=1}^{n} w_i} \sum_{i=1}^{n} w_i \left(1 - \frac{|p_1 f_i - p_2 f_i|}{\max Value\,(i) - \min Value\,(i)}\right) \qquad (1)$$

5.3   Performance Estimation Algorithm

CBE (Algorithm 1) enables a computationally inexpensive and accurate evaluation of potential configurations provided by the Mission Manager. Evaluating each potential configuration in the simulators can be expensive. Thus, CBE allows the Mission Manager to provide feedback about which configurations should be evaluated in more detail. To begin with, CBE receives a set of problems *Probs* representing possible missions under consideration from the Mission Manager (MM). For each problem $p \in Probs$, it retrieves a similar case C from case base CB using *SimilarCase(p, λ, CB)*, which examines all cases in CB that are above similarity $λ$ to p using from Eq. 1 and returns the most similar case with a known O_h (i.e., preference is given to cases with more known values). If no above-threshold cases have a known O_h, the most similar case is returned. If no cases are above threshold, a null value is returned. If a case is retrieved, CBE uses it. Otherwise, CBE evaluates p using the evaluation function (i.e., computing O_e) and creates a new case (the values for all other outcomes are set to null). The retrieved or created case is then added to a set of cases to be further evaluated.

Algorithm 1: Case-Based Estimator (CBE)

Inputs: $probs = \{p_1, p_2, \ldots, p_n\}$
Returns: CB'_E // Performance of problems in MM filtered subset of cases
Legend:
   CB // The (entire) case base
   $CB_E \leftarrow \{\emptyset\}$ // Subset of cases to be evaluated

Function: $FindBestProblem(probs)$ **returns** CB'_E
foreach $p \in probs$ **do**
   $C \leftarrow SimilarCase(p, \lambda, CB)$;
   if $C = \emptyset$ **then**
      $O_e \leftarrow EvaluationFunction(p)$;
      $C_{new} \leftarrow \langle p, O_e, \emptyset, \emptyset, \emptyset, \emptyset \rangle$;
      $CB \leftarrow CB \cup C_{new}$;
      $CB_E \leftarrow CB_E \cup C_{new}$;
   else
      $CB_E \leftarrow CB_E \cup C$;
   $CB_E \leftarrow ComputeEstimates(CB_E, CB)$;
   $CB'_E \leftarrow MissionManagerFilter(CB_E)$;
   $CB'_E \leftarrow RunMASONAndReviseCases(CB'_E)$;
   $CB'_E \leftarrow ComputeEstimates(CB'_E, CB)$;
return CB'_E;

Function: $ComputeEstimates(CB_E, CB)$
foreach $C \in CB_E$ **do**
   if $C.O_l = \emptyset$ **then**
      $C.O_{l_{estimate}} \leftarrow EstimateMASON(C, CB)$;
   if $C.O_h = \emptyset$ **then**
      $C.O_{h_{estimate}} \leftarrow EstimateAMASE(C, CB)$;
return CB_E

If the problem's MASON and AMASE values are not known (i.e., the problem has never been evaluated in the simulators), CBE then estimates the MASON and AMASE values (i.e., $O_{l_{estimate}}$ and $O_{h_{estimate}}$). The resulting cases are then sent to the Mission Manager for filtering because it is best able to choose what problems and metrics to optimize over given the overall mission context.

The Mission Manager returns a subset of cases (CB'_E) for further evaluation. For each of these cases, if the actual MASON outcome values are not known, it is run in the MASON simulator and its corresponding O_l values are revised. Afterward, the estimation routine is run again to generate new estimations for the AMASE outcome values and the resulting subset is returned to the Mission Manager. Not shown in Algorithm 1 is once the Mission Manager has filtered the set of problems a subset of these are picked to be deployed based on the Mission Manager's criteria. The resulting AMASE outcome values are subsequently stored in the case (i.e., as O_h) if the case did not previously have AMASE outcome values. If AMASE outcome values already exist (e.g., for a repeated surveillance task), rather than ignore the data a new case is created from the current problem. Its MASON outcome values are also computed to ensure no cases have AMASE outcome values without MASON outcome values.

The functions *EstimateMason(C, CB)* and *EstimateAMASE(C, CB)* are implemented using a linear regression algorithm for each outcome attribute. For MASON, the linear regression function takes the form $(p, O_e) \rightarrow O_{l_{estimate}}$. Similarly, the AMASE regression function is of the form $(p, O_l) \rightarrow O_{h_{estimate}}$. These regression algorithms are trained using cases that are above similarity θ to the current case and have known O_l (for MASON regression) or O_h (for AMASE regression) values. This similarity threshold ensures the regression functions are generated using only data from similar problems, helping isolate problems into clusters.

This is an online learning algorithm for estimation, with data acquired every time the estimation system is run. For each problem, a new case can potentially be generated. As the Operator selects problems to evaluate further, the MASON and AMASE outcome metrics are added to the cases. As more values are known, the algorithm will have more data to use for regression and should increase estimation accuracy.

6 Empirical Study

We empirically tested the following hypotheses:

H1: CBE's estimate of a problem's outcome approaches the actual outcome when evaluated using the high fidelity simulation over time.

H2: CBE provides more accurate estimates than the evaluation function and low fidelity simulator.

H3: CBE is more computationally efficient than the low fidelity simulator as the number of cases increases.

H4: CBE's filtered regression approach yields more accurate predictions than a non-filtered regression.

In the following sections we describe the evaluation methods, algorithms tested and metrics used.

6.1 Empirical Method

An objective of these tests is to verify that CBE accurately predicts the performance of a configuration when run on a high fidelity simulation. Thus, our ideal performance baseline is provided by the high-fidelity AMASE.

Problem sets were generated using a custom PostGIS system (Roberts et al. 2015). Each problem was formed by choosing a random airport from the OpenStreetMaps data set (Geofabrik 2014) and finding five random buildings within a 3–5 km radius of the airport. Each of these six locations is given a buffer region of 300 meters around their perimeters and the result is the search area to use in a given problem. Each search area is also assigned a random wind speed between 0–20 meters per second. The six search areas (i.e., airport and five nearby buildings) are stored as a search and rescue (S&R) problem. We repeated this 100 times to obtain 100 S&R problems. Problem features (e.g., Distance to AOI, Wind Direction) are derived from these problems at run time.

Each S&R problem was used to create a problem set that contains potential vehicle assignments for the problem. In CBE, only one vehicle assignment was used (i.e., the STUAS with default camera configuration). Future work will evaluate other vehicles and configurations as parameters, such as the use of static cameras and multiple vehicles. All problems in the problem set were run in AMASE to obtain ground truth data (i.e., how well that vehicle will perform when assigned to search a specific region).

A problem set run consists of giving an entire problem set to CBE and comparing its estimates to the known ground truth. Because we cannot know what the Mission Manager seeks at this level of abstraction, since it could be a human or intelligent subsystem, our method for selecting a subset of cases to run on MASON randomly selects 4 of the 6 cases. Similarly, when the final estimates are returned, 2 randomly selected problems (among the 4 selected) will be evaluated in AMASE.

A test run consists of randomly ordering the 100 problem sets and sequentially giving them as input to CBE, simulating 100 sequential uses of CBE. At the end of the 100 runs there will be 200 fully evaluated cases. We repeated this process 50 times and aggregated the results. All regression calculations were computed using WEKA's linear regression implementation (Hall et al. 2009).

6.2   Algorithms and Baselines Tested

We used the following algorithms and baselines to evaluate CBE. Each was run using the same evaluation problems presented in identical orderings.

- **CBE:** We set $\theta = 0.75$ to allow discrimination between building searches and airport searches. We set $\lambda = 0.99$ so that cases are reused only when they are highly similar to the problem.
- **Func:** Results obtained from running the estimation function on each problem.
- **MASON:** Results obtained by running MASON on each problem.
- **FuncReg:** Results obtained using linear regression to estimate AMASE's outcomes using problem features and the function estimate. This uses all available AMASE data for regression (i.e., not only data from similar problems, as with the CBR approach).
- **MASONReg:** Similar to *FuncReg*, but problem features and MASON outcomes are used to estimate AMASE outcomes. This also uses all available AMASE data for regression.

FuncReg and *MASONReg* use all data that is available to perform linear regression. For example, for the 100th input problem set 198 cases are used (since two AMASE outcomes are determined from each of the previous 99 input sets) while the 1st problem set will have no known AMASE outcomes. When any algorithm is unable to predict an AMASE outcome (i.e., no data to perform regression) a default error of 200 % is used. The *Func* and *MASON* baselines serve to show that using the values from only the lower fidelity simulators is inferior to using a mapping function such as the regression approach of CBE.

6.3   Results and Analysis

We now describe whether our results support our hypotheses:

H1: Fig. 2 displays results showing support for H1. It graphs the mean error of CBE over 50 runs after it has calculated its kth AMASE outcome estimate (100 problem sets with two estimates per set). As the number of problems evaluated increases, the error decreases and eventually converges to approximately 20% error, which is an improvement over our low fidelity estimate as shown in H2. The graph shows the error when predicting Duration. Although not shown, Energy converges similarly.

H2: Fig. 2 also displays the performance of baselines *MASON* and *Func*. CBE consistently outperforms the evaluation function and eventually outperforms *MASON* giving support to H2. Table 2 confirms this; using a paired t-test we found that CBE significantly outperforms *Func* for Energy and Duration. It also significantly outperforms *MASON* overall for Energy and for Duration over the final 75 % of problems (i.e., after learning).

H3: CBE requires fewer problems to be evaluated than if every problem is evaluated in MASON (only 4 of 6 are evaluated in MASON, so 67 % of the evaluations). Reducing necessary simulations runs is a large reduction in run time considering that, for an example run of an 18 min real world mission, the MASON simulation can take 35 s while AMASE takes 95 s. Additionally, setting an appropriate value for λ can influence how often MASON is used by CBE. If there is a case that is similar to an input problem, that problem does not need to be evaluated in MASON if the case has a recorded MASON outcome. In the current evaluation about 23 cases per test run were found to be similar enough to an input problem to be reused. If the Mission Manager wanted evaluations for all of those problems (i.e., they were among the 4 of 6 selected for MASON evaluation), that would result in 23 fewer MASON evaluations out of 400. However, the Mission Manager may not require evaluations for any of those problems and would instead evaluate other problems, resulting in no additional improvements. Reducing λ to 0.98 increases this to 107 reuses. This hypothesis has some support, but further exploration of parameters is warranted to find optimal values for this domain.

H4: Support of H4 is shown in Fig. 3 which graphs the results of CBE versus the two full regression approaches (i.e., they use *all* the cases), which begin with no data for the first two AMASE estimates and as such default to 200 % error. This accounts for the highest error in the first two AMASE estimates. In contrast, CBE uses the estimation function and MASON estimates, accounting for a lower initial error. For the remaining AMASE estimates, all three algorithms converge towards approximately 20 % error. Although CBE does not appear to converge faster, its errors are never as large as the initial regression models. This could be due to case reuse or the filtering of non-similar cases in the regression calculation.

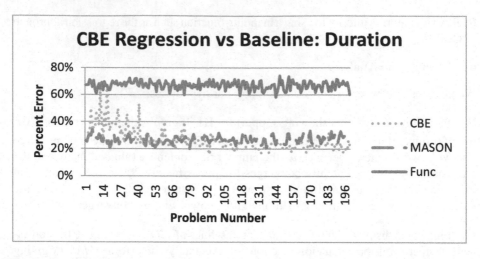

Fig. 2. Graph plotting percent error for CBE and the baseline algorithms across the 200 problems on 50 runs.

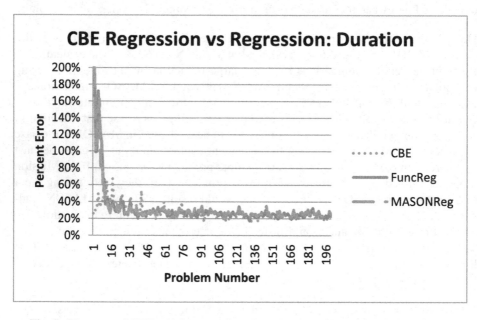

Fig. 3. The error of CBE and the regression algorithms for 200 problems over 50 runs.

Shown in Fig. 3, CBE's performance is better than or equal to the other algorithms. In the comparison of the 50 test run averages, CBE significantly outperformed both alternatives using a paired t-test; see Table 2. Table 3 displays the mean number of times, across all 50 runs, that CBE recorded lower error than the other algorithms. Additionally, it shows the mean reduction in absolute error when using CBE across all test runs.

Table 2. Results of t-tests ($p <$ value) showing CBE's improvement vs. other algorithms

CBE vs	Duration	Energy
Func	0.00000	0.00000
MASON	0.34228 (0.0 after 25% of the cases)	0.00000
FuncReg	0.00137	0.00446
MASONReg	0.00062	0.00130

Table 3. Improvement of CBE versus the regression algorithms

	Duration		Energy	
CBE vs	**# Improvements**	**Mean Error Reduction**	**# Improvements**	**Mean Error Reduction**
FuncReg	54.69%	5.05%	54.05%	2.75%
MASONReg	55.09%	5.33%	55.03%	2.97%

7 Discussion

The results in Sect. 6 clearly indicate the benefits of CBE, which recorded a 5 % reduction in Duration error and an almost 3 % reduction in Energy error. The reason for Energy's lower improvement could be differences with how the low and high fidelity simulators are modelling recharging. In MASON a vehicle is supposed to remain still while recharging, while AMASE (which was built to model fixed wing aircraft) does not restrict movement as much while recharging. Future versions of these simulators will address these discrepancies and also implement a procedure for returning to base and landing to increase scenario realism.

We expect that further improvements to performance will be found as more discriminating problem features are identified. For example, another type of vehicle would yield entirely new data clusters as Energy burn rates, and flight profiles would differ. As more data is collected over time the accuracy of the algorithms should increase and require fewer simulation runs.

8 Conclusion

Most CBR systems that estimate functions, such as cost, attempt to find a similar case and adapt their solution. We report on a novel online hybrid algorithm that can reuse prior learned values from similar problems and creates new estimates for others. For the scenarios in the domain we examine, the Case-Based Estimator (CBE) produced more accurate estimates from less data than two other regression algorithms.

One of the next steps from these results is to combine the benefits of this approach with the parameter selection approach from our previous investigation (Auslander et al. 2014). Benefits may include improving suggested solutions by estimating their actual outcomes. Beyond this there are many ways to improve the CBE algorithm. One of the most promising directions would be the exploration of non-linear regression models.

It is likely that some of our problem features are not independent, and a model that considers co-variance information may return more accurate results. We expect there to be tradeoffs in performance with these new models (e.g., additional computational resources required for increased training samples).

Over time the amount of data in the case base will become sufficiently large to necessitate the use of case-base maintenance techniques. While in general the accuracy of regression algorithms will increase given more data, improvements may also accrue by removing anomalous cases. In addition, as more data is obtained it may be possible to be more discriminative in case selection by increasing λ and θ. Future research may identify ways to scale these parameters with the data.

Acknowledgements. Thanks to OSD ASD (R&E) for sponsoring this research.

References

Abi-Zeid, I., Yang, Q., Lamontagne, L.: Is CBR applicable to the coordination of search and rescue operations? A feasibility study. In: Althoff, K.-D., Bergmann, R., Branting, L. (eds.) ICCBR 1999. LNCS (LNAI), vol. 1650, pp. 358–371. Springer, Heidelberg (1999)

Aha, D.W., Salzberg, S.L.: Learning to catch: applying nearest neighbor algorithms to dynamic control tasks. In: Proceedings of the Fourth International Workshop on Artificial Intelligence and Statistics, Ft. Lauderdale, FL, pp. 363–368 (1993, Unpublished)

Atkeson, C., Moore, A., Schaal, S.: Locally weighted learning. Artif. Intell. Rev. **11**(1–5), 11–73 (1997a)

Atkeson, C., Moore, A., Schaal, S.: Locally weighted learning for control. Artif. Intell. Rev. **11**(1–5), 75–113 (1997b)

Auslander, B., Apker, T., Aha, D.W.: Case-based parameter selection for plans: coordinating autonomous vehicle teams. In: Lamontagne, L., Plaza, E. (eds.) ICCBR 2014. LNCS, vol. 8765, pp. 32–47. Springer, Heidelberg (2014)

Borck, H., Karneeb, J., Alford, R., Aha, D.W.: Case-based behavior recognition in beyond visual range air combat. In: Proceedings of the Twenty-Eighth Florida Artificial Intelligence Research Society Conference, pp. 379–384. AAAI Press, Hollywood, FL (2015)

Cleveland, W.S., Devlin, S.J.: Locally weighted regression: an approach to regression analysis by local fitting. J. Am. Stat. Assoc. **83**(403), 596–610 (1988)

Duquette, M.: Effects-level models for UAV simulation. In: Proceedings of the AIAA Modeling and Simulation Technologies Conference. AIAA, Chicago, IL (2009)

Forbus, K.D., Gentner, D., Law, K.: MAC/FAC: A model of similarity-based retrieval. Cogn. Sci. **19**, 141–205 (1995)

Gabel, T., Riedmiller, M.: An analysis of case-based value function approximation by approximating state transition graphs. In: Weber, R.O., Richter, M.M. (eds.) ICCBR 2007. LNCS (LNAI), vol. 4626, pp. 344–358. Springer, Heidelberg (2007)

Geofabrik: OpenStreetMap data extracts (2014). http://download.geofabrik.de/index.html

Gillespie, K., Molineaux, M., Floyd, M.W., Vattam, S.S., Aha, D.W.: Goal reasoning for an autonomous squad member. In: Aha, D.W. (ed.) Goal Reasoning: Papers from the ACS Workshop, Atlanta, GA (2015). www.cc.gatech.edu/~svattam/goal-reasoning

Hall, M., Frank, E., Holmes, G., Pfahringer, B., Reutemann, P., Witten, I.H.: The WEKA data mining software: an update. SIGKDD Explor. **11**(1), 10–18 (2009)

Jaidee, U., Muñoz-Avila, H., Aha, D.W.: Case-based goal-driven coordination of multiple learning agents. In: Delany, S.J., Ontañón, S. (eds.) ICCBR 2013. LNCS, vol. 7969, pp. 164–178. Springer, Heidelberg (2013)

Likhachev, M., Kaess, M., Arkin, R.: Learning behavioral parameterization using spatio-temporal case-based reasoning. In: Proceedings of the International Conference on Robotics and Automation, pp. 1282–1289. IEEE Press, Washington, DC (2002)

Luke, S., Cioffi-Revilla, C., Panait, L., Sullivan, K., Balan, G.: MASON: a multiagent simulation environment. Simulation **81**(7), 517–527 (2005)

Martinson, E., Apker, T., Bugajska, M.: Optimizing a reconfigurable robotic microphone array. In: Proceedings of the International Conference on Intelligent Robots and Systems, pp. 125–130. IEEE Press, San Francisco, CA (2011)

Molineaux, M., Aha, D.W., Moore, P.: Learning continuous action models in a real-time strategy environment. In: Proceedings of the Twenty-First Florida Artificial Intelligence Research Conference, pp. 257–262. AAAI Press, Coconut Grove, FL (2008)

Muñoz-Avila, H., Aha, D., Breslow, L., Nau, D.: HICAP: an interactive case-based planning architecture and its application to noncombatant evacuation operations. In: Proceedings of the Ninth National Conference on Innovative Applications of Artificial Intelligence, pp. 879–885. AAAI Press, Orlando, FL (1999)

Muñoz-Avila, H., Aha, D., Nau, D., Weber, R., Breslow, L., Yaman, F.: SiN: integrating case-based reasoning with task decomposition. In: Proceedings of the Seventeenth International Joint Conference on Artificial Intelligence, pp. 999–1004. Morgan Kaufmann, Seattle, WA (2001)

O'Connor, C.: Foreign humanitarian assistance and disaster-relief operations: lessons learned and best practices. Nav. War Coll. Rev. **65**(1), 152–160 (2012)

Roberts, M., Apker, T., Johnson, B., Auslander, B., Wellman, B., Aha, D.W.: Coordinating robot teams for disaster relief. In: Proceedings of the Twenty-Eighth Florida Artificial Intelligence Research Society Conference, pp. 366–371. AAAI Press, Hollywood, FL (2015)

Ros, R., Arcos, J., Lopez de Mantaras, R., Veloso, M.: A case-based approach for coordinated action selection in robot soccer. Artif. Intell. **173**, 1014–1039 (2009)

Wilson, M., Molineaux, M., Aha, D.W.: Domain-independent heuristics for goal formulation. In: Proceedings of the Twenty-Sixth Florida Artificial Intelligence Research Society Conference, pp. 160–165. AAAI Press, St. Pete Beach, FL (2013)

Case-Based Policy and Goal Recognition

Hayley Borck[1]($\boxtimes$), Justin Karneeb[1], Michael W. Floyd[1],
Ron Alford[2,3], and David W. Aha[3]

[1] Knexus Research Corporation, Springfield, VA, USA
{hayley.borck,justin.karneeb,michael.floyd}@knexusresearch.com
[2] ASEE Postdoctoral Fellow, Washington, DC, USA
[3] Navy Center for Applied Research in Artificial Intelligence,
Naval Research Laboratory (Code 5514), Washington, DC, USA
{ronald.alford.ctr,david.aha}@nrl.navy.mil

Abstract. We present the Policy and Goal Recognizer (PaGR), a case-based system for multiagent keyhole recognition. PaGR is a knowledge recognition component within a decision-making agent that controls simulated unmanned air vehicles in Beyond Visual Range combat. PaGR stores in a case the goal, observations, and policy of a hostile aircraft, and uses cases to recognize the policies and goals of newly-observed hostile aircraft. In our empirical study of PaGR's performance, we report evidence that knowledge of an adversary's goal improves policy recognition. We also show that PaGR can recognize when its assumptions about the hostile agent's goal are incorrect, and can often correct these assumptions. We show that this ability improves PaGR's policy recognition performance in comparison to a baseline algorithm.

Keywords: Policy recognition · Intelligent agents · Goal reasoning · Air combat

1 Introduction

The World War I and World War II dogfighting style of air combat is mostly obsolete. The modern capabilities of aircraft instead facilitate a longer range style of combat deemed *Beyond Visual Range* (BVR) combat. In this paper, we focus on how an Unmanned Aerial Vehicle (UAV) may participate in BVR combat operations. In BVR combat, significant time between maneuvers allows a UAV more time to plan its reactions. Accurately recognizing the actions and plans of other agents greatly reduces the complexity for planning in adversarial multi-agent systems [1]. However, long-range aerial observations are generally restricted to the basic telemetry data returned by radar (i.e., position and speed). This makes accurately recognizing an opponent's actions difficult [2].

In previous work, we described a reasoning system called the *Tactical Battle Manager* (TBM), which uses goal reasoning [3,4], policy recognition [5], and automated planning techniques to control a UAV in a simulated BVR environment. The existing policy recognition system, like most, uses observations to

© Springer International Publishing Switzerland 2015
E. Hüllermeier and M. Minor (Eds.): ICCBR 2015, LNAI 9343, pp. 30–43, 2015.
DOI: 10.1007/978-3-319-24586-7_3

directly predict the tactics of the opposing agents. However, the choice of tactic (or plan) for an agent is driven by its over-arching mission (or goal), which cannot be directly observed. For example in a defensive mission, maintenance of stealth and prevention of losses may drive the enacted policies of the agent. In contrast, aggressiveness and firing missiles at opponents may be more important in an offensive mission.

In this paper, we introduce the Policy and Goal Recognizer (PaGR; Sect. 4), which leverages agent goals in an attempt to improve policy recognition accuracy. PaGR is given a mission briefing, which includes the details of the upcoming mission such as the expected capabilities of the hostiles and their expected goal. PaGR's cases contain the goal, observations, and policy of a hostile agent. During the mission, the goal and current observations are used in the similarity calculation to retrieve a newly-observed agent's predicted policy. Our experiments in a simulated BVR environment (Sect. 5), involving 2v2 scenarios (i.e., 2 allied agents vs. two hostile agents), show that case retrieval is significantly more accurate when goal information is employed.

However, mission briefings may not be accurate, and our experiments also show that making incorrect assumptions about an adversary's goal may negatively impact retrieval performance. To counteract inaccurate goal assumptions, PaGR leverages past observations to recognize when there is a discrepancy between the goals provided by the mission briefing and the actual goals of the hostiles. Once a discrepancy is recognized, PaGR then attempts to predict a goal that more correctly describes, and thus predicts, hostile agent intentions. Our experiments show that goal revision can reduce most of the negative impacts from incorrect briefings.

We begin with a discussion of related work (Sect. 2), followed by a description of the TBM (Sect. 3). We then describe PaGR (Sect. 4) and our empirical study of it (Sect. 5), and finish with a discussion of future work (Sect. 6) and conclusions (Sect. 7).

2   Related Work

Plan, activity, and intent recognition has been an active field of research in AI in recent years [6]. Our system endeavors to recognize the tactics of opposing agents. We define an agent's tactics as a sequence of actions it performs as part of a mission scenario, with the possibility it will switch tactics during a scenario. In this sense, an adversarial agent uses a plan to perform a tactic while having the autonomy to change tactics as necessary (e.g., because a previous tactic was unsuccessful, as countermeasures against another agent, it changed its goals).

Vattam et al. [7] use action sequence graphs to perform case-based plan recognition in situations where actions are missing or noisy. Ontanón et al. [8] use case-based reasoning (CBR) to model human driving vehicle control behaviors and skill level to reduce teen crash risk. Their CBR system predicts the next action the driver would take given the current environment state. However, these systems cannot identify plans that were previously identified incorrectly and refine the plan recognition process.

Fagundes et al. [9] focus on determining when it is appropriate or necessary to interact with agents to gain more information about their plan. Their system implements symbolic plan recognition via feature decision trees to determine if an agent should interact with other agents in multiagent scenarios. Unlike the BVR domain, they assume full observability of other agents. Similar to Fagundes et al., we have shown previously that acting intelligently via automated planning significantly reduces the time it takes to classify agent behaviors in single-agent scenarios [10]. Although we do not perform active recognition in this paper, we extend its recognition system by adding an agent goal to the case.

Laviers and Sukthankar [11] present an agent that learns team plan repair policies by recognizing the plans of the opposing American football team. Molineaux et al. [12] also study plan recognition in this domain; their system uses recognized plans to aid in case-based reinforcement learning. Similarly, human activity recognition [13] has been used to recognize a senior citizen's activity and adapt an agent to be more helpful when assisting the senior. These efforts are similar to our own in that they attempt to revise the model of other agents. However, unlike our agent in the BVR domain, these recognition agents have access to perfect information.

Single-agent keyhole plan recognition can be expanded to multiagent domains, but has been shown to be a much more difficult problem [14]. Zhou and Li [15] show that multi-agent plan recognition can be performed in a partially observable environment by representing team plans as a weighted maximum satisfiability problem. However, their algorithm's runtime is proportional to the amount of missing information which could prove problematic in a real-time combat domain like BVR. As such, we instead determine the probability of each tactic being used. Our system can then assume that a specific tactic is being used or perform exploratory actions that will help differentiate between possible tactics.

Plan recognition can also assist in behavior or intent recognition. Geib and Goldman [16] describe the need for plan and intent recognition in intrusion detection systems. Their system operates in a partially observable environment with a single adversarial agent. They use a probabilistic plan recognizer that reasons about observed actions, generates hypotheses about the adversary's goals, and retrieve plans from a plan library. Their system operates under the assumption that an attacker targets the agent's system. In our work, we instead use plan recognition to identify an adversary's plans and the target of those plans (e.g., is the hostile attacking the agent or one of its teammates).

Corchado et al. [17] present an agent that encodes the belief-desire-intention (BDI) model in a CBR context. Their agent uses its current beliefs and desires (the problem) to select its intentions (the solution). This is similar to our own work in that it explicitly encodes the desires of an agent in cases. However, our cases are created from observations of adversarial agents, so they may contain erroneous or incomplete information about beliefs, desires, or intentions.

Uncertainty about the quality (e.g., due to sensing errors) or completeness (e.g., due to partial observability and hidden internal information) of observed information has been a focus of CBR systems that learn by observation [18–20]. These systems learn to perform a task by observing an expert, either with or without the expert's knowledge and assistance. In our system, we are not trying to replicate an agent's behavior but instead use observations to identify the agent's behavior and use this to guide our agent's decision making process.

Traces of a user's behavior have been used as cases in trace-based reasoning [21] and episodic reasoning [22] systems. These systems store a sequence of the user's past actions, or both past actions and past environment states. This allows the systems to reason over the current problem, the problem's evolution over time, and the solutions that have been applied to the problem. These systems differ from our own in that we do not have direct knowledge of the observed agent's action but must instead infer them from observable data.

3 Tactical Battle Manager

PaGR will serve as a component in the Tactical Battle Manager (TBM; Fig. 1), a system we are developing for collaborative pilot-UAV interaction and autonomous UAV control. During experimentation and testing the pilot is modeled by using a static policy (see Sect. 5). In its current state, the TBM contains, among other components, a goal reasoning component (*Goal Management System*), a planning and prediction component (*Planning System*), and a policy recognition component (*Policy Recognizer* in *Knowledge Updater*).

The goal reasoning (GR) component chooses between competing goals. In the TBM a goal is a list of weighted desires. Each desire has an agitation function that is updated based on the world state at each update cycle. The more a desire

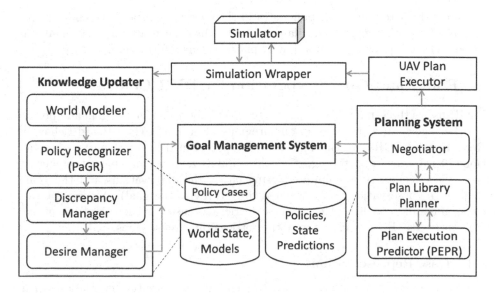

Fig. 1. The Tactical Battle Manager (TBM) architecture.

is agitated, the more urgently the agent wants to alleviate the symptoms (i.e., perform actions that reduce agitation). The GR component is not the focus of this paper. In our evaluations, a static goal is used by the UAV. In the future, the UAV will use observations, recognized hostile policies, and goals provided by PaGR to revise the current goal.

Our planner is an extension of a simple plan library planner [23]. The Planning System chooses a policy from a fixed set of eight candidates, which are either one of four basic movements (listed immediately below) or one of these four movements where the agent also fires a missile.

- **Pure Pursuit**: An agent flies directly at a target.
- **Drag**: An agent flies directly away from a target.
- **Beam**: An agent flies at a 90° offset to a target.
- **Crank**: An agent flies at the maximum offset but tries to keep its target observable in radar.

The missile-firing variants of these policies are: *ShootPursuit*, *ShootDrag*, *ShootBeam*, and *ShootCrank*. While these policies are simple, they encompass a range of real-world BVR combat tactics as described by our subject matter experts.

The planner uses the Plan Execution Predictor (PEPR) [24] to generate potential future states for all eight policies (i.e., what would happen if the UAV used each of those policies). These policies are then ranked by the weighted desire agitation over the duration of the prediction (i.e., how well the UAV's goals are met), and the highest-ranked policy is returned. For example, if the current goal has a highly weighted safety desire, the planner will favor policies that do not put the agent in danger (i.e., Drag). If the safety desire is only somewhat agitated, the planner may choose a safe policy that also satisfies its other desires, such as ShootBeam.

Since the outcome of our policies (and thus desire agitation) depends on the actions of the other agents, the planner needs to have accurate information about the policies of opposing agents. PaGR provides this information.

4   Policy and Goal Recognition for BVR Combat

PaGR uses a three-part case structure to effectively recognize the goal and policy of other agents in a scenario. In our previous work on the Case-Based Behavior Recognizer (CBBR) system [2], cases were partitioned into two parts: observations of the agent and the agent's policy. PaGR adds a third part, the agent's goal. Case similarity is computed using either the observations and the goal, or just the observations (i.e., when trying to revise the goal). A typical CBR cycle generally consists of four steps: retrieval, reuse, revision, and retention. Currently, PaGR implements only the retrieval and reuse steps.

4.1   Case Representation

Each case in PaGR represents a hostile agent and its potential interactions with one agent of the friendly team in the scenario. During case authoring, we created

Table 1. Example Case within PaGR

Component	Representation
Goal	A list of weighted desires
	- 0.25 Aggressive Posture
	- 0.50 Safety
	- 0.25 Fire Weapon
Observations	A set of features discretized from the observable world state
	- 1.00 FacingTarget
	- 0.45 TargetRange
	- 0.15 InDanger
Policy	A set of ungrounded actions
	- Fire Missile
	- Fly Drag

a case for each pair of friendly and hostile aircraft in the scenario based on their interactions. We will later use these cases to determine the most likely target of a hostile agent.

A case $C = \langle G, O, \Pi \rangle$ has three components (see Table 1). First, the *goal* G is a list of the weighted desires of the agent. Second, the *observations* O of the position and heading of the hostile agent are recorded. These observations are then projected onto features such as FACINGTARGET (the relative bearing to the target), TARGETRANGE (a normalized distance between the agent and its target), and INDANGER (whether the agent is in the weapon range of the target). Each feature, normalized to between 0 and 1, represents how much the feature is currently exhibited by the observations. For example, FACINGTARGET will be 1 if the hostile agent is directly facing the target but closer to 0.5 if it is facing the target at a 90° angle. Finally, a *policy* Π is recorded as a sequence of ungrounded actions that define an air combat tactic. We use policies rather than plans, which have grounded actions, to help PaGR overcome the missing information that is prevalent in this domain. As stated previously, which parts of the case represent the problem and which represent the solution change depending on PaGR's current task (i.e., policy identification or updating goal assumptions). The possible observational features are listed below:

1. Facing Target
2. Closing on Target
3. Target Range
4. Within a Target's Weapon Range
5. Has Target within Weapon Range
6. Is in Danger
7. Is moving with respect to Target

4.2   Similarity and Retrieval

The similarity and retrieval steps of PaGR are relatively straightforward. During the recognition of a policy, similarity is calculated between the goal and observations of the current query q and those of each case c in the case base CB.

The similarity of two observations is defined as the average distance of the matching features (Eq. 1), where $\sigma(w_f, q_f, c_f)$ is the weighted distance between two values for feature f and N is the set of time step features.

$$sim_o(q, c) = \frac{\sum_{f \in N} \sigma(w_f, q_f, c_f)}{|N|} \tag{1}$$

The similarity of two goals is defined as the distance of the weights of the desires that define the goal. If a desire is missing (e.g., q has desire $f \in D$ but c does not) then it is treated as having a weight of 0 (Eq. 2).

$$sim_g(q, c) = \frac{\sum_{f \in D} (w_f, q_f, c_f)}{|D|} \tag{2}$$

The similarity between two cases is then defined as the weighted average of the observation and goal similarities (Eq. 3).

$$sim(q, c) = w_o \times sim_o(q, c) + w_g \times sim_g(q, c), \tag{3}$$

where $w_o + w_g = 1$. Using this equation for case similarity, PaGR retrieves a set of cases C_q, where each case in the set has a similarity greater than a given threshold parameter τ_r. If no cases were retrieved over the threshold, the policy is marked as unknown and system continues until the next timestep to repeat the retrieval process with more information. After it retrieves C_q, PaGR returns a normalized ratio of the policies in the cases of C_q that represents their relative frequencies with respect to their case weights generated during pruning. We interpret this as a probability distribution, so that given a policy p with a ratio of 0.7, there is a 70 % chance that this is the policy being executed.

4.3   Pruning Strategy

We perform a limited amount of case base maintenance (CBM) on our case library. In particular, we prune cases in a manner that is similar to the strategy we used in our previous work [10], although now designed for our revised three-part case structure. As we have done previously, our pruning algorithm merges similar cases that have the same policy while maintaining their overall importance. That is, whenever two cases with the same policy have a similarity over a certain threshold τ_π, we merge them and increment an associated counter. After all similar cases have been merged, an additional sweep is made over the (now pruned) case base. Each case has its weight normalized with respect to all cases sharing its policy. This prevents over-abundant cases from overpowering less common ones during retrieval. The resulting case base is smaller than the original but preserves the differences between prominent and atypical cases.

4.4   Using PaGR in BVR Mission Scenarios

Before each simulated 2v2 scenario is run, a mission briefing is given to the UAV for use by PaGR. This mission briefing includes details on the expected goal of the hostiles, and a case base (described in Sect. 4.1). For example, one of the mission briefings used in our experiments informs the UAV that all encountered enemy aircraft will have highly aggressive goals and do not care about their own safety.

During the simulation, PaGR can be configured to use the expected goal in one of three ways: (1) ignore it (a baseline approach), (2) use the given expected goal during case retrieval, or (3) perform goal revision and, during retrieval, use whatever goal it thinks is most likely being pursued.

PaGR performs goal revision by searching for the goal that best explains all prior observations. For each such observation, PaGR retrieves all the cases C_q whose similarity is above τ_r (ignoring their goals). Just as with policy retrieval, PaGR creates a probability distribution over the retrieved goals. These distributions are averaged over the duration of the scenario. During goal revision, the goal with the highest average probability is used as the new mission-briefing goal and is used for retrieving future policies.

5   Empirical Study

In our empirical study we examine the following hypotheses:

H1: Using an agent's goal during retrieval will increase PaGR's policy recognition accuracy.

H2: PaGR can effectively determine a goal discrepancy through observation.

H3: If given an incorrect goal, PaGR can revise it to increase policy recognition accuracy.

5.1   Scenarios

We created three scenarios and five mission briefings to test PaGR. Creation of the scenarios and briefings were guided by subject matter experts. These briefings specify goals for the opposing agents aircraft with varying levels of aggressiveness, namely:

- HIGHLYAGGRESSIVE: Approach directly and fire.
- AGGRESSIVE: Approach indirectly and fire.
- SAFETYAGGRESSIVE: Fire at range and leave.
- OBSERVE: Approach indirectly but do not fire.
- SAFETY: Do not approach.

All our 2v2 scenarios involve two *blue* allied aircraft flying into an airspace with two *red* hostile aircraft approaching from various angles (Fig. 2). The blue aircraft are governed by a static HIGHLYAGGRESSIVE policy; they fly directly

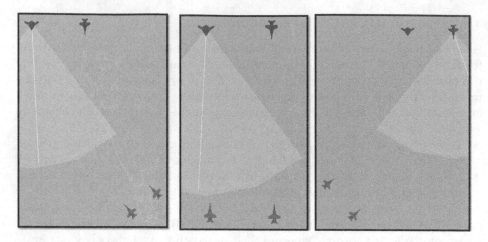

Fig. 2. Base 2v2 scenarios with friendlies in blue and hostiles in red. Lightened areas represent the radar observation cones of a single aircraft (other cones are not shown) (Color figure online).

at their target and attempt to shoot a missile when in range. We run PaGR on only *one* of the blue aircraft (representing the UAV). We chose to not use a full TBM to control the second blue aircraft while testing PaGR so that the results would not be influenced by the TBM's other components. The red aircraft are controlled by a simplified version of the TBM; they are given a static goal from the mission briefing to use with the plan library planner, and do not invoke PaGR. The red aircraft mark all opposing agents as using a basic policy that continues the current heading and speed for the purposes of prediction. Thus, in a given mission, the red aircraft will react to the blue aircraft using the goal weights specified in the mission briefing.

To create a large pool of test scenarios, we applied a random perturbation 33 times to each of the 15 ⟨scenario, mission briefing⟩ pairs, which yields a total of 495 variants. This scenario perturbation step modifies the headings and positions for each agent individually in the scenario within bounds to generate random but valid scenarios. By "valid", we mean that one of the agents will eventually come within radar range of another agent. The mission briefings and scenarios were created to encompass a set of real-world scenarios as described by our subject matter experts.

5.2   Empirical Method

We used a stratified k-fold cross validation strategy, where $k = 10$, and split the variants evenly across each fold. For each (test) fold, a case base was created using the other (training) folds and pruned as described in Sect. 4.3. Finally, all variants contained within the current testing fold were run for evaluation purposes. Section 5.3 describes the experiments and algorithms used to test our hypotheses.

To test hypothesis **H1**, that using goals improves PaGR's policy recognition accuracy, we tested PaGR in the following conditions.

1. **Correct Goal**: This was the ideal case for PaGR; it was given a correct goal from the mission briefing and never tried to change it. Additionally, when performing case retrieval it weighted goal similarity at 20 % ($w_g = 0.2$) and observations at 80 % ($w_o = 0.8$). Preliminary tests showed that the specific value for goal weighting had little impact on overall recognition as long as $w_g \in [15\%, 25\%]$.
2. **No Goal**: In this condition, PaGR did not use goals during recognition. That is, during case retrieval, it weighted goal similarity at 0 %, meaning it used only observations to recognize policies.
3. **Incorrect Goal**: This tests PaGR in its worst-case scenario. This configuration was identical to the Correct Goal condition except it was always given an *incorrect* goal.
4. **Revised Goals**: We used this condition to test **H2** and **H3**. PaGR was given an incorrect goal, but in addition to the normal policy retrieval recognition, PaGR ran an additional retrieval using only observations. However, rather than returning a policy, it returns a probability distribution of goals. PaGR averages this probability across the entire scenario executed so far and checks whether the current goal is the most probable. If not, it reports a discrepancy that triggers goal recognition of a recognized goal.

To calculate the results of these runs, we computed the percentage of recognition attempts that PaGR returned the actual policy being enacted by the target agent. We averaged these values for each policy across all variants.

5.3 Results

Figure 3 displays the average results for each of the 8 policies and their average (*Overall*). The results support **H1**; in all cases where PaGR was given a correct goal, it performed at least as well as when it was run without using goals and in many instances it performed better. In the two cases where the systems performed similarly (Drag and ShootDrag), the policies are easily recognizable given the observations as they are the only plans that involve completely facing away from the target. We conducted an independent samples t-test to compare the results for the Correct Goal and No Goal conditions for all the policies. This yields p values in the range of 10^{-5} to 10^{-9} for all policies except Drag and ShootDrag. When run in the Incorrect Goal condition, PaGR performs far worse than the No Goal condition.

The results for our second experiment are displayed in Fig. 4, where the blue bars denote PaGR's ability to recognize that a given goal is incorrect and the red bars denote that it also identifed the hostile agent's true goal (i.e., the most probable goal was the true goal). PaGR's discrepancy recognition is high; it almost always correctly identifies when the current goal is incorrect given the observations. This lends some support to **H2**. Figure 4 also shows that some true goals are easier for PaGR to recognize than others.

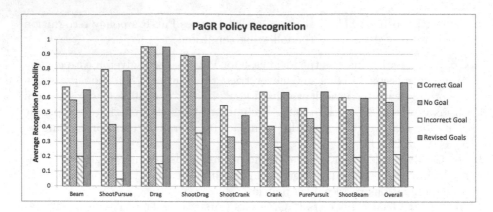

Fig. 3. Average policy recognition accuracy results from our first experiment showing the policy recognition accuracy of PaGR tested in three conditions

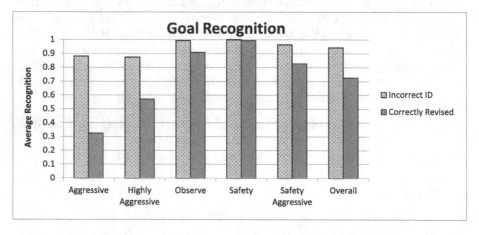

Fig. 4. Goal recognition accuracy results. The mission briefing goal is recognized as being incorrect (blue bars) if goal revision selects any other goal to replace it. The goal is correctly revised (red bars) if PaGR selects the true goal (Color figure online).

Finally, to test hypothesis **H3** we re-ran the Revised Goals condition used to test **H2**, except this time we allowed PaGR to replace the current goal with the most probable goal based on its averaged observations. Figure 3 displays the Revised Goals results. PaGR's recognition probability when using goal revision is almost equivalent to when using the Correct Goal condition. At first glance these results may seem confusing because the Correct Goal recognition scores are not nearly high enough to always revise the goal correctly. Often times it seems that PaGR confuses two similar goals, such as HIGHLYAGGRESSIVE and AGGRESSIVE. However, because these two goals can often lead to similar policies, PaGR's overall correct recognition is high. A t-test comparing the Revised Goals and No Goal results yields p scores in the same range as we found for hypothesis **H1**. Therefore, this lends some support for hypothesis **H3**.

6   Future Work

In Beyond Visual Range air combat a hostile agent will often try and disguise their actions or true intent. Therefore, in future work we plan to extend PaGR to recognize deceptive agents in BVR scenarios. We also plan to integrate PaGR into the TBM. Leveraging PaGR's ability to accurately predict hostile policies and goals should increase the TBM's ability to effectively reason about goals and plans. We also plan to investigate how active techniques for policy recognition could impact PaGR's results in comparison to our prior work on active behavior recognition [10]. Finally, we plan to show that using PaGR allows for overall improvements to mission performance in longer and more complex scenarios, compared to using no recognition system (or an alternative recognition system).

7   Conclusions

In this paper we presented the Policy and Goal Recognizer (PaGR). We tested PaGR in simple 2v2 Beyond Visual Range air combat scenarios, where one of the two friendly vehicles uses PaGR. We found that using PaGR's prediction of a hostile agent's goal increases its policy recognition accuracy (as compared to when using only its observations). In our scenarios, PaGR can effectively determine whether the mission briefing provided the correct goal for a hostile agent. We also found that, when using the same case structure and information used for policy recognition, PaGR can revise incorrect goal information and use goals that better fit the observations, which increases policy recognition accuracy.

Acknowledgements. Thanks to OSD ASD (R&E) for supporting this research. Thanks also to our subject matter experts for their many contributions and to the reviewers for their helpful comments.

References

1. Carberry, S.: Techniques for plan recognition. User Model. User-Adap. Inter. **11**(1–2), 31–48 (2001)
2. Borck, H., Karneeb, J., Alford, R., Aha, D.W.: Case-based behavior recognition in beyond visual range air combat. In: Proceedings of the Twenty-Eighth International Florida Artificial Intelligence Research Society Conference. AAAI Press (2015)
3. Muñoz-Avila, H., Jaidee, U., Aha, D.W., Carter, E.: Goal-driven autonomy with case-based reasoning. In: Bichindaritz, I., Montani, S. (eds.) ICCBR 2010. LNCS, vol. 6176, pp. 228–241. Springer, Heidelberg (2010)
4. Molineaux, M., Klenk, M., Aha, D.W.: Goal-driven autonomy in a navy strategy simulation. In: Proceedings of the Twenty-Fourth AAAI Conference on Artificial Intelligence. AAAI Press (2010)
5. Borck, H., Karneeb, J., Alford, R., Aha, D.W.: Case-based behavior recognition to facilitate planning in unmanned air vehicles. In: Vattam, S.S., Aha, D.W., eds.: Case-Based Agents: Papers from the ICCBR Workshop, Technical report. University College Cork, Cork, Ireland (2014)

6. Sukthankar, I.G., Goldman, R., Geib, C., Pynadath, D., Bui, H.: An introduction to plan, activity, and intent recognition. In: Sukthankar, I.G., Goldman, R., Geib, C., Pynadath, D., Bui, H. (eds.) Plan, Activity, and Intent Recognition. Elsevier (2014)
7. Vattam, S.S., Aha, D.W., Floyd, M.: Case-based plan recognition using action sequence graphs. In: Lamontagne, L., Plaza, E. (eds.) ICCBR 2014. LNCS, vol. 8765, pp. 495–510. Springer, Heidelberg (2014)
8. Ontañón, S., Lee, Y.-C., Snodgrass, S., Bonfiglio, D., Winston, F.K., McDonald, C., Gonzalez, A.J.: Case-based prediction of teen driver behavior and skill. In: Lamontagne, L., Plaza, E. (eds.) ICCBR 2014. LNCS, vol. 8765, pp. 375–389. Springer, Heidelberg (2014)
9. Fagundes, M.S., Meneguzzi, F., Bordini, R.H., Vieira, R.: Dealing with ambiguity in plan recognition under time constraints. In: Proceedings of the International Conference on Autonomous Agents and Multi-Agent Systems, pp. 389–396. ACM Press (2014)
10. Alford, R., Borck, H., Karneeb, J., Aha, D.W.: Active behavior recognition in beyond visual range combat. In: Proceedings of the Third Conference on Advances in Cognitive Systems, Cognitive Systems Foundation (2015)
11. Laviers, K., Sukthankar, G.: A real-time opponent modeling system for Rush Football. In: Proceedings of the Twenty-Second International Joint Conference on Artificial Intelligence, pp. 2476–2481. AAAI Press (2011)
12. Molineaux, M., Aha, D.W., Sukthankar, G.: Beating the defense: using plan recognition to inform learning agents. In: Proceedings of the Twenty-Second International Florida Artificial Intelligence Research Society Conference, pp. 337–343. AAAI Press (2009)
13. Levine, S.J., Williams, B.C.: Concurrent plan recognition and execution for human-robot teams. In: Twenty-Fourth International Conference on Automated Planning and Scheduling. ACM Press (2014)
14. Banerjee, B., Lyle, J., Kraemer, L.: The complexity of multi-agent plan recognition. Auton. Agent. Multi-Agent Syst. **29**(1), 40–72 (2015)
15. Zhuo, H.H., Li, L.: Multi-agent plan recognition with partial team traces and plan libraries. In: Proceedings of the Twenty-Second International Joint Conference on Artificial Intelligence, pp. 484–489. AAAI Press (2011)
16. Geib, C.W., Goldman, R.P.: Plan recognition in intrusion detection systems. In: Proceedings of the DARPA Information Survivability Conference, pp. 46–55. IEEE Press (2001)
17. Corchado, J.M., Pavón, J., Corchado, E., Castillo, L.F.: Development of CBR-BDI agents: a tourist guide application. In: Funk, P., González Calero, P.A. (eds.) ECCBR 2004. LNCS (LNAI), vol. 3155, pp. 547–559. Springer, Heidelberg (2004)
18. Ontañón, S., Mishra, K., Sugandh, N., Ram, A.: Case-based planning and execution for real-time strategy games. In: Weber, R.O., Richter, M.M. (eds.) ICCBR 2007. LNCS (LNAI), vol. 4626, pp. 164–178. Springer, Heidelberg (2007)
19. Rubin, J., Watson, I.: On combining decisions from multiple expert imitators for performance. In: Proceedings of the Twenty-Second International Joint Conference on Artificial Intelligence, pp. 344–349. AAAI Press (2011)
20. Floyd, M.W., Esfandiari, B., Lam, K.: A case-based reasoning approach to imitating RoboCup players. In: Proceedings of the Twenty-First International Florida Artificial Intelligence Research Society Conference, pp. 251–256. AAAI Press (2008)

21. Zarka, R., Cordier, A., Egyed-Zsigmond, E., Lamontagne, L., Mille, A.: Similarity measures to compare episodes in modeled traces. In: Delany, S.J., Ontañón, S. (eds.) ICCBR 2013. LNCS, vol. 7969, pp. 358–372. Springer, Heidelberg (2013)
22. Sánchez-Marré, M., Cortés, U., Martínez, M., Comas, J., Rodríguez-Roda, I.: An approach for temporal case-based reasoning: episode-based reasoning. In: Muñoz-Ávila, H., Ricci, F. (eds.) ICCBR 2005. LNCS (LNAI), vol. 3620, pp. 465–476. Springer, Heidelberg (2005)
23. Borrajo, D., Roubíčková, A., Serina, I.: Progress in case-based planning. ACM Comput. Surv. 47(2), 1–39 (2015)
24. Jensen, B., Karneeb, J., Borck, H., Aha, D.: Integrating AFSIM as an internal predictor. Technical report AIC-14-172, Naval Research Laboratory, Navy Center for Applied Research in Artificial Intelligence, Washington, DC (2014)

Adapting Sentiments with Context

Flávio Ceci[1(✉)], Rosina O. Weber[2], Alexandre L. Gonçalves[1],
and Roberto C.S. Pacheco[1]

[1] Federal University of Santa Catarina, Florianopolis, SC, Brazil
flavio.ceci@unisul.br, a.l.goncalves@ufsc.br,
pacheco@egc.ufsc.br
[2] Drexel University, Philadelphia, PA, USA
rosina@drexel.edu

1 Introduction and Background

Sentiment analysis is a valuable application of text classification because of the high volume of crowdsourced online content [1]. The typical texts targeted by sentiment analysis systems consist of opinions about an entity (e.g., individual, product, service).

The value of sentiment analysis goes beyond classifying the polarity of a document; it resides in providing the *aspects* of the entity being reviewed and the sentiment associated with these aspects, which is known as aspect-level sentiment analysis (e.g., [2, 3]). For example, the value of the analysis of the opinion "*The flash recovery time is ridiculously slow, but that I can live with, the 4 out of every 5 pictures that come out blurry I cannot*" is not that its overall sentiment is negative, but it is that *flash recovery time* and *blurry pictures* are both negative aspects. Consumers are interested in the sentiment associated to aspects of multiple opinions to make decisions about products, organizations or people [1, 3]. To be useful, systems should thus aggregate information contained in multiple opinions. Furthermore, the literature [1] tells us that users prefer structured visualizations that summarize opinions over textual summaries.

Figure 1 shows the application context where a user is interested in opinions about a product x. A filter $F(x)$ produces n opinions p_i on product x ($i = 1,.., n$) and submits each opinion p_i as new case c_i. Cases are pairs P, S where the problem P is an opinion p_i and a solution S is a triple $S(t_i, a_{ij}, C)$ where t_i are sentiment trees, a_{ij} are m ($j = 1,.., m$) polarized aspects, and C is the global polarization of the opinion. The case-based sentiment analysis module produces individual solutions that are aggregated into one resulting structured summary in response to the user's request. Figure 2 shows an example of a structured review produced from ten sentiment trees. In this paper, we limit our discussion to the case-based module.

Despite vast amounts of training data available for document-level analysis, training data for phrase- or aspect-level analysis is usually not available and needs to be manually annotated (e.g., [4]). Document-level training data comes from the web (e.g., Epinions.com, Amazon.com) where customers write reviews and select a score in a five-star scale. The requirement to analyze individual aspects and the lack of training data at the aspect-level are both limitations of supervised methods.

© Springer International Publishing Switzerland 2015
E. Hüllermeier and M. Minor (Eds.): ICCBR 2015, LNAI 9343, pp. 44–59, 2015.
DOI: 10.1007/978-3-319-24586-7_4

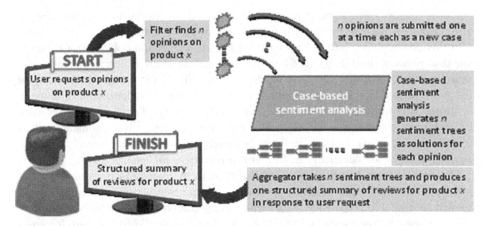

Fig. 1. User request is answered with a structured summary of reviews for product *x*.

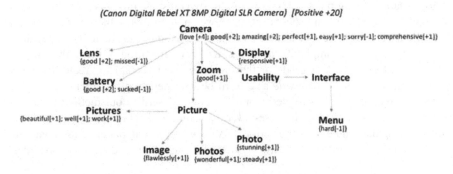

Fig. 2. Structured summary from ten reviews for Canon Digital Rebel XT.

Unsupervised methods typically use phrases or terms that bear sentiment. The usual approach to determine document-level polarity is to compare positive versus negative sentiment bearing excerpts, phrases, or words and assign the polarity of the document based on the majority or a threshold. There are variations of how to identify these excerpts or phrases but they commonly rely on lexicons of sentiment bearing (or opinion) words [2, 5], such as adjectives (e.g., great) and adverbs (e.g., effectively). A sentiment lexicon (we use the one from [5] in this paper) and an entity-specific sentiment ontology [6] provide all necessary elements to create a visual summary of the aspects of opinions.

Wilson et al. [4] refer to the polarity of a term in a lexicon as its *prior polarity*, suggesting that it may vary with the context and thus a *contextual polarity* may need to be verified. Contextual polarity may be domain-dependent. For example, the word *unpredictable* can be used positively in the domain of movie reviews but negatively to describe car handling [1]. It may also be domain-independent, such as the expression *perfect mess* that replaces the polarity of the positive *perfect* with negative [7]; this form uses context from adjacent words and creates a contextual lexicon that permanently defines terms like *perfect mess* as bearing negative sentiment.

In this paper, we explore an impermanent form of contextual polarity that is purely context-dependent, and thus needs to be assessed in each different context (i.e., opinion), and does not permanently change the polarity of the terms. The polarity of these words depends solely on the author's discourse and are independent of domain. We target two forms of impermanent contextual polarity. The first form includes regular expressions that may be used with either positive or negative sentiment, such as the adjective *expensive* that can be used positively as in *without the need to buy an expensive camera*, or negatively as in *it is too expensive for what it gives*.

The second form targets expressions referred to as *thwarted-expectations*, a linguistic pattern where authors use terms with sentiment that is opposite to the overall intended sentiment to prepare the reader with an expectation that is suddenly contrasted, creating emphasis. See example from [1] p. 13, *"This film should be brilliant. It sounds like a great plot, the actors are first grade, and the supporting cast is good as well, and Stallone is attempting to deliver a good performance. However, it can't hold up."*

Our proposed treatment to contextual polarity for thwarted-expectations is to revert the prior polarity of sentiment bearing words for two reasons. First because this leads to the correct interpretation, and second because we contend that authors never intend to include the aspects of the expectation for any other reason but to create the expectation. In other words, in the example above, it was never the intention of the author to point out that the actors were first grade or the plot was great, except to create the expectation. We thus interpret *first grade* as a term bearing negative sentiment.

The next section presents related work. Section 3 describes our method to analyze aspect-level sentiments within the case-based reasoning (CBR) methodology including the use of singular value decomposition (SVD) to represent patterns for contextual polarity. Section 4 evaluates it. Section 5 discusses negation, and Sect. 6 concludes.

2   Related Work

The main resource used in sentiment analysis is a sentiment lexicon [2, 5]. Lexicons are usually domain-independent and can be expanded to fit a target domain, which can be done automatically by searching for domain-specific confirmation of polarity via linguistic constructs on a corpus [8]. For example, assuming that the word *elegant* bears positive sentiment, and we do not know the sentiment of the word *light* in a domain, a phrase describing an aspect as *light and elegant* suggests *light* in this domain is positive, whereas a phrase *light but elegant* would suggest *light* is negative. Lexicons can also be expanded from a set of domain-specific seed words using sources like WordNet [1].

Ontologies are used in aspect-level sentiment analysis (e.g., [9]) as sources of associations between aspects and target products. They retain information such as that a part of a camera is the lens and a case is an accessory. They sometimes need to be learned or expanded from data (e.g., [6]) or sources like WordNet (e.g., [9]).

The work in this paper uses opinions from two domains (Fig. 3), cameras and movies (more in Sect. 4.1). The ontology for cameras is expanded from the ontology in [6]. The ontology for the domain of movies is based on movieontology.org [21].

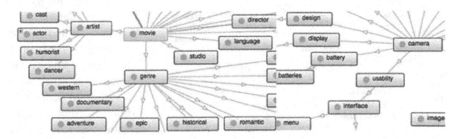

Fig. 3. Partial view of ontologies for the camera and movie domains used in this paper.

Contextual polarity modifies prior polarities recorded in sentiment lexicons based on context. This may require manual annotation of single phrases [4, 10]. Like in this paper, [7] uses document-level polarization for learning contextual polarities.

Singular value decomposition (SVD) [11] represents a semantic space revealing strongly associated terms based on their occurrence in a corpus. SVD has been used before for cross-domain sentiment classification [3, 12] but not for contextual polarity.

Supervised methods such as Naïve Bayes and support vector machines (SVM) are considered top performers in topical text classification [13]. These methods do not perform as accurately in sentiment classifications because of its subjectivity, but they are still among the most accurate [3] together with variations of deep learning [14]. For this reason, we use Naïve Bayes and SVM as references of accuracy and use unsupervised methods that enable the generation of a structured summary [3].

Most recent work in CBR and sentiment analysis typically uses opinions to produce recommendations of the reviewed products [15, 16, 18, 19]. Authors in [17] use information from reviews to help users write comprehensive reviews. More recently, [15] explored other sources of data to improve the quality of their resulting recommendations. Chen et al. [18] also explore aspect-level analysis for recommendation, ranking aspects and user preferences. CBR was proposed for cross-domain sentiment analysis [20] where each case includes resources for analyzing sentiment in each domain.

3   Case-Based Sentiment Analysis

Case-based sentiment analysis is an unsupervised algorithm to generate sentiment trees for textual opinions. We incorporate sentiment analysis within the CBR methodology so previous solutions, when available, are reused, adapted, and learned. Cases are pairs (P, S) where the problem P is an opinion p_i and a solution S is a triple S (t_i, a_{ij}, C) composed of a sentiment tree t_i, a_{ij} polarized aspects, and a global

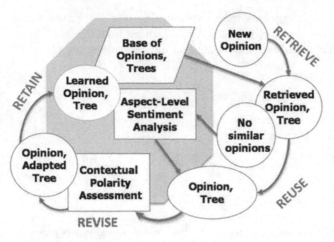

Fig. 4. Case-based sentiment analysis (solution is abbreviated as tree).

polarization C. Figure 4 depicts how the sentiment analysis module is incorporated into the CBR methodology. For a new opinion (i.e., query), if there is an existing solved problem (i.e., opinion) with a similarity score above a threshold θ, its solution (i.e., sentiment tree with polarized aspects and global polarization) is reused; when no similar cases are available, a new sentiment tree is produced through aspect-level sentiment analysis. Solved cases are adapted for contextual polarity (i.e., that may revert the polarity of some aspects, adapting the tree and its polarity) and then retained for further reuse.

3.1 Text Processing

New cases are textual opinions such as, *"This case will protect your camera but has no place for spare battery or even a extra memory card. It only has a spring clip to fasten it to a belt or bag there is no belt loop or other handle strap. There are other cases available with better features for less money."* This text is processed for (1) to generate a new tree (Sect. 3.2) and (2) to reuse a tree (Sect. 3.3). For both purposes, the text is subject to the steps of tokenization, part-of-speech (POS) tagging, selection of aspects, and of sentiment words, in this order of precedence, as they all use the results from the immediately preceding step. Tokenization and POS tagging produce the tokens in Fig. 5 for the last sentence of the example above.

```
there (Existential there)
are (Verb, non-3rd person singular present)
other (Adjective)
cases (Noun, plural)
available (Adjective)
with (Preposition or subordinating conjunction)
better (Adjective, comparative)
features (Noun, plural)
for (Preposition or subordinating conjunction)
less (Adjective, comparative)
money (Noun, singular or mass)
```

Fig. 5. Last sentence of the example opinion after tokenization and POS tagging.

Now we describe the selection of aspects and of sentiment words. Aspects are mentioned in opinions as nouns. To identify them, we use target-specific ontologies (Sect. 2) and select the noun tokens that occur in the ontologies as aspects of each product (e.g., an aspect of a camera is lens). Thus, this step recognizes the token *cases* as a concept because *cases* is in the ontology as an accessory of camera.

To identify sentiment words, the input consists of the vector's terms that occur in the chosen sentiment lexicon [5]. The polarization of the sentiment words is transferred to the aspects by association. Aspects are associated to sentiment words by proximity in the sentence. In the example above, the concept *cases* is associated with sentiment words *better* and *available*, both having positive sentiment. If an aspect has no sentiment word in close proximity then it receives no polarization and is removed from the vector. Only aspects that have a sentiment word associated remain in the vector. The resulting vector from the example above consists of the tokens: *other, cases, bag, extra, battery, spare, camera, features, only, even, available*. These tokens and their respective labels are the input to generate new trees and queries for similarity assessment.

3.2   Generating a New Tree

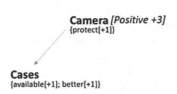

Fig. 6. New tree (e.g., solution) generated from the tokens for example opinion from Fig. 5.

When opinions are initially acquired and there are no existing cases to reuse, a new sentiment tree is generated from scratch to each opinion. Once a set of cases is acquired, new sentiment trees are generated.

The generation of a new tree uses the labeled tokens as exemplified above that result from the processing described in the previous section. The new tree uses the associations between the aspects and target object. The aspects receive a polarization based on the sum of polarities of the sentiment words associated with them, where positive polarity is +1 and negative is -1. The global polarization of the target object is given by the simple majority of the sum of polarities of sentiment words. Figure 6 shows the resulting tree for the example above with global polarity *positive*, which does not coincide with the original polarity from the author's score.

3.3   Similarity Assessment, Retrieval, and Reuse

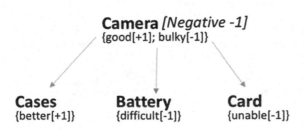

Fig. 7. Reused solution from most similar opinion.

The queries (i.e., case problems) are the tokens resulting from the processing described in Sect. 3.1; they are represented in inverse document frequency vectors without labels. The similarity is computed by the cosine between the vectors.

A candidate case is retrieved if there is one with similarity score equal or higher than a threshold θ; the highest scoring case is retrieved when more than one satisfies the criterion.

The value of the similarity threshold is computed for each set of opinions. The threshold θ is defined as the average of the similarity scores obtained by the most similar case when submitting all opinions by removing one at a time in a leave-one-out fashion. The values used in this paper are 0.43 for cameras and 0.18 for movies.

When a similar case is available, its complete solution is reused, that is, its tree with its polarized aspects and global polarization. The most similar case to the opinion of the example above has a similarity score of 0.47. This is above the threshold of 0.43 for the domain of cameras, and thus its solution is reused. Figure 7 shows the reused solution that has a *negative* global polarization, which does coincide with the original polarity from the author's score. This example illustrates how the reuse of a similar solution can improve overall accuracy (see Sect. 4.4, H2, Table 4) despite introducing noise.

3.4   Contextual Polarity Assessment

To assess contextual polarity, first the patterns of the data are learned to capture the context of sentiment words in positive and negative opinions. Our approach assumes that words with one prior polarity, say negative, are expected to occur in negative opinions, that is, in opinions whose global polarity is the same as the prior polarity of the word. We refer to the sentiment words that occur in opinions with both global polarities as *ambivalent*; these sentiment words are candidates for contextual polarity.

Four term-document matrices are created, two for each domain, one for positive and one for negative opinions. Ambivalent words are in rows and sentiment words in columns. The value associating a sentiment and an ambivalent word describes the number of times the sentiment term occurs in the context of an ambivalent word. This context is given by a window of five terms to the right and five terms to the left from a vector of all and only the sentiment words occurring in an opinion. The SVD process generates, respectively, two vectors for each ambivalent word, one for positive and one for negative opinions. These vectors represent the space where an ambivalent word occurs in each context. To determine the contextual polarity of an ambivalent word, we compare the context of sentiment words where the ambivalent word occurs and compare it to the positive and negative context vectors. The ambivalent word is assigned the polarity from the vector with the majority of overlapping words from the comparison.

Fig. 8. Sentiment tree before and after adaptation via contextual polarity assessment.

Figure 8 shows the example from the opinion, *"Sure my canon sd-30 is super cute and I get lots of compliments on how small and sexy it is, but heaven forbid you want to take pictures with it! This camera cannot take pictures in low light at all! Every picture I've taken at parties has been out of focus, which adds up to a lot of hazy photos of great memories. I took it back to the camera shop and the guy said I had the setting right but "yeah, they don't work well in low light" Well if I needed a bright sunny day for every shot I could make a pinhole camera out of a cardboard box! The pictures it takes in daylight are nice. But anyone who wants a tiny little camera like this is planning on putting it in their pocket and taking it to parties! I got the Sony Cybershot 10.1 and it takes GREAT pictures in low light. Later, Canon."* The words of this opinion after processing (Sect. 3.1) are in Fig. 8, where the right-side shows, in red, the terms that change polarity. These changes resulted from the comparison between the vectors as described above. For example, all the words in both vectors (1) and (2) for the word *cute* (below), produced from SVD, overlap with words of the context where *cute* occurs.

The negative vector has more overlapping words and thus the final polarity is negative.

$$\overrightarrow{Cute}_{\text{Positive vector}} = \{\text{like, work, great, nice, well}\}. \tag{1}$$

$$\overrightarrow{Cute}_{\text{Negative vector}} = \{\text{like, work, bright, right, great, nice, heaven, super, well}\}. \tag{2}$$

This result is expected because, as stated earlier, we contend that the author of this opinion did not intend to mention the terms *cute* and *sexy* positively but merely to create a thwarted-expectation.

4   Experimental Design

Although our proposed methods improve aspect-level sentiment analysis, we demonstrate our results using document-level accuracy because improvements in aspect-level analysis ultimately impacts accuracy of documents.

4.1   Data

The proposed method and its contributions are not intended to be domain-independent, so we use two domains given that accuracy varies depending on domain [1]. We use data from the domains of cameras and movies, which are sufficiently distinct to demonstrate the difficulty of sentiment analysis in different domains. The domain of cameras is easier than the domain of movies. Opinions on cameras are objective [3] because cameras have the tangible purpose of taking pictures, mentioning facts about aspects such as lenses and picture quality. Opinions on movies, contrarily, include subjective expressions from emotions [3], given it is an entertainment product. The latter poses particular challenge to prior polarity of sentiment words [7].

The opinions from both domains come from the Multi-Domain Sentiment Dataset [12]. There are 1443 camera opinions, 726 positive and 717 negative. The 1991 opinions about movies refer to DVD movies, 996 positive and 995 negative. Positive opinions are the ones original authors assigned a 4 or 5 in a five-star scale. Analogously, negative opinions received a score of 1 or 2 stars. We present the results for each subset of positive and negative opinions also separately. This facilitates our analyses because methods impact each subset differently, as it will be shown.

4.2   Evaluation and Metrics

We compute average accuracy via leave-one-out cross validation (LOOCV) (e.g., [22]). Statistical significance is tested with McNemar's [23] with $\rho = 0.05$. Significant results are shown in bold. TP, TN, FP, and FN stand for true and false positives and negatives; average accuracy for negative and positive opinions defined as:

$$Accuracy\ Negative = \frac{TN}{TN + FP} \quad Accuracy\ Positive = \frac{TP}{TP + FN}$$

4.3   Methods for Comparison

We compare the case-based approach (Sect. 3) against two methods. We define as baseline the results from the processing and generation of a new tree reuse or adaptation. Our goal is to improve aspect-level sentiment analysis from these results. Table 1 presents the performance of the baseline on the two domains of cameras and movies.

Table 1. Average accuracy of baseline for domains of cameras and movies.

	Positive opinions	Negative opinions	All
Camera	0.915	0.537	0.727
Movies	0.807	0.670	0.739

As references of accuracy we implement Naïve Bayes and SVM using LightSide (http://lightsidelabs.com). We used as input the same texts used in the processing (Sect. 3.1) from the two data sets (Sect. 4.1), and the polarities from authors' scores. We used the default settings that include unigrams, and a rare threshold of five (i.e., minimum required occurrences of a term in the corpus). The tool does not provide the representation or code of the classifier, just its functionality to produce results, which are summarized in Table 2.

Performance of the baseline and supervised methods shown in Tables 1 and 2 demonstrate the importance of analyzing results separately for positive and negative opinions, and for each domain. Aggregating the results may conceal important weaknesses of methods under study. Next we present proposed studies.

Table 2. Average accuracy of Naïve Bayes and SVM for domains of cameras and movies.

Domain	Method	Positive opinions	Negative opinions	All
Camera	Naïve Bayes	0.854	0.837	0.845
	SVM	0.820	0.872	0.845
Movies	Naïve Bayes	0.825	0.748	0.787
	SVM	0.788	0.783	0.786

4.4   Studies

We demonstrate four hypotheses, comparing the case-based approach against the baseline and supervised methods. Two hypotheses for positive and two for negative sets.

H1. Positive opinions polarized with CBR will produce statistically significant higher average accuracy than polarization obtained by the baseline for both domains.

H1 is disproved because CBR produced higher average accuracy with statistical significance only in the domain of movies (Table 3). The difference produced in the domain of cameras is negligible. In the domain of movies, CBR shows an increase of 7 % in true positives. Detailed analysis reveals that this was caused mostly by adaptation, where 90 cases had their global polarity swapped moving them into the class of true positives.

Table 3. Average accuracy for baseline and CBR for positive opinions for both domains.

Method	Positive opinions	
	Camera	Movies
Baseline	0.915	0.807
CBR	0.931	**0.864**

The statistically significant increase in average accuracy and the observation that it occurs during adaptation suggests the benefit of our approach for contextual polarity. It is reasonable that contextual polarity produces more significant impact in the domain of movies given its more subjective nature.

H2. Negative opinions polarized with CBR will produce statistically significant higher average accuracy than polarization obtained by the baseline for both domains.

H2 is demonstrated in both domains (Table 4). In the domain of cameras, CBR increases the number of correctly polarized opinions in 25 %, with 95 cases receiving correct polarization. Detailed analysis reveals that this improvement was not caused by adaptation, but from reuse. The example in Sect. 3 illustrates this situation where a correct polarization has a potentially problematic side effect of producing structured summaries with artificial aspects that did not belong in the original opinion.

In the domain of movies, the 14 % increase in correctly polarized opinions are caused mostly by adaptation via contextual polarity. This confirms the result in positive opinions for the domain of movies, where contextual polarity is expected to impact given the subjectivity of this domain.

Table 4. Average accuracy for baseline and CBR for negative opinions for both domains.

Method	Negative opinions	
	Camera	Movies
Baseline	0.537	0.670
CBR	**0.669**	**0.767**

H3. Positive opinions polarized with CBR will produce statistically significant higher average accuracy than polarization obtained by Naïve Bayes and SVM for both domains.

H3 is demonstrated in both domains (Table 5). All results are statistically significant. For the domain of cameras, CBR produce respectively 9 % and 14 % more true positives than Naïve Bayes and SVM, both results are statistically significant. In the domain of movies, differences are small but significant. These results meet our goal of reaching the levels of accuracy of supervised methods while keeping the ability to build a structured summary with the polarity of aspects.

Table 5. Average accuracy of Naïve Bayes and SVM compared to CBR for positive opinions.

Method	Positive opinions	
	Camera	Movies
Naïve Bayes	0.854	0.825
SVM	0.820	0.788
CBR	**0.931**	**0.864**

H4. Negative opinions polarized with CBR will produce statistically significant higher average accuracy than polarization obtained by Naïve Bayes and SVM for both domains.

In the domain of cameras, CBR is inferior to both supervised methods at the 20 % level, with statistical significance. In the domain of movies, there are no statistically significant differences (Table 6).

Table 6. Average accuracy for Naïve Bayes and SVM compared to CBR for negative opinions.

Method	Negative opinions	
	Camera	Movies
Naïve Bayes	**0.837**	0.748
SVM	**0.872**	0.783
CBR	0.669	0.767

4.5   Discussion

The studies show that our goal to improve accuracy of a sentiment analysis approach that can produce the structured summary with polarities of individual aspects exemplified in Fig. 1 was attained. We demonstrate statistically significant increments in accuracy at the document-level through aspect-level analysis. These improvements in accuracy originate from reuse and from adaptation through contextual polarity.

The total number of textual opinions that moved from the false into the positive categories over the baseline were 376. The total number of opinions whose classification became correct through reuse was 142, 38 %; through adaptation was 234 opinions, 62 %. When only looking at the reuse step, 80 % of improvements occurred in the domain of cameras against 20 % in the domain of movies. The improvements from adaptation were 86 % in the domain of movies and 14 % in the domain of cameras.

Solution reuse, as implemented (i.e., roughly $^{1}/_{3}$ of the improvements), may change aspects of an opinion introducing a form of noise in the final structured summary, which we do not currently evaluate. Further studies are needed to determine how worthwhile this is. This is important because it is in the core of the value of using CBR. An alternative would be to create a tree that adds aspects to the existing tree instead of replacing it. This is aligned with the intuition that authors may not express all aspects that are relevant about a product. This is a limitation of online reviews also explored by [15].

Improvements originating from adaptation via contextual polarity do not impose any limitations. The fact that contextual polarity has produced more improvements in the domain of movies confirms its suitability to subjective domains where polarity can be context-dependent. Adaptation through contextual polarity is currently implemented as a step that is divorced from reuse, bringing up the question of whether it needs to be incorporated into CBR. In the form it is implemented now, the main advantage of implementing it within CBR is that contextual polarities can be learned and this may benefit future cases, although this was not evaluated.

When compared to Naïve Bayes and SVM, CBR was more accurate in positive opinions in both domains. The reasons for this performance are likely to originate from the focus on aspects. It is important to note that the case-based approach can be implemented in an unsupervised fashion, but the results shown in this paper use supervised methods both when defining the threshold for similarity assessment and when learning patterns for contextual polarity.

The inferiority of CBR in negative opinions in the domain of cameras may be partly explained by the fact that it does not treat negation whereas the supervised methods indirectly do. We will return to the discussion of negation in the next section.

5   Negation

The studies in Sect. 4 demonstrate contributions to sentiment analysis via CBR. Our ultimate goal is a method that can produce a structured aspect-level summary (Fig. 1) with accuracy comparable to supervised methods. For this reason, we used the same data as input to the supervised methods (i.e., SVM and Naïve Bayes) that we used in the case-based method. In order to use exactly the same data, we chose not to treat negation in our CBR approach. Neither did we expand the ontology or lexicon used.

Negation is consensually described as an important issue to address in sentiment analysis (e.g., [1, 3, 4]). We did not include it because these methods are typically *ad hoc* and could be misconstrued as a change in the data leading to biased results.

We believe that lack of treatment for negation can explain the difficulty of the case-based approach to produce better accuracy in the subsets of negative opinions. This is an assumption that relies on the notion that negative opinions may include more negative expressions, that is, expressions that include polarity shifters through negation such as *It wasn't really funny*, or *This product doesn't add any benefit*.

Next, we show the results we obtain when we treat negation. We identify terms that we consider as sentiment shifters such as *don't* and *isn't*. Then, we search for sentiment words in a window of four words in each direction from the shifter. If a sentiment word is found, then we change its polarity from negative to positive and vice versa.

When we compare performance without and with treating negation, results in positive subsets are negligible; we will not thus show results for positive subsets. The following Tables 7 and 8 show comparisons of results without treating versus treating negation for subsets of negative opinions.

Table 7. Average accuracy for the domain of cameras without and with negation.

Cameras	Negative opinions	
Method	Without negation	With negation
Baseline	0.537	0.671
CBR	**0.669**	**0.748**

In the domain of cameras, CBR is more accurate when negation is treated than when it is not treated to polarize negative opinions with statistical significance (Table 7). The average accuracy increases about 11 %, producing 56 new correct polarizations.

In the domain of movies, when negation is treated, CBR produces a small but statistically significant increment in average accuracy (Table 8) than when negation is not treated. The average accuracy increases about 3 %, producing 23 new correct polarizations.

Table 8. Average accuracy for the domain of movies without and with negation.

Movies	Negative opinions	
Method	Without negation	With negation
Baseline	0.670	0.696
CBR	**0.767**	**0.790**

It is important to note that these results shall be interpreted cautiously in comparison to supervised methods because this treatment of negation is not implemented when adopting Naïve Bayes or SVM. This is because we treat negation after identifying sentiment bearing words whose polarity is swapped. These supervised methods do not allow the same opportunity.

6 Conclusions, Limitations, and Future Work

In this paper, we described a case-based sentiment analysis approach that targets aspect-level methods in the CBR methodology. The main contribution of this paper is the approach to assess contextual polarity in the adaptation step of the CBR methodology. The results show statistically significant improvements in document-level accuracy obtained by enhancements in aspect-level sentiment analysis. Data and other resources used in this paper are available to be reproduced at https://dl.dropboxusercontent.com/u/3025380/iccbr2015/site/index.html.

Our approach to contextual polarity differs from previous works (e.g., [4]) in that we do not need to manually annotate phrases; we rely on the author-generated scores. It is also different from [7] because our approach is impermanent in that it allows one expression to change polarity in each different context whereas theirs builds a permanent lexicon for a contextual 1-word window.

As an implementation of the CBR methodology, this work has not yet explored the potential of various similarity measures. Our studies have shown that the reuse of solutions is sometimes more accurate than producing new solutions from scratch, but this may introduce noise in the final structured summary that is the ultimate application, as shown in Fig. 1.

Within the sentiment analysis application, our work has room for improvements in expanding ontologies and sentiment lexicons, in using other POS from the opinions such as verbs, in including compound expressions. Our example in Sect. 3, for example, fails to recognize *heaven forbid* as one expression. The implemented approach is not exploring sub-aspects of batteries such as battery weight and life. We do not incorporate, for example, a way to derive aspects from adjectives such as the occurrence of the adjective *expensive* leading to the creation of the aspect *price* [3]. We do not address comparative opinions [3] or sarcasm [2, 3], nor do we analyze the potential of our proposed SVD method to address those forms.

Although our work is not supervised, it does contain two elements that rely on some form of training. One is the contextual polarity approach that relies on the polarization of opinions. The other is the definition of the retrieval threshold.

In future work, within CBR, we will explore other measures of similarity assessment, and the implementation of contextual polarity within reuse. We will also implement negation so it does not interfere with contextual polarity.

We evaluated our contributions using author-generated document-level scores even if our contributions targeted aspect-level sentiment. We shall next evaluate our work with excerpts that have been annotated for aspects and sentiments. In terms of the data, we only used two domains. We shall test generalizability of our proposed contributions for more domains. We also plan to better cleanse the data to remove duplicates. Finally, given that our approach to contextual polarity relies on assessing similarity, we shall investigate the benefits of improving its quality using contexts as cases.

References

1. Pang, B., Lee, L.: Opinion mining and sentiment analysis. Found. Trends Inf. Retrieval **2**, 1–135 (2008)
2. Feldman, R.: Techniques and applications for sentiment analysis. Commun. ACM **56**(4), 82–89 (2013)
3. Liu, B.: Sentiment analysis and opinion mining. Synth. Lect. Hum. Lang. Technol. **5**(1), 1–167 (2012)
4. Wilson, T., Wiebe, J., Hoffmann, P.: Recognizing contextual polarity: an exploration of features for phrase-level sentiment analysis. Comput. Linguis. **35**(3), 399–433 (2009)
5. Hu, M., Liu, B.: Mining and summarizing customer reviews. In: Gehrke, J., DuMouchel, W. (eds.) Proceedings of the Tenth ACM SIGKDD International Conference on Knowledge Discovery and Data Mining, pp. 168–177. ACM, New York (2004)
6. Wei, W., Gulla, J.A.: Sentiment learning on product reviews via sentiment ontology tree. In: Hajic, J. (ed.) Proceedings of the 48th Annual Meeting of the ACL, pp. 404–413. Association for Computational Linguistics, Stroudsburg (2010)
7. Weichselbraun, A., Gindl, S., Scharl, A.: Extracting and grounding context-aware sentiment lexicons. IEEE Intell. Syst. **28**(2), 39–46 (2013)
8. Hatzivassiloglou, V., McKeown, K.: Predicting the semantic orientation of adjectives. In: Cohen, P.R., Wahlster, W. (eds.) Proceedings of the 35th Annual Meeting of the ACL and Eighth Conference of the European Chapter of the ACL, pp. 174–181. Association for Computational Linguistics, Stroudsburg (1997)
9. Carenini, G., Ng, R., Pauls, A.: Multi-document summarization of evaluative text. In: McCarthy, D., Wintner, S. (eds.) Proceedings of the European Chapter of the Association for Computational Linguistics, pp. 305–312. Association for Computational Linguistics, Stroudsburg (2006)
10. Ghazi, D., Inkpen, D., Szpakowicz, S.: Prior and contextual emotion of words in sentential context. Comput. Speech Lang. **28**(1), 76–92 (2014)
11. Deerwester, S., Dumais, S.T., Furnas, G.W., Landauer, T.K., Harshman, R.: Indexing by latent semantic analysis. J. Am. Soc. Inf. Sci. **41**(6), 391–407 (1998)
12. Blitzer, J., Dredze, M., Pereira, F.: Biographies, bollywood, boom-boxes and blenders: domain adaptation for sentiment classification. In: Bosch, A., Zaenen, A. (eds.) Proceedings of the 45th Annual Meeting of the Association of Computational Linguistics, pp. 440–447. Association for Computational Linguistics, Stroudsburg (2007)
13. Pang, B., Lee, L., Vaithyanathan, S.: Thumbs up? Sentiment classification using machine learning techniques. In: Hajic, J., Matsumoto, Y. (eds.) Proceedings of the ACL-02 Conference on Empirical Methods in Natural Language Processing (EMNLP 2002), vol. 10, pp. 79–86. Association for Computational Linguistics, Stroudsburg, PA (2002)
14. Bespalov, D., Qi, Y., Bai, B., Shokoufandeh, A.: Sentiment classification with supervised sequence embedding. In: Flach, P.A., De Bie, T., Cristianini, N. (eds.) ECML PKDD 2012, Part I. LNCS, vol. 7523, pp. 159–174. Springer, Heidelberg (2012)
15. Dong, R., O'Mahony, M.P., Smyth, B.: Further experiments in opinionated product recommendation. In: Lamontagne, L., Plaza, E. (eds.) ICCBR 2014. LNCS, vol. 8765, pp. 110–124. Springer, Heidelberg (2014)
16. Dong, R., Schaal, M., O'Mahony, M.P., McCarthy, K., Smyth, B.: Opinionated product recommendation. In: Delany, S.J., Ontañón, S. (eds.) ICCBR 2013. LNCS, vol. 7969, pp. 44–58. Springer, Heidelberg (2013)

17. Dong, R., Schaal, M., O'Mahony, M.P., McCarthy, K., Smyth, B.: Mining features and sentiment from review experiences. In: Delany, S.J., Ontañón, S. (eds.) ICCBR 2013. LNCS, vol. 7969, pp. 59–73. Springer, Heidelberg (2013)
18. Chen, Y.Y., Ferrer, X., Wiratunga, N., Plaza, E.: Sentiment and preference guided social recommendation. In: Lamontagne, L., Plaza, E. (eds.) ICCBR 2014. LNCS, vol. 8765, pp. 79–94. Springer, Heidelberg (2014)
19. Vasudevan, S.R., Chakraborti, S.: Enriching case descriptions using trails in conversational recommenders. In: Lamontagne, L., Plaza, E. (eds.) ICCBR 2014. LNCS, vol. 8765, pp. 480–494. Springer, Heidelberg (2014)
20. Ohana, B., Delany, S.J., Tierney, B.: A case-based approach to cross domain sentiment classification. In: Agudo, B.D., Watson, I. (eds.) ICCBR 2012. LNCS, vol. 7466, pp. 284–296. Springer, Heidelberg (2012)
21. Bouza, A.: MO – the Movie Ontology. http://movieontology.org
22. Devroye, L., Wagner, T.J.: Distribution-free inequalities for the deleted and hold-out error estimates. IEEE Trans. Inf. Theor. **25**(2), 202–207 (1979)
23. McNemar, Q.: Note on the sampling error of the difference between correlated pro-portions or percentages. Psychometrika **12**(2), 153–157 (1947)

Aspect Selection for Social Recommender Systems

Yoke Yie Chen[1]([✉]), Xavier Ferrer[2,3], Nirmalie Wiratunga[1], and Enric Plaza[2]

[1] IDEAS Research Institute, Robert Gordon University, Aberdeen, Scotland
{y.y.chen,n.wiratunga}@rgu.ac.uk
[2] Artificial Intelligence Research Institute (IIIA-CSIC),
Spanish National Research Council (CSIC),
Campus UAB, Bellaterra, Catalonia, Spain
[3] Universitat Autònoma de Barcelona, Bellaterra, Catalonia, Spain
{xferrer,enric}@iiia.csic.es

Abstract. In this paper, we extend our previous work on social recommender systems to harness knowledge from product reviews. By mining product reviews, we can exploit sentiment-rich content to ascertain user opinion expressed over product aspects. Aspect aware sentiment analysis provides a more structured approach to product comparison. However, aspects extracted using NLP-based techniques remain too large and lead to poor quality product comparison metrics. To overcome this problem, we explore the utility of feature selection heuristics based on frequency counts and Information Gain (IG) to rank and select the most useful aspects. Here an interesting contribution is the use of top ranked products from Amazon to formulate a binary classification over products to form the basis for the supervised IG metric. Experimental results on three related product families (Compact Cameras, DSLR Cameras and Point & Shoot Cameras) extracted from Amazon.com demonstrate the effectiveness of incorporating feature selection techniques for aspect selection in recommendation task.

Keywords: Social recommenders · Online reviews · Feature selection

1 Introduction

Recommender systems provide a ranked list of products to assist user purchase needs. With content-based systems, products similar to those that have been liked by the user are ranked higher [7]. Central to this is the ability to establish similarity between the target 'liked' product and the rest. How to best represent products to achieve effective product comparison is an area of interest to Case-Based Reasoning (CBR) in the context of recommender systems [12]. Increasing effort is being focused on incorporating knowledge from product reviews into product representation. In particular, the rich information embedded in product reviews permits recommender systems to learn implicit preferences of

© Springer International Publishing Switzerland 2015
E. Hüllermeier and M. Minor (Eds.): ICCBR 2015, LNAI 9343, pp. 60–72, 2015.
DOI: 10.1007/978-3-319-24586-7_5

users by considering product aspects (also called features) mentioned in product reviews [1].

Our previous work proposed a social recommender system using two social media knowledge sources: online product reviews and purchase preferences. As a result, recommendation was improved by the combination of aspect based sentiment analysis with preference knowledge [2]. More importantly, we showed that recommendations generated based on aspect-based sentiment analysis to be far superior to one that is agnostic of aspects. However, most NLP-based aspect extraction techniques rely on POS tagging and syntactic parsing which are known to be less robust when applied to informal text [10]. As a result, it is not unusual to have a large numbers of spurious content to be extracted incorrectly as aspects. Methods to infer aspect importance and thereafter rank them for selection are needed to achieve a manageable aspect subset size.

Feature selection is known to enhance accuracy in supervised learning tasks such as text classification by identifying redundant and irrelevant features [15]. In this paper, we address the problem of selecting important aspects using feature selection heuristics. Specifically, we explore two feature selection approaches to evaluate aspect usefulness: Information Gain (IG) and aspect frequency. In our solution, we capitalise on top ranked products from Amazon to formulate a binary classification over products to form the basis for the supervised IG metric. In addition, we investigate the transferability of selected aspects from a particular product family (e.g. Compact Cameras) to other related product families (e.g. DSLR Cameras and Point & Shoot Cameras).

The rest of the paper is organised as follows: In Sect. 2 we present related research. Next we describe the process of aspect extraction and feature selection heuristics in Sect. 3. Finally, evaluation results are presented in Sect. 4 followed by conclusions in Sect. 5.

2 Related Works

Recent work in social recommender systems utilise sentiment analysis as key features for product representation. An interesting idea here is to compare products not simply on the basis of sentiment polarity (i.e. positive or negative sentiment scores) but on the basis of similar sentiment over product aspects. This then requires aspects to be extracted from the product reviews before they can be associated with polarity scores [4,6]. Fundamental to this comparison is the relevance of the product aspects extracted from online reviews.

Frequency of aspects is commonly used as a heuristic to select genuine aspects from product reviews [4,5]. This frequency score can further be combined with sentiment scores to bias these rankings when the task involves opinionated content [16]. Similarly, frequency can also be combined with similarity knowledge whereby aspects that contribute most to product similarity computations are considered more relevant than those that do not [11].

Unlike frequency-based heuristics, supervised selection heuristics have been successfully employed to reduce dimensionality and achieve significant gains in accuracy for text classification [14]. In this paper we explore how the supervised Information Gain (IG) heuristic can be adopted in the context of social recommenders to reduce the dimensionality of product aspects. Whilst Vargas-Govea et al. [13] have also used a supervised selection method in the context of semantic based restaurant recommender systems, they did so to identify influential contextual features using user rating values as the class label. Unlike with typical classification tasks where class labels are explicitly defined, in our work the notion of class and its boundaries need to be considered carefully to enable the application of IG for aspect selection.

3   Review Based Product Recommendation

Central to a social recommender system is the source of opinionated content in the form of product reviews. As depicted in Fig. 1, this source can be harnessed to generate a product ranking using the following three steps:

1. Extract product aspects from reviews and quantify the strength of sentiment over these aspects within the range of [-1,1];
2. Select aspects according to a selection heuristic; and
3. Generate recommendations using evidence from sentiment based strategies.

3.1   Aspect Extraction from Product Reviews

Grammatical extraction rules [8] are used to identify a set of candidate aspect phrases from sentences. These rules operate on dependency relations in parsed

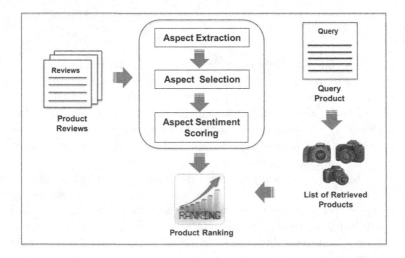

Fig. 1. Overview of social recommender process.

DP = set of dependency pattern rules
{

   $dp_1 : amod(N, A) \rightarrow \langle N, A \rangle,$
   $dp_2 : acomp(V, A) + nsubj(V, N) \rightarrow \langle N, A \rangle,$
   $dp_3 : cop(A, V) + nsubj(A, N) \rightarrow \langle N, A \rangle,$
   $dp_4 : dobj(V, N) + nsubj(V, N') \rightarrow \langle N, V \rangle,$
   $dp_5 : \langle h, m \rangle + nn(h, N) \rightarrow \langle N + h, m \rangle,$
   $dp_6 : \langle h, m \rangle + nn(N, h) \rightarrow \langle h + N, m \rangle$

}

Fig. 2. Extraction rules.

sentences[1]. Figure 2 lists the rules that we have employed in this work. The rule conclusions contain the constructs that form the extracted aspect following rule activation. Here N is a noun, nn is a compound-noun, A an adjective, V a verb, h a head term, m a modifier. Candidate phrases include $\langle h, m \rangle$, $\langle N, A \rangle$, $\langle N, V \rangle$, $\langle h + N, m \rangle$ and $\langle N + h, m \rangle$. For each candidate, non-noun (N) words are eliminated and the remainder forms the set of aspects. See [2] for the detailed definition of grammatical relations and its application.

3.2 Aspect Selection

Aspects extracted are not equally important and therefore are subjected to selection-based dimensionality reduction. Let $\mathcal{P}$ be a set of products, $\mathcal{A}$ a set of aspects that appear in online reviews, $\mathcal{R}$. A product p is represented as $\overrightarrow{x} = \{x_1, ... x_{|\mathcal{A}|}\}$ where x is binary valued and corresponds to the presence or absence of an aspect $a \in \mathcal{A}$. The aim of feature selection is to reduce $|\mathcal{A}|$ to a smaller aspect subset size n by selecting aspects according to the score assigned by the feature selection technique. The selected aspects then form a new aspect vector $\overrightarrow{x'}$ and a corresponding reduced aspect set $\mathcal{A}'$ for product p, where $\mathcal{A}' \subset \mathcal{A}$ and $|\mathcal{A}'| \leq |\mathcal{A}|$. The algorithm used to rank aspects for selection is shown in Algorithm 1. Here $S = \{p_1, ..., p_w\}$ denotes the sample of products for training purpose.

Algorithm 1. Aspect Selection

---

   n = aspect subset size
   for each $a \in \mathcal{A}$ **do**
      Calculate aspect score using S
   end for
   Sort aspects based on frequency or IG scores
   $\mathcal{A}' = \{a_1, ..., a_n\}$
   return $\mathcal{A}'$

---

[1] Sentences are parsed using the Stanford Dependency parser [3].

Aspect Selection by Frequency. Frequency of extracted aspects is calculated according to the number of times an aspect occurs over the set of reviews. Accordingly aspects are ranked based on their FREQUENCYRANK scores computed as follows:

$$\text{FREQUENCYRANK}(a_i) = \frac{f(a_i)}{\sum_{j=1}^{\mathcal{A}} f(a_j)} \tag{1}$$

where FREQUENCYRANK(a_i) returns the relative frequency of an aspect a_i appearing in reviews $\mathcal{R}$. Here frequent occurrence of aspects in online reviews is perceived as important. However, frequency based approaches have a tendency to select general aspects such as "camera" and "quality", which fail to provide sufficient context for product comparisons (see Table 3). Instead of relying simply on frequency, what is required here is a heuristic that can identify aspects that have strong discriminative power. One such strategy is discussed next which measure the discriminative power of an aspect in discriminating between top ranked and non-top ranked products.

Aspect Selection with Information Gain. Features that are able to discriminate between classes are considered important in text classification [15]. Using this same principle, here aspects which are able to discriminate between top-ranked and non-top ranked products are deemed important for recommendation. In the absence of predefined class labels, we use a product ranking benchmark to derive class labels whereby a rank position is used as a class boundary to separate top ranked products from the rest of the products. Here we use a binary class such that c is either 0 meaning that the product is in the top ranked set; or is 1 meaning it is not among the top ranked products. In this way, each product in the product sample, S, can be assigned a binary label. Accordingly we rewrite the product notation as a pair $(\overrightarrow{x}, c)$ where c is a binary class label for p. Essentially, increasing the rank position that derives the class boundary for products will lead to a skewed class distribution; resulting in a decrease of the number of products belonging to $c = 0$ whilst increasing the size of class $c=1$. Given this supervised context, the discriminative power of an aspect a given the classes is computed as follows:

$$\text{IG}(X, C) = \sum_{x \in 0,1} \sum_{c \in 0,1} P(X = x, C = c).log_2 \frac{P(X = x, C = c)}{P(X = x).P(C = c)} \tag{2}$$

3.3 Aspect Sentiment Scoring

Given a target query product and its set of similar products we can rank these based on their sentiment (positive and negative) scores. Essentially such a product score is an aggregation of sentiment scores over the selected subset of aspects.

$$score(p_i) = \frac{\sum_{j=1}^{|\mathcal{A}'|} SentiScore(p_i, a_j)}{|\mathcal{A}'|} \tag{3}$$

Where the sentiment of the product p_i is associated with individual aspects a_j and $|\mathcal{A}'|$ is the aspect set for product p_i. Here, $SentiScore$ of an aspect is derived from product reviews using SmartSA [9] and is computed as:

$$SentiScore(p_i, a_j) = \frac{\displaystyle\sum_{m=1}^{|\mathcal{R}_j^i|} SentiScore(r_m)}{|\mathcal{R}_j^i|} \tag{4}$$

where $\mathcal{R}_j^i$ is a set of reviews for product p_i related to aspect a_j and $r_m \in \mathcal{R}_j^i$.

4   Evaluation

The primary aim of our evaluation is to study the impact of aspect selection on recommendation quality. To do this, we evaluate how well the recommendation system works in practice on Amazon.com data. We conveniently use Amazon's product *Star-Ratings* as the benchmark ranking to derive a comparison metric based on rank improvement. A secondary aim is to explore the transferability of the aspects learned from a particular product family (e.g. Compact cameras) to other related product families (e.g. DSLR cameras and Point & Shoot cameras).

4.1   Amazon Datasets

We crawled 1179 Amazon products during September 2014 from three different Amazon Digital Cameras categories: Compact (COMPACT), DSLR (DSLR) and Point & Shoot (PAS). The products extracted contain more than 100,000 different user generated reviews. Since we are not focusing on the cold-start problem, we use 1st January 2010 and less than 15 reviews as the pruning factor for the three product families. Finally, any synonymous products are united leaving us data for 98 COMPACT, 102 DSLR and 93 PAS products (see Table 1).

The aspect extraction algorithm described in Sect. 3.1 extracted 300–450 unique aspects for COMPACT, DSLR and PAS. On average, each product is defined by 220 different aspects, with standard deviations of 110, 115 and 86 aspects for COMPACT, DSLR and PAS cameras respectively. Importantly, more than 50 % of the products shared at least 100 different aspects with other products of the same family, whilst almost 30 % shared more than 150 aspects (more than 200 for COMPACT) on average. The fact that there are many shared aspects between products of the same family is reassuring for product comparison.

4.2   Evaluation Metrics

In the absence of a manual qualitative estimate of recommendation or access to user specific purchase trails, we derived approximations from the Amazon data we have crawled. For this purpose, using a *leave-one-out* methodology, the average gain in rank position of recommended products over the left-out query

Table 1. Statistics of Amazon compact, DSLR and PAS camera datasets

Category	Compact	DSLR	PAS
No. of Products	98	102	93
No. of Reviews	6349	7451	11,202
Aspects Mean (Std. Dev.)	267.25 (110.28)	226.78 (115.40)	186.39 (86.72)
No. of Different aspects	424	438	308

product is computed relative to a benchmark product ranking for each of the three product families.

$$RankImprovement\%(RI) = \frac{\sum_{i=1}^{n=3} benchmark(P_q) - benchmark(P_i)}{n * |\mathcal{P} - 1|} \quad (5)$$

where n is the size of the retrieved set and $benchmark$ returns the position on the benchmark list. Greater average gains in rank position over the query product would lead to a higher RI. For instance, assume the query product is ranked 40th on the benchmark list of 81 unique products $\mathcal{P}$, and the recommended product is ranked 20th on this list, then the recommended product will have a relative benchmark RI of 25 %.

We generated three benchmark lists according to Amazon's *Star-Ratings* of the three product families we crawled. In cases where two or more products had the same *Star-Ratings*, the products were ordered by the number of comments.

4.3 Ranking Strategies

The retrieval set of a query product consists of products that share a similar number of k aspects such that higher values of k denote lower number of products retrieved. This retrieval set is ranked using the sentiment-based recommendation strategies presented in Sect. 3.3. Central to this strategy is the selection of aspects using the following feature selection methods:

- BASE: recommend using aspect sentiment analysis with all aspects (see Eq. 3);
- FREQUENCYRANK (FR): same as BASE but only considering a subset of aspects selected by FR (see Eq. 1);
- INFORMATIONGAIN (IG): same as BASE but only considering a subset of aspects selected by IG (see Eq. 2).

The experiments were performed using 5 fold cross validation. To assess the transferability of the important aspects learned from different product families, we apply the selected aspects learned from a particular family to other two related product families when ranking the products.

4.4   Recommendation Performance Using IG

The objective of using feature selection techniques is to exploit important aspects generated by these techniques to rank products. We assess the effect of IG on recommendation performance by manipulating class sizes and aspect subset sizes. Figure 3 shows the performance of each product family in terms of average RI on benchmark *Star-Rating* with increasing class size. Here, the average RI is computed using different k shared aspects where k ranges from 0 to 240 (Note from Sect. 3.2 that class size relates to the top products rank position being used to create a class boundary separation). The results show that a small class size leads to better performance. For instance, Fig. 3 shows the performance of COMPACT improves from 5 % to 10 % but starts to fall after 10 %. Similar observations can be made on DSLR and PAS where their performance starts to drop after 15 %.

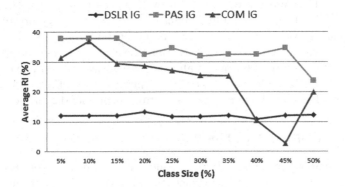

Fig. 3. Average RI for all products at different class size

Figure 4 presents the average RI for all product families when selecting aspects at different aspect subset size using IG at class size 10 %. In general, the average RI of all product families is at its best when 90 aspects were selected and remains constant for $n > 90$. It is interesting to note that when $n < 90$, products are compared using a smaller number of aspects. For example, only 40 % of PAS and COMPACT contain more than 25 aspects in the aspect subset size of 50. This explains the fluctuations in average RI for both families when considering low values of n. Based on the observations in both experiments, from this point onwards we use fixed aspect subset size $n = 90$ and a class size of 10 % for the rest of the experiments.

4.5   Comparison of Feature Selection Techniques

The graphs in Figs. 5, 6 and 7 illustrate the results of our comparison using RI at increasing k number of shared aspects. An overall view of these graphs shows that IG performs best for all three product families. However, we observed that

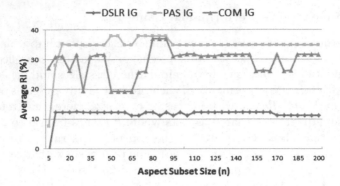

Fig. 4. Average RI for all products at different aspect subset size

the RI of IG is 15 % more than BASE on average, obtaining an absolute RI of more than 40 % for PAS family. This means that for every query product over a set of 90 products, we are able to recommend a better product ranked 40 positions higher on average. It is also worth pointing out that the performance of FR improves the recommendations of all three categories at 5 % on average compared to BASE. These results show that selecting a subset of aspects which are important provides a significant improvement on recommendation performance.

4.6   Similarity of Product Families

In Table 2, we studied the similarity of product aspects between the three related product families by computing the Jaccard similarity coefficient between the sets of aspects of each family. Furthermore, we created a ranking of frequent aspects for each family (see Table 3) and applied Spearman rank correlation coefficient to compare those ranked lists of aspects. As it can be observed, DSLR and COMPACT share a similar set of aspects with a 0.72 Jaccard coefficient (even higher when considering top 20 products), whilst the set of aspects used in

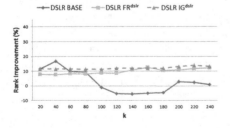

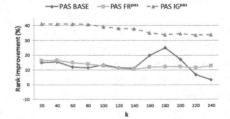

Fig. 5. RI with aspect selection on DSLR.

Fig. 6. RI with aspect selection on PAS.

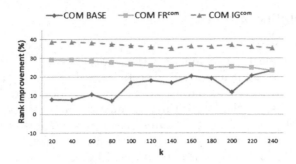

Fig. 7. RI with aspect selection on COMPACT.

Table 2. Aspect similarity for different camera families

	All Aspects		
	DSLR+Comp	DSLR+PAS	Comp+PAS
Jacc.	0.72	0.58	0.61
Spear.	0.87	0.75	0.76
	Top 20 Aspects		
Jacc.	0.81	0.60	0.66
Spear.	0.80	0.62	0.64

Table 3. Top 10 most frequent aspects by product families

Compact	DSLR	PAS
Camera	Camera	Camera
Lens	Use	Picture
Use	Lens	Use
Focus	Picture	Photo
Picture	Video	Video
Quality	Focus	Quality
Image	Time	Zoom
Time	Shoot	Battery
Photo	Image	Time
Shoot	Quality	Shot

PAS is slightly different (with a Jaccard coefficient of 0.58 between DSLR and PAS, and 0.61 between COMPACT and PAS). Furthermore, the Spearman rank correlation coefficient value shows that the aspects shared between families have similar frequency values. For instance, aspect *lens* in COMPACT is the second most frequent aspect whilst it occupies the third position in DSLR (both families have a 0.87 Spearman rank correlation).

4.7   Transferability of Aspects

In Table 2, we observed that the product aspects from three related product families have some degree of similarity. Here, we assess the transferability of the aspects by observing if important aspects learned from one product family are able to improve recommendation performance on other product families. Figures 8, 9 and 10 show the RI for three product families using FR and IG in aspect selection. Here PAS FR^{DSLR} indicates that the results presented correspond to the FR strategy for PAS using DSLR selected aspects. Similarly, PAS IG^{DSLR} indicates PAS results using DSLR aspects selected by IG.

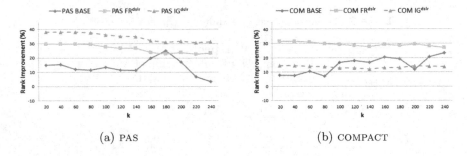

(a) PAS   (b) COMPACT

Fig. 8. RI using transferred aspects from DSLR

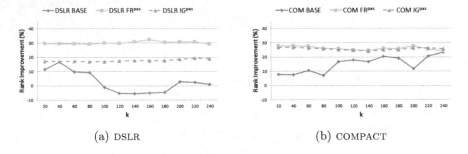

(a) DSLR   (b) COMPACT

Fig. 9. RI using transferred aspects from PAS

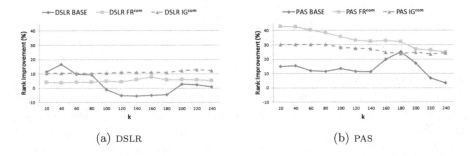

(a) DSLR   (b) PAS

Fig. 10. RI using transferred aspects from COMPACT

The benefit of aspects transferability can be observed when FR is used in aspect selection. For instance, Figs. 9a and 10b show FR provides significant improvements in recommendation for DSLR and PAS respectively. Furthermore, we observed that COMPACT FR^{DSLR} (Fig. 8b) obtains similar RI to COMPACT FR^{COMPACT} (Fig. 7), indicating the selected aspects of both families are similar. This result is expected given high aspects correlation between the frequent aspects of DSLR and COMPACT. The result obtained using IG is mixed. One explanation for its poor performance is that the product families do not share similar subset of aspects, resulting in a drop in RI of IG in Figs. 8b and 10a. This indicates that aspects selected by IG are domain-dependent as they provide little benefits to other product families.

The high transferability of the aspects using FR suggests that general aspects are suitable to be used in recommending cameras products. However, we observed that not all product families benefit from the transfer of aspects. For instance, COMPACT does not benefit from aspects learned from other product families. Essentially, best results are achieved when domain-dependent aspects are learned using IG in PAS and COMPACT (see Figs. 6 and 7).

5   Conclusion

In this paper we extended our previous work on social recommender systems to harness knowledge from product reviews, and explored the utility of the frequency based approach and supervised Information Gain to rank and select the most useful aspects for recommendation. The benefits are demonstrated in a realistic recommendation setting using benchmarks generated from *Star-Rating*. We confirmed that aspect selection using feature selection techniques help improve recommendations of the three datasets; the best results are obtained using Information Gain when considering only a small subset of aspects. On the other hand, we presented how the aspects selected by the frequency based aspect selection technique is transferable between product families and that leads to better recommendation performance. However, better results are achieved when using domain-dependent aspects. Our results show that Information Gain is promising in identifying important aspects and improve recommendations, but further work is needed to explore other feature selection techniques such as mutual information and the Chi-squared statistic.

Acknowledgments. This research has been partially supported by AGAUR Scholarship (2013FI-B 00034) and NASAID (CSIC Intramural 201550E022).

References

1. Chen, L., Chen, G., Wang, F.: Recommender systems based on user reviews: the state of the art. User Model. User-Adap. Interact. **25**(2), 99–154 (2015)
2. Chen, Y.Y., Ferrer, X., Wiratunga, N., Plaza, E.: Sentiment and preference guided social recommendation. In: Lamontagne, L., Plaza, E. (eds.) ICCBR 2014. LNCS, vol. 8765, pp. 79–94. Springer, Heidelberg (2014)

3. De Marneffe, M., MacCartney, B., Manning, C., et al.: Generating typed dependency parses from phrase structure parses. In: Proceedings of Language Resources and Evaluation Conference, pp. 449–454 (2006)
4. Dong, R., Schaal, M., OMahony, M., McCarthy, K., Smyth, B.: Opinionated product recommendation. In: International Conference on Case-Based Reasoning (2013)
5. Hu, M., Liu, B.: Mining and summarising customer reviews. In: Proceedings of ACM SIGKDD International Conference on Knowledge Discovery and Data Mining, KDD 2004, pp. 168–177 (2004)
6. Huang, J., Etzioni, O., Zettlemoyer, L., Clark, K., Lee, C.: Revminer: an extractive interface for navigating reviews on a smartphone. In: Proceedings of the 25th Annual ACM Symposium on User Interface Software And Technology, pp. 3–12. ACM (2012)
7. Lops, P., De Gemmis, M., Semeraro, G.: Content-based recommender systems: State of the art and trends. In: Ricci, R., Rokach, L., Shapira, B., Kantor, P.B. (eds.) Recommender Systems Handbook, pp. 73–105. Springer US, New York (2011)
8. Moghaddam, S., Ester, M.: On the design of lda models for aspect-based opinion mining. In: Proceedings International Conference on Information and Knowledge Management, CIKM 2012 (2012)
9. Muhammad, A., Wiratunga, N., Lothian, R., Glassey, R.: Contextual sentiment analysis in social media using high-coverage lexicon. In: Bramer, M., Petridis, M. (eds.) Research and Development in Intelligent Systems, pp. 79–93. Springer, Switzerland (2013)
10. Owoputi, O., O'Connor, B., Dyer, C., Gimpel, K., Schneider, N., Smith, N.A.: Improved part-of-speech tagging for online conversational text with word clusters. In: Association for Computational Linguistics (2013)
11. Ronen, R., Koenigstein, N., Ziklik, E., Nice, N.: Selecting content-based features for collaborative filtering recommenders. In: Proceedings of the 7th ACM Conference on Recommender Systems, pp. 407–410. ACM (2013)
12. Smyth, B.: Case-based recommendation. In: Brusilovsky, P., Kobsa, A., Nejdl, W. (eds.) Adaptive Web 2007. LNCS, vol. 4321, pp. 342–376. Springer, Heidelberg (2007)
13. Vargas-Govea, B., González-Serna, G., Ponce-Medellın, R.: Effects of relevant contextual features in the performance of a restaurant recommender system. ACM RecSys 11 (2011)
14. Wiratunga, N., Koychev, I., Massie, S.: Feature selection and generalisation for retrieval of textual cases. In: Funk, P., González Calero, P.A. (eds.) ECCBR 2004. LNCS (LNAI), vol. 3155, pp. 806–820. Springer, Heidelberg (2004)
15. Yang, Y., Pedersen, J.O.: A comparative study on feature selection in text categorization. In: ICML, vol. 97, pp. 412–420 (1997)
16. Zha, Z.-J., Jianxing, Y., Tang, J., Wang, M., Chua, T.-S.: Product aspect ranking and its applications. IEEE Trans. Knowl. Data Eng. **26**(5), 1211–1224 (2014)

Music Recommendation: Audio Neighbourhoods to Discover Music in the Long Tail

Susan Craw[✉], Ben Horsburgh, and Stewart Massie

School of Computing Science and Digital Media,
Robert Gordon University, Aberdeen, UK
{s.craw,s.massie}@rgu.ac.uk
http://www.rgu.ac.uk/dmstaff/craw-susan-massie-stewart

Abstract. Millions of people use online music services every day and recommender systems are essential to browse these music collections. Users are looking for high quality recommendations, but also want to discover tracks and artists that they do not already know, newly released tracks, and the more niche music found in the 'long tail' of on-line music. Tag-based recommenders are not effective in this 'long tail' because relatively few people are listening to these tracks and so tagging tends to be sparse. However, similarity neighbourhoods in audio space can provide additional tag knowledge that is useful to augment sparse tagging. A new recommender exploits the combined knowledge, from audio and tagging, using a hybrid representation that extends the track's tag-based representation by adding semantic knowledge extracted from the tags of similar music tracks. A user evaluation and a larger experiment using Last.fm user data both show that the new hybrid recommender provides better quality recommendations than using only tags, together with a higher level of discovery of unknown and niche music. This approach of augmenting the representation for items that have missing information, with corresponding information from similar items in a complementary space, offers opportunities beyond content-based music recommendation.

Keywords: Recommender systems · Novelty and serendipity · Knowledge extraction · CBR similarity assumption

1 Introduction

Long tail marketing techniques have a sales model based upon promoting less popular products in the 'long tail' as shown in Fig. 1. It is most effectively employed by online retailers, so is very relevant for online music services. Therefore music recommenders should not overlook or ignore recommendations in this 'long tail'. A track that is not often listened to may be a niche recommendation that offers serendipity and an opportunity to discover new music. These recommendations encourage sales in this important area of the online music market.

Query-by-example music recommenders have access to different representations for items: audio representations like texture (timbre), harmony, rhythm;

© Springer International Publishing Switzerland 2015
E. Hüllermeier and M. Minor (Eds.): ICCBR 2015, LNAI 9343, pp. 73–87, 2015.
DOI: 10.1007/978-3-319-24586-7_6

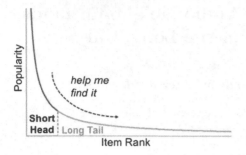

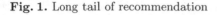

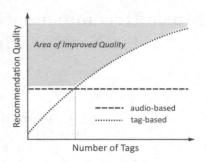

Fig. 1. Long tail of recommendation

Fig. 2. Recommendation with sparse tags

meta-data such as track title, artist, year, etc; and semantic information such as social tagging from on-line music services. Many state-of-the-art recommender systems make use of social tagging [1]. These tags can provide useful semantic information for recommendation including genres, topics, opinions, together with social, contextual and cultural information. However, not all tracks within a collection are tagged equally: popular tracks tend to have more tags describing them, and niche tracks may have no tags at all.

Figure 2 illustrates the effect of tag sparseness on recommenders. When the query has few tags the tag-based recommendations have poor quality, and an audio recommender that is not affected by tags provides better recommendations. Conversely, tag-based recommenders cannot reliably identify good recommendations if they have few tags. This diagram motivates the idea for a hybrid recommender that combines tag and audio representations. Hybrid recommenders typically merge representations, or combine the processes of sub-recommenders. Our approach is different, because it augments existing tags when necessary. It exploits the similarity assumption of case-based reasoning to extract additional tag knowledge from audio neighbourhoods. It extends the notion of recommendation, that "similar tracks will be good recommendations", to "similar tracks will have useful tagging". It injects novelty and serendipity into its recommendations, since it is not biased against sparsely tagged tracks in the 'long tail'.

In the rest of this paper we first review relevant literature in music recommendation and serendipity. Section 3 introduces our music dataset and describes the tag and audio representation for tracks. Our new recommender that combines knowledge extracted from audio neighbourhoods with existing tagging is presented in Sect. 4. Its performance for accuracy and novelty in a user trial and a system-centric evaluation is discussed in Sects. 5 and 6.

2   Related Work

One advantage that content-based recommendation has over collaborative filtering is that it does not suffer from 'cold start' where user data is not available, nor from the 'grey sheep' problem [2], where users with niche tastes are excluded because no similar users exist. Instead, content-based recommendation has the potential to recommend any track within a collection to a user. The main disadvantage of content-based recommenders is that they rely entirely on the strength of each track's representation. If the representation is weak, then it is difficult to define meaningful similarity, and the quality of recommendations will be poor. The two core approaches used to represent music tracks are audio content and tags. However, audio content representation is weak, and does not provide high quality recommendations. When tag-based features are used, high quality recommendations can be made, and these have been shown to provide better quality recommendations than collaborative filtering methods [3].

Tags come directly from users and are most commonly generated socially, via user collaboration, and so a wealth of social and cultural knowledge is available to describe tracks. However, this also means that tagging is not evenly distributed, and a popularity bias in music listening habits further skews the distribution of tags [4]. New and niche tracks are in the popularity 'long tail' of Fig. 1, so few people are listening to them, so few/no people are tagging them, so these tracks have few/no tags, so tag-based recommenders do not recommend them, so few people are listening to them, etc.

The Million Song Dataset [5] includes a Last.fm contribution containing a tagged dataset[1] and the tagging of this reference dataset is typical. It contains almost 950 k tracks tagged with more than 500 k unique tags, and on average each tagged track has 17 tags, but 46 % of tracks do not have any tags at all. The 25 k most tagged tracks each has 100 tags, but this number very quickly drops off in a 'long tail' similar to Fig. 1. Halpin et al. found similar tagging 'long tails' in the various del.icio.us sites they investigated [6].

Auto-tagging is designed to overcome sparse social tagging [7]. A popular approach learns tags that are relevant to a track from a Gaussian Mixture Model of the audio content [8]. While this approach may guarantee a certain degree of tagging throughout a collection, humans are not involved with the association of tags with tracks, and thus it is likely that erroneous tags will be propagated to many tracks. It is also easy to learn common tags which co-occur often, but runs the risk of excluding more niche tags, which may be most appropriate for tracks with few tags. Track similarity has also been used for auto-tagging style and mood [9]. Here tag vectors of similar tracks are aggregated, and the most frequently occurring tags are propagated. The advantage of this method is that there is no attempt to correlate content directly with tags, or presume that tagging must fit any prior distribution. Instead it exploits consensus of human tagging. We take inspiration for our pseudo-tagging from this approach, but the way we use pseudo-tags recognises that they are not 'real' tags [10].

Hybrid representations that combine tag and audio representations can also cope with sparse tagging. Levy & Sandler [11] create a code-book from clustered audio content vectors, and these muswords are used as the audio equivalent of tags. Concepts are extracted using Latent Semantic Analysis (LSA) from the

---

[1] http://labrosa.ee.columbia.edu/millionsong/lastfm.

combined representation of tags and muswords. In previous work we concatenated tag and texture representations, before extracting latent concepts [12].

Taste in music is highly subjective, and so generating novel and serendipitous recommendations is particularly important, and challenging. Kaminskas & Bridge's [13] exploration of serendipity notes the trade-off in standard recommender approaches between quality and serendipity. The Auralist [14] and TRecS [15] hybrid recommenders address this trade-off by amalgamating sub-recommenders with differing priorities including quality, serendipity and novelty. In both systems, special novelty and serendipity recommenders influence choice.

3   Music Collection

Our music collection was created from a number of CDs that contain different genres, a range of years, and many compilation CDs to keep the collection diverse [16]. This dataset includes 3174 tracks by 764 separate artists. The average number of tracks per artist is 4, and the most common artist has 78 tracks. The tracks fall into 11 distinct super-genres: Alternative (29%), Pop (25%), Rock (21%), R&B (11%); and Dance, Metal, Folk, Rap, Easy Listening, Country and Classical make up the remaining 14% of the collection. We now describe two standard music representations applied to this dataset: one based on the tagging of Last.fm users; and a texture representation built from audio files.

Music tracks often have tag annotations on music services. Last.fm is used by millions of users, and their tagging can be extracted using the Last.fm API[2]. When a user listens to a track, they may decide to tag it as 'rock'. Each time a unique user tags the track as 'rock', the relationship of the tag to the track is strengthened. A track's tag vector $t = < t_1\ t_2\ \ldots, t_m >$ contains these tag frequencies t_i, and m is the size of the tag vocabulary. Last.fm provides normalised frequencies for the tags assigned to each track, with the most frequent tag for a track always having frequency 100. A total of $m = 5160$ unique tags are used for our music collection in the Last.fm tagging. On average each track has 34 tags with a standard deviation of 24.4, and the most-tagged track has 99 tags. The tagging is realistically sparse: 3% of the tracks have no tags at all; there are 24% with fewer than 10 tags; and 42% with fewer than 20 tag.

Texture (timbre) is one of the most powerful audio-based representations for music recommendation [17]. We use the MFS Mel-Frequency Spectrum texture [18], available through the Vamp audio analysis plugin system[3]. MFS is a musical adaptation of the well-known Mel-Frequency-Cepstral-Coefficients (MFCC) texture [19]. Figure 3 illustrates the main stages in transforming audio tracks into MFS (and MFCC) vectors, and demonstrates the relationship between MFS and MFCC. Audio waveforms, encoded at 44.1 kHz, are first split into windows of length 186 ms, and each window is converted into the frequency domain using a Discrete Fourier Transform (DFT). Each frequency spectrum computed has a maximum frequency of 22.05 kHz, and a bin resolution of 5.4 Hz. Next, each

---

[2] www.last.fm/api.
[3] www.vamp-plugins.org/download.html

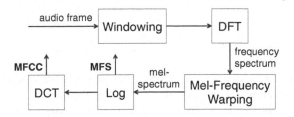

Fig. 3. Extraction of MFS and MFCC

window is discretised into a feature vector, based on the mel-scale [20]. We use 40 mel filters, the granularity found to be best for aggregation-based recommender models [18]. A mean feature vector MFS is computed for each track and these are used to construct a track-feature matrix. Latent Semantic Indexing (LSI) is used to discover musical texture concepts, and each track is projected into this texture space to create its MFS-LSI texture vector.

4   Hybrid Recommenders

Query-by-track recommender systems *Tag* and *Audio* may be defined using standard vector cosine similarity with these tag-based and texture vector representations. However, each individually can be problematic. Tag can give good recommendations but cannot recognise recommendations that have few or no tags, and cannot retrieve good recommendations for poorly tagged queries. Audio does not suffer this problem because all tracks have audio data, but does not offer the same performance as Tag with well-tagged tracks. Our two new hybrid recommenders are designed to reduce the semantic gap between audio content and tags, and allow recommendation quality to be improved when tracks are under-tagged. They take advantage of tagging, but also exploit similarity neighbourhoods in the audio space to learn pseudo-tags. These hybrid query-by-track recommenders are defined by standard tag-based representations and cosine similarity retrieval.

4.1   Learning Pseudo-Tags

Pseudo-tagging is different from other hybrid representations that combine tag- and audio-based representations. Instead, pseudo-tags are extracted from the tags of tracks that have similar audio content, and these pseudo-tags are used within a tag-based representation.

The first step to generating pseudo-tags for a track is to find tracks that are similar to this track. A k nearest-neighbour retrieval using cosine similarity in the musical texture MFS-LSI space identifies the K most similar tracks. A rank-based weighted sum of the tag vectors $t(1)$ ... $t(K)$ for these K retrieved tracks are used to learn the pseudo-tag vector $p = \ <p_1 \ p_2 \ ... \ p_m>$:

$$p_i = \sum_{k=1}^{K}(1 - \frac{k-1}{K})t_i(k) \tag{1}$$

where $t_i(k)$ is the frequency of the ith tag in the tag vector of the kth nearest neighbour track[4] Retrieved tracks from lower positions have less influence and so the retrieval list is restricted to $K = 40$ neighbours for our experiments.

Our *Pseudo-Tag* recommender retrieves tracks using cosine similarity of these pseudo-tag vectors. The pseudo-tag representation reduces sparsity in tag-based representations because audio neighbourhoods of tracks are unlikely to be uniformly sparsely tagged. The advantage of using pseudo-tags over audio content directly is that factors such as context and opinions will also be present in the pseudo-tag representation, inherited from the neighbourhoods.

4.2   Augmenting Tags with Pseudo-Tags

Pseudo-tag vectors are useful when a track has few tags, but can influence the representation too much if the track is already well-tagged. In particular, the pseudo-tag vector has ignored any tag information that may be associated with the track itself, and includes all tags that are associated with any of the track's neighbours. Our *Hybrid* recommender uses a tag representation that augments any existing tags for a track by merging the track's learned pseudo-tag vector p. with its tag vector t.

A pseudo-tag vector p is much less sparse (fewer zero frequencies) than a tag vector t because p has been aggregated from tag vectors belonging to a number of tracks in the neighbourhood of t's track. The first step in creating the hybrid tag/pseudo-tag representation selects the number of pseudo-tags P to be included, so that it balances the number of existing tags T; i.e. non-zero frequencies in t. We experimented with different values of $P = 0, 10, 20, ...100$. The solid dark line in Fig. 4 shows the best performing number of pseudo-tags P for tracks with different numbers of tags T grouped into tag buckets of size 10. Under-tagged tracks need higher numbers of pseudo-tags, and well-tagged tracks use fewer; this is consistent with intuition. The dark dashed line is the line-of-best-fit through these data points. We select the number of pseudo-tags retained based on an approximation of this line: $P = 100 - T$ for our dataset.

The vector of selected pseudo-tags $\tilde{p}$ is created by retaining the P highest frequencies in p and zeroing the rest. Next, an influence weighting α determines the influence of the selected pseudo-tags $\tilde{p}$ on the hybrid vector h[4]:

$$h_i = \alpha\tilde{p}_i + (1 - \alpha)t_i \tag{2}$$

Experiments similar to those for P, alter the weighting α from 0 to 0.5 in steps of 0.1. The grey lines and secondary axis in Fig. 4 show the best weighting and dashed line-of-best-fit, estimated as $\alpha = 0.5 * (1 - T/100)$.

---

[4] All tag-based vectors t, $t(k)$, p, $\tilde{p}$, and h are routinely normalised as unit vectors before use. For clarity, normalisation has been omitted from Eqs. (1) and (2).

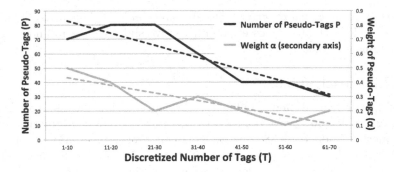

Fig. 4. Number of pseudo-tags and weighting for hybrid

The Hybrid recommender uses representation h to retrieve tracks. For well-tagged tracks, the tag vector dominates h, and the Hybrid recommender benefits from the strengths of tag-based recommendation. Weakly-tagged tracks are augmented by the inclusion of pseudo-tags in h. The Pseudo-Tag representation is a variant of Hybrid, where the weighting α is 1, and all pseudo-tags are used.

5   User Evaluation

A user evaluation was undertaken to test the quality of recommendations with real users, but also to measure the level of discovery of new tracks in the recommendations. The two new hybrid recommenders Pseudo-Tag and Hybrid are included in the experiments to see the effect of replacing or augmenting tags with learned pseudo-tags. The Tag recommender is also included as a baseline.

5.1   Design of User Evaluation

The selection and presentation of the query and recommendations are designed to avoid bias, and the screen provides the same information for every query. The user is shown a query track and the top five recommended tracks from a single recommender. Each track has its title, artist, and a play button that allows the user to listen to a 30 s mid-track sample. The recommender is chosen randomly, and the top five recommendations are presented in a random order. Each query track is selected at random from either a fixed pool or the entire collection, with 50:50 chance. The pool contains 3 randomly selected tracks for each of the 11 genres in the collection. The 33 pool tracks will be repeated more frequently, whereas the other tracks are likely to be used at most once. Users evaluate as many queries as they choose, without repetition.

A user gives feedback on the quality of each of the recommendations by moving a slider on a scale between very bad (0) to very good (1). Each slider is positioned centrally on the scale initially, and records feedback in $\frac{1}{1000}$ ths. To capture feedback on each track's novelty, the user also selects from 3 options:

knows artist & track; *knows artist only*; or *knows neither*. When feedback for a query is complete, the user presses submit to save slider values and novelties for its 5 recommendations.

5.2   User Participation

The on-line user evaluation was publicised through social media and mailing lists. It was available for 30 days and a total of 132 users took part, evaluating a total of 1444 queries. There were 386 queries where all 5 recommendations scored 500, suggesting that the user clicked submit without moving any of the sliders. These were discarded, and the remaining 1058 valid queries provide explicit feedback on their recommendations. On average users evaluated recommendations for 6.24 queries, and the most active user scored 29 queries.

Prior to providing feedback, each user completed a questionnaire to indicate their gender, age, daily listening hours, and musical knowledge: *none* for no particular interest; *basic* for lessons at school, reads music magazine/blogs, etc.; or *advanced* for play instrument, edit music on computer, professional musician, audio engineer, etc. Each user also selects any genres they typically listen to. Figure 5 contains a summary of the questionnaire data, showing there is a good spread across age, gender, and knowledge, and that the musical interests align well with the genres in the pool and collection overall.

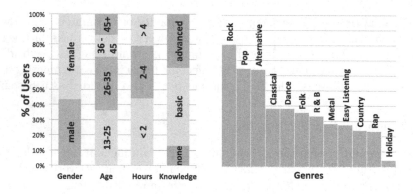

Fig. 5. Profile of user group

5.3   Results for Recommendation Quality

We calculate recommendation quality Q for a query-recommendation pair q, r by aggregating the individual scores by a user u, across all users U providing feedback for this query's recommendations. We then use a $Q@N$ average of the top N recommendations r_n to evaluate the recommendations for query q.

$$Q(q, r) = \frac{1}{|U|} \sum_{u \in U} \mathrm{score}_u(q, r) \qquad Q@N(q) = \frac{1}{N} \sum_{n=1}^{N} Q(q, r_n) \qquad (3)$$

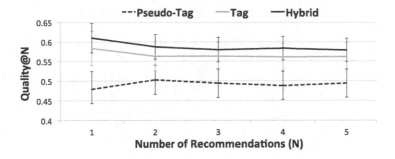

Fig. 6. Recommendation quality from user evaluation

Figure 6 shows the $Q@N$ values averaged across all pool queries in the user evaluation. We focus on pool queries, approximately 47 % of all queries, since non-pool queries are typically evaluated by only a single user. The error bars indicate 95 % confidence and are included to give a sense of separation for the graphs. They show quite high variability because of the following: a user provides feedback on the recommendations of a single recommender for each of their queries, there are a relatively small number of queries, and each user gives feedback on only a subset of these.

The Hybrid recommender provides higher quality recommendations than Tag and Pseudo-Tag by augmenting existing tags used by Tag with some of the pseudo-tags from Pseudo-Tag. The small improvement of Hybrid over Tag shows that augmenting the tags with pseudo-tags does not damage recommendation. Sparsely tagged tracks gain from additional pseudo-tags, although pseudo-tags on their own are not so good for recommending. It appears that the adaptive balancing of existing tags with pseudo-tags from the audio neighbourhood is helpful. The equivalent figure for all queries is similar, but it has the Hybrid and Tag graphs slightly closer together, and a larger separation from Pseudo-Tags.

In general, the $Q@N$ drops slowly as N increases, so tracks later in the recommender's list gradually decrease in quality as expected. Remember that the recommendations are presented in a random order so there is no user bias towards tracks higher up the list. It is not clear why the top recommendation by Pseudo-Tag is poorer than those that are ranked lower, but possibly users are not good at ranking accurately recommendations that are generally poor.

5.4 Results for Discovery with Quality

We are interested in recommenders that offer new and niche tracks as serendipitous recommendations whilst retaining the all-important quality of recommendation. Here we explore the novelty of recommendations in the user evaluation by analysing the user replies about knowing the track.

One interesting observation from the user evaluation is the confirmation that users give higher feedback to recommendations that they know, and slightly higher ratings to tracks where they know the artist. Figure 7 shows the average

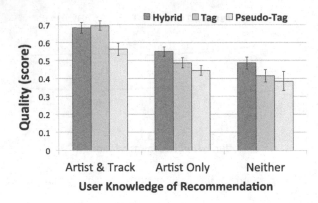

Fig. 7. Quality by what is known

score for recommendations from users according to the user's knowledge of the artist and track.

Figure 8 captures the quality and novelty of all recommendations in the user evaluation. The location and spread of the clusters for Hybrid (black), Tag (grey) and Pseudo-Tag (white) demonstrate well the trade-off between quality and novelty. Good quality recommenders are higher; and those suggesting more recommendations that are unknown are towards the right, so best recommenders that combine novelty with quality are towards the top right. The individual points in the clusters show the score@N and % unknown tracks for different $N = 1..5$. For hybrid the top point with highest quality is $N = 1$; larger Ns have increasingly lower quality. For Tag and Pseudo-Tag the isolated point to the left is $N = 1$; the other 4 Ns are very tightly clustered.

Hybrid achieves quality recommendations and is able to suggest unknown tracks; it recommends novel tracks 50–55 % of the time. Although Tag has comparable quality it is significantly poorer for novel recommendations. Only 30–40 % of its recommendations are unknown because Tag tends to recommend well-tagged tracks and these are also often well-known. Hybrid and Pseudo-Tag are comparable for novelty since they each exploit the tags inherited from neighbouring tracks. However, quality is also important, and Hybrid gives significantly better recommendations – despite the users' quality bias towards known tracks!

6   Evaluation Using Last.fm User Data

A larger system centric evaluation has also been undertaken using leave-one-out testing on the whole music collection. We use the socialSim score that defines the recommendation quality Q as the association between the numbers liking and listening to tracks q and r:

$$Q(q,r) = \text{socialSim}(q,r) = \frac{\text{likers}(q,r)}{\text{listeners}(q,r)} =_{est} \frac{\text{likers}(q,r)}{\text{listeners}(q) \cdot \text{listeners}(r)} \qquad (4)$$

where likers(q, r) and listeners(t) are available through the Last.fm API (see [21] for details). This evaluation uses socialSim@N averaged over all tracks q in the collection. Notice that tag data used in the recommenders is distinct from user data underpinning socialSim, although both are extracted from Last.fm.

Figure 9 contains the quality results for Hybrid, Tag and Pseudo-Tag as in the user evaluation, now for $N = 1..10$ recommendations. Results for an Audio recommender based on MFS-LSI texture are also included as a purely audio-based baseline; it was omitted from the user evaluation, to reduce the number of very poor recommendations presented to users for feedback. The 95 % error bars are much more compressed now because of the very large set of queries from leave-one-out testing, and the combined opinions of very many Last.fm users.

The overall findings confirm those from the user evaluation: Hybrid and Tag are comparable, with Hybrid having a tendency to give higher quality recommendations. Pseudo-Tag is significantly poorer and, as expected, Audio is much poorer still. Compared to Fig. 6, the recommendation quality drops more quickly for all three recommenders, and continues decreasing as N increases. With users, later recommendations did not dilute the quality of earlier ones, but since a user rated all 5 recommendations at the same time perhaps less variation between a query's recommendations is natural. Also, we have seen that whether a track is known or not affects a user's score, and the system-centric evaluation does not suffer the effect of individual subjectivity. The placing of the Hybrid and Tag graphs is slightly higher than with users, and there is a significantly increased gap between Pseudo-Tag and Tag. However, exact values are not really comparable. There is a prevalence of zeros in the socialSim score when there is no evidence of likers in user data, but users may give less pessimistic ratings for

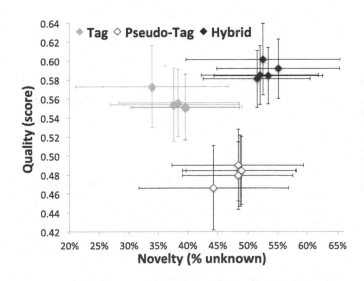

Fig. 8. Balance between quality and novelty from user evaluation

poor recommendations when responding to real queries, and Q is unlikely to generate 0; i.e. all users scoring 0 for a recommendation.

The notion of novelty is difficult to capture from user data. Instead we introduce an artificial measure that exploits the link between tracks that are well-known and the level of tagging. Recommendations with few tags will be classed as novel and the % of novel recommendations will measure novelty. Figure 10 has quality replicated from Fig. 9, and novelty is the % of recommendations with fewer than 30 tags; i.e. those whose tags have been augmented with 70–100 pseudo-tags and 35–50 % weighting with tags. Again the advantage from combined quality and novelty for Hybrid over Tag is clear. The points in each cluster, showing quality and novelty with differing numbers of recommendations, are for $N = 1..10$ with $N = 1$ being the top point, with larger N strictly in order below. Hybrid gives a better level of discovery of tracks with relatively few tags although this tendency is not significant for $N \leq 5$.

What happens with a more demanding criterion for novelty than 30 tags? The quality-novelty scatter for discovery involving fewer tags is a little more overlapping on the novelty axis, as shown in Fig. 11 for the discovery of tracks with fewer than 20 tags. Hybrid and Tag are now comparable for novelty, with a tendency for Hybrid to be better for $N > 5$.

In Figs. 10 and 11, only the novelty values change for different levels of tagging, and the heights of the points on the quality scale are identical in both figures. Hybrid and Tag are clearly recommending different tracks because of the significant quality gains for Hybrid's use of pseudo-tags in these figures. With an overall frequency in the dataset of 54 % for < 30 tags, and 42 % for < 20 tags, the discovery rates in these figures indicate fair treatment of the 'long tail'.

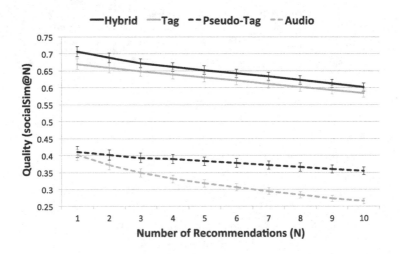

Fig. 9. Recommendation quality from last.fm user data

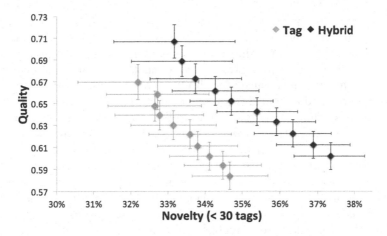

Fig. 10. Quality and discovery from system-centric evaluation (30 tags)

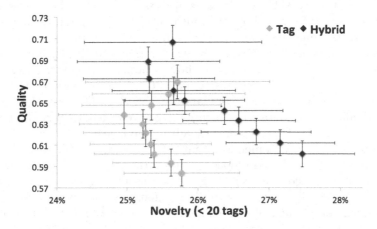

Fig. 11. Quality and discovery from system-centric evaluation (20 tags)

7   Conclusions

Tags introduce a blur between the classical notion of a recommender being either content-based or collaborative filtering. In the strict sense tags are content, but because they are created collaboratively, recommendations made based on tags are influenced by other users, which classical content-based systems are not. As a result tag-based recommenders offer some of the advantages of collaborative filtering, but also suffer some of their disadvantages. Collaborative user tagging provides semantics including contextual, social and cultural information that allow tag-based recommenders to take advantage of this information when making recommendations. However, this also means that tag-based recommenders are affected by under-tagging of new and niche tracks that are in the 'long tail' of music tracks. There is a reinforcement loop whereby tracks that are not listened to often, do not get tagged frequently, so have relatively few tags, and so do not get recommended. The combination of the 'long tail' of music tracks and the popularity bias of tagging means that there is also a 'long tail' of tagging of tracks. These are precisely the tracks that one wishes to include in a recommender that will introduce serendipity and novelty for users.

We have developed a Hybrid recommender that learns pseudo-tags for tracks with fewer tags so that the tagging 'long tail' is removed, and a tag-based recommender does not face the sparseness of user tagging. Pseudo-tags are related to audio since they are learned from tracks that are similar in the audio space, but they also capture semantics that users have given to neighbouring tracks. Further the weighting and selection of pseudo-tags allows only the most popular tags to be inherited from the musical neighbourhood. Finally the balancing of tags with pseudo-tags ensures that user-generated tags are used, and are most influential, whenever they are available.

The user trial and larger off-line evaluation demonstrate that Hybrid is effective in bridging the semantic gap between user tagging and audio. The semantic knowledge extracted from audio neighbourhoods is useful in improving the quality of Hybrid recommendations over those from the Tag recommender. These evaluations also explored the novelty of recommendations. Importantly the user trial results were based on responses about whether the track, or track and artist, were unknown, and there is a significant separation between the Hybrid cluster for 'novelty with quality' compared to Tag. The off-line evaluation gave consistent findings, but its tagging criterion for novelty is artificial, and for more sparsely tagged tracks, Hybrid's novelty advantage is less.

Our interest is in recommenders that offer serendipity whilst maintaining good recommendations. Hybrid indeed achieves this, without introducing a specialised serendipity recommender. Augmenting pseudo-tags has even increased recommendation quality. This approach of augmenting a weak representation with equivalent knowledge from neighbourhoods in a complete representation may be useful in related recommendation tasks; e.g. extracting pseudo-captions to improve image retrieval, learning pseudo-ratings for collaborative filtering. We have focused on pseudo-tags to improve recommendation but it could be interesting to understand inconsistencies between tags and pseudo-tags that indicate possible malicious tagging and shilling attacks.

References

1. Nanopoulos, A., Rafailidis, D., Symeonidis, P., Manolopoulos, Y.: MusicBox: personalized music recommendation based on cubic analysis of social tags. IEEE Trans. Audio Speech Lang. Process. **18**(2), 407–412 (2010)
2. Barragáns-Martínez, A.B., Costa-Montenegro, E., Burguillo, J.C., Rey-López, M., Mikic-Fonte, F.A., Peleteiro, A.: A hybrid content-based and item-based collaborative filtering approach to recommend TV programs enhanced with singular value decomposition. Inf. Sci. **180**(22), 4290–4311 (2010)
3. Firan, C.S., Nejdl, W., Paiu, R.: The benefit of using tag-based profiles. In: Proceedings of Latin American Web Conference, pp. 32–41 (2007)

4. Celma, O., Cano, P.: From hits to niches?: Or how popular artists can bias music recommendation and discovery. In: Proceedings of 2nd Netflix-KDD Workshop, pp. 1–8 (2008)
5. Bertin-Mahieux, T., Ellis, D.P., Whitman, B., Lamere, P.: The million song dataset. In: Proceedings 12th International Society for Music Information Retrieval Conference, pp. 591–596 (2011)
6. Halpin, H., Robu, V., Shepherd, H.: The complex dynamics of collaborative tagging. In: Proceedings of 16th International Conference on World Wide Web, pp. 211–220 (2007)
7. Bertin-Mahieux, T., Eck, D., Mandel, M.: Automatic tagging of audio: the state-of-the-art. In: Wang, W. (ed.) Machine Audition: Principles, Algorithms and Systems, pp. 334–352. IGI Global, Hershey (2010)
8. Turnbull, D., Barrington, L., Torres, D., Lanckriet, G.: Semantic annotation and retrieval of music and sound effects. IEEE Trans. Audio Speech Lang. Process. **16**(2), 467–476 (2008)
9. Sordo, M., Laurier, C., Celma, O.: Annotating music collections: How content-based similarity helps to propagate labels. In: Proceedings of 8th International Conference on Music Information Retrieval (ISMIR) (2007)
10. Horsburgh, B., Craw, S., Massie, S.: Learning pseudo-tags to augment sparse tagging in hybrid music recommender systems. Artif. Intell. **219**, 25–39 (2015)
11. Levy, M., Sandler, M.: Music information retrieval using social tags and audio. IEEE Trans. Multimedia **11**(3), 383–395 (2009)
12. Horsburgh, B., Craw, S., Massie, S., Boswell, R.: Finding the hidden gems: recommending untagged music. In: Proceedings of 22nd International Joint Conference in Artificial Intelligence, pp. 2256–2261. AAAI Press (2011)
13. Kaminskas, M., Bridge, D.: Measuring surprise in recommender systems. In: Proceedings of ACM RecSys Workshop on Recommender Systems Evaluation: Dimensions and Design (2014)
14. Zhang, Y.C., Séaghdha, D.Ó., Quercia, D., Jambor, T.: Auralist: introducing serendipity into music recommendation. In: Proceedings of the 5th ACM International Conference on Web Search and Data Mining, pp. 13–22 (2012)
15. Hornung, T., Ziegler, C.N., Franz, S., Przyjaciel-Zablocki, M., Schatzle, A., Lausen, G.: Evaluating hybrid music recommender systems. In: Proceedings of IEEE/WIC/ACM International Joint Conferences on Web Intelligence and Intelligent Agent Technology, vol. 1, pp. 57–64. IEEE (2013)
16. Horsburgh, B.: Integrating content and semantic representations for music recommendation. PhD thesis, Robert Gordon University (2013)
17. Celma, O.: Music Recommendation and Discovery: The Long Tail, Long Fail, and Long Play in the Digital Music Space. Springer, Heidelberg (2010)
18. Horsburgh, B., Craw, S., Massie, S.: Music-inspired texture representation. In: Proceedings of the 26th AAAI Conference on Artificial Intelligence, pp. 52–58. AAAI Press (2012)
19. Mermelstein, P.: Distance measures for speech recognition, psychological and instrumental. Pattern Recogn. Artif. Intell. **116**, 91–103 (1976)
20. Stevens, S., Volkmann, J., Newman, E.: A scale for the measurement of the psychological magnitude pitch. J. Acoust. Soc. Am. **8**, 185–190 (1937)
21. Craw, S., Horsburgh, B., Massie, S.: Music recommenders: User evaluation without real users? In: Proceedings of the 24th International Joint Conference in Artificial Intelligence, pp. 1749–1755. AAAI Press (2015)

Goal-Driven Autonomy with Semantically-Annotated Hierarchical Cases

Dustin Dannenhauer$^{(\boxtimes)}$ and Héctor Muñoz-Avila

Department of Computer Science and Engineering,
Lehigh University, Bethlehem 18015, USA
dtd212@lehigh.edu

Abstract. We present *LUiGi-H* a goal-driven autonomy (GDA) agent. Like other GDA agents it introspectively reasons about its own expectations to formulate new goals. Unlike other GDA agents, *LUiGi-H* uses cases consisting of hierarchical plans and semantic annotations of the expectations of those plans. Expectations indicate conditions that must be true when parts of the plan are executed. Using an ontology, semantic annotations are defined via inferred facts enabling *LUiGi-H* to reason with GDA elements at different levels of abstraction. We compared *LUiGi-H* against an ablated version, *LUiGi*, that uses non-hierarchal cases. Both agents have access to the same base-level (i.e. non-hierarchical plans), while only *LUiGi-H* makes use of hierarchical plans. In our experiments, *LUiGi-H* outperforms *LUiGi*.

1 Introduction

Goal-driven autonomy (GDA) is a goal reasoning method in which agents introspectively examine the outcomes of their decisions and formulate new goals as-needed. GDA agents reason about their own expectations of actions by comparing the state obtained after executing actions against an expected state. When a discrepancy occurs, GDA agents formulate an explanation for the discrepancy and based on this explanation, new goals are generated for the agent to pursue.

Case-based reasoning (CBR) has been shown to be an effective method in GDA research. CBR alleviates the knowledge engineering effort of GDA agents by enabling the use of episodic knowledge about previous problem-solving experiences. In previous GDA studies, CBR has been used to represent knowledge about the plans, expectations, explanations and new goals (e.g. [1–3]). A common trait of these works is a plain (non-hierarchical) representation for these elements. In this work we propose the use of episodic GDA knowledge in the form of hierarchical plans that reason on stratified expectations and explanations modeled with ontologies. We conjecture that hierarchical representations enable modeling of stronger concepts thereby facilitating reasoning of GDA elements beyond object-level problem solving strategies on top of the usual (plain) plan representations.

To test our ideas we implemented a new system, which we refer to as *LUiGi-H* and compared it against a baseline that uses plain GDA representations: *LUiGi*.

© Springer International Publishing Switzerland 2015
E. Hüllermeier and M. Minor (Eds.): ICCBR 2015, LNAI 9343, pp. 88–103, 2015.
DOI: 10.1007/978-3-319-24586-7_7

Crucially, both *LUiGi-H* and its baseline *LUiGi* include the same primitive plans. That is, they have access to the same space of sequences of actions that define an automated player's behavior. Hence, any performance difference between the two is due to the enhanced reasoning capabilities; not the capability of one performing actions that the other couldn't. For planning from scratch, HTN planning has been shown to be capable of expressing strategies that cannot be expressed in STRIPS planning [4]. But in this work, plans are not generated from scratch (our systems don't even assume STRIPS operators); instead, plans are retrieved from a case library so those expressiveness results do not apply here.

It is expected that *LUiGi-H* will require increased computation time to reason over the state in order to compute higher level expectations. We test the performance of both *LUiGi-H* and *LUiGi* on the real-time strategy game: Starcraft. Hence, both systems experience a disadvantage if the computation time during reasoning (i.e. planning, discrepancy detection, goal-selection, etc.) is too large. Increased computation time manifests as a delay in the issuing of macro-level strategy (i.e. changing the current plan) to the game-interfacing component of the agent. This will become more clear in Sect. 5 where we discuss the architecture of both agents. In our results *LUiGi-H* outperforms *LUiGi* demonstrating that it can take advantage of the case-based hierarchical knowledge without incurring periods of inactivity from running time overhead.

2   Example

We present an example in the real-time strategy game Starcraft. In Starcraft, players control armies of units to combat and defeat an opponent. In our example and experiments we concentrate on macro-level decisions; low-level management is performed by the underlying default game controller.

Figure 1 shows a hierarchical plan or h-plan used by *LUiGi-H*. This plan, and every plan in the case base, is composed of the primitive actions found in Table 1 at the lowest level of the h-plan (we refer to the lowest level as the 0-level plan). This h-plan achieves the *Attack Ground Surround* task. For visualization purposes we divide the h-plan into two bubbles A and B. Bubble A achieves the two subtasks *Attack Ground Direct* (these are the two overlapping boxes) while Bubble B achieves the *Attack Units Direct* task. For the sake of simplicity we don't show the actual machine-understandable representation of the tasks. In the representation the two *Attack Ground Direct* tasks would only differ on the parameters (one is attacking region A while the other one is attacking region B as illustrated in Fig. 2).

Bubble A contains the two Attack Ground Direct tasks, each of which is composed of the actions: Produce Units, Move Units, and Attack Units. Bubble B contains the task Attack Units Direct which is composed of the actions: Move Units, Attack Units. This h-plan generates an Attack Ground Surround plan for each region surrounding the enemy base. In the example on the map shown in Fig. 2, this happens to be two regions adjacent to the enemy base, therefore the plan contains two Attack Ground Direct that are executed concurrently.

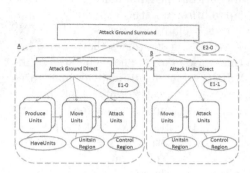

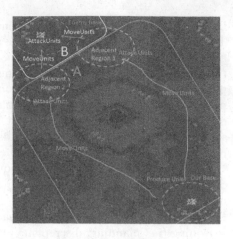

Fig. 1. High Level Plan: AttackGround-Surround

Fig. 2. AttackGroundSurround on map Volcanis

Once the execution of both Attack Ground Direct tasks are completed, the agent's units will be in regions adjacent to the enemy base. At this point, the next task Attack Units Direct is executed, which moves the units into the enemy base and attacks. Reasoning using a more abstract plan such as this one requires representing the notion of surrounding. This is only possible because of *LUiGi-H*'s use of more complex expectations. Specifically, the expectation labeled *E1-0* in Fig. 1 represents the condition that all regions that were attacked are under control (Table 2). In the ontology, the explicit notion of Region Surrounded can be inferred for a region if all of that region's adjacent regions are controlled by the agent (represented by Control Region). In this example there are only two Attack Ground Direct because there are only two adjacent regions to the enemy base). In Fig. 1 each ellipse contains the expectation for its corresponding task. For the primitive tasks or actions, the expectations are as shown in Table 1. For the expectations of tasks at higher levels in the plan, such as for *Attack Ground Direct*, the expectation indicates that our units are successfully located in regions adjacent to the enemy base. Only after this expectation is met, then the agent proceeds to *Attack Units Direct* task (denoted by B in Fig. 1).

3   Goal Driven Autonomy

Goal-driven autonomy consists of a four step cycle. First, a goal is selected by the Goal Manager and sent to the planner. While the plan is being executed, the Discrepancy Detector checks to see if the plan's expectations are met before and after actions are being executed. If a discrepancy is found, the discrepancy is sent to the Explanation Generator and the system comes up with an explanation, which is then sent to the Goal Formulator to create new goal(s). Finally those goals are sent to the Goal Manager and the cycle is repeated again.

Table 1. Primitive Actions and Corresponding Expectations

Action	Pre-expectations	Post-expectations
Produce units	1. Control home base	1. Our player has the given units requested
Move units	1. Control home base	1. Our units are within a given radius of the destination
Attack units	None	1. We control the given region
Attack worker units	None	1. We control the given region

Table 2. High Level Expectations used in Attack Ground Surround

Expectation	Description
E1-0	Control all of the regions from Attack Ground Direct
E1-1	Control region from Attack Direct
E2-0	Control same region as in E1-1

Discrepancy detection plays an important role as the GDA cycle will not begin the process of considering a new goal unless an anomaly occurs. In the domain of Starcraft, the state is very large (on the order of thousands of atoms). The baseline *LUiGi* system solved the problem of mapping expectations to primitive plan actions such as Produce Units, Move Units, and Attack Units by using an ontology. Using the ontology, abstract concepts are inferred from the current state and are used as expectations (i.e. the enemy controls region A). The state is updated as the agent pperceives it (i.e. taking into account fog of war: partial observability of the state). This is done to restrict the size of the state while still maintaining the ability to infer necessary concepts which serve as higher level expectations. The ontology is discussed in more detail in Sect. 5.3.

The Discrepancy Detector uses the ontology to reason over the state and compares current inferred facts to the expectations of the current plans' actions to determine if there is a discrepancy. The Explanation Generator provides an explanation for the discrepancy. The Goal Formulator generates a new goal based on the explanation. The Goal Manager manages which goals will be achieved next.

The crucial difference between *LUiGi* and *LUiGi-H* is that *LUiGi* performs the GDA cycle on level-0 plans. That, is on the primitive tasks or actions such as Produce Units and their expectations (e.g. *Have Units*). In contrast, *LUiGi-H* reasons on expectations at all echelons of the hierarchy. The next sections describe details of the inner workings of *LUiGi-H*.

4   Representation Formalism and Semantics of h-plans

LUiGi-H maintains a library of h-plans. h-plans have a hierarchical structure akin to plans used in hierarchical task network (HTN) planning but, unlike plans in HTN planning, h-plans are annotated with their expectations. In HTN planning only the concrete plan or level-0 plan (i.e. the plan at the bottom of the hierarchy) has expectations as determined by the actions' effects. This tradition is maintained by existing goal-driven autonomy systems that use HTN planners. For example, [5] uses the actions' semantics of the level-0 plans to check if the plans' expectations are met but does not check the upper layers. Our system *LUiGi-H* is the first goal-driven autonomy system to combine expectations of higher echelons of a hierarchical plan and case-based reasoning.

These h-plans encode the strategies that *LUiGi-H* pursues (e.g. the one shown in Fig. 1). Each case contains one such an h-plan including its corresponding expectations. We don't assume the general knowledge needed to generate HTN plans from scratch. Instead, we assume a CBR solution, whereby these h-plans have been captured in the case library. For example, they are provided by an expert as episodic knowledge. This raises the question about how we ensure the semantics of the plans are met; HTN planners such as SHOP guarantee that HTN plans correctly solve the planning problems but require the knowledge engineer to provide the domain knowledge indicating how and when to decompose tasks into subtasks (i.e. methods). In addition, the STRIPS operators must be provided. In our work, we assume that the semantics of the plans are provided in the form of expectations for each of the levels in the h-plan and an ontology Ω that is used to define these expectations.

We define a task to be a symbolic description of an activity that needs to be performed. We define an action or primitive task to be a code call to some external procedure. This enables us to implement actions such as "scorched earth retreat U to Y"(telling unit U to retreat to location Y while destroying any bridge or road along the way) and the code call is implemented by a complex procedure that achieves this action while encoding possible situations that might occur without worrying about having to declare each action's expectations as *(preconditions, effects)* pairs. This flexibility is needed for constructing complex agents (e.g. an Starcraft automated player) where a software library is provided with such code calls but it would be time costly and perhaps infeasible to declare each procedure in such library as an STRIPS operator. We define a compound task as a task that it is not defined through a code call (e.g. compound tasks are decomposed into other tasks, each of which can be compound or primitive).

Formally, an h-plan is defined recursively as follows.

Base Case. A level-0 plan π_0 consists of a sequence of primitive tasks. Each primitive task in a level-0 plan is annotated with an expectation. *Example:* In Fig. 1 the level-0 plan consists of 8 actions: the produce, move, attack sequence is repeated twice (but with different parameters; parameters are not shown for simplicity) followed by the move and attack actions. Each task (shown as a rectangle) has an expectation (shown as an ellipse).

The base case ensures that the bottom level of the hierarchy consists exclusively of primitive tasks and hence can be executed.

Recursive Case. Given a plan π_k of level k (with $k \geq 0$), a level-$k + 1$ plan, π_{k+1} consists of a sequence of tasks such for each task t in π_{k+1} either:

(d1) t is a task in π_k, or

(d2) t is decomposed into a subsequence $t_1...t_m$ of tasks in π_k. *Example:* In Fig. 1, the task Attack Ground Direct is decomposed into the produce, move, attack primitive tasks.

Conditions (d1) and (d2) ensure that each task t in level $k + 1$ either also occurs in level k or it is decomposed into subtasks at level k.

Finally, we require that each task t in the π_{k+1} plan to be annotated with an expectation e_t such that:

(e1) if t meets condition (d1) above, then t has the same expectation e_t for both π_k and π_{k+1}.

(e2) if t meets condition (d2) above, then t is annotated with an expectation e_t such that $e_t \models_\Omega e_m$, where e_m is the expectation for t_m. That is, e_m can be derived from e_t using the ontology Ω or loosely speaking, e_t is a more general condition that e_m. *Example:* The condition *control region* can be derived from condition **E1-0** (Table 2).

An h-plan is a collection $\pi_0, \pi_1, ..., \pi_n$ such that for all k with $(n-1) \geq k \geq 1$, then π_{k+1} is a plan of level $(k+1)$ for π_k. *Example:* the plan in Fig. 1 consists of 3 levels. The level-0 plan consists of 8 primitive tasks starting with *produce units*. The level-1 plan consists of 3 compound tasks: *Attack ground direct* (twice) and *attack unit direct*. The level-2 plan consists of a single compound task: *Attack Ground surround*.

A case library consists of a collection $hp_1, hp_2, ..., hp_m$ where each hp_x is an h-plan.

GDA with h-plans. Because *LUiGi-H* uses h-plans the GDA cycle is adjusted as follows: discrepancies might occur at any level of the hierarchy of the h-plan. Because each task t in the h-plan has an expectation e_t, then the discrepancy might occur at any level-k plan. Thus the cycle might result in a new task at any level k. This in contrast to systems like HTNbots-GDA where discrepancies can only occur at level-0 plans. When a discrepancy occurs for a task t in a level k-plan, an explanation is generated for that discrepancy, and a new goal is generated. This new goal repairs the plan by suggesting a new task repairing t while the rest of the k-level plan remains the same. At the top level, say n, this could mean retrieving a different h-plan. This provides flexibility to do local repairs (e.g. if unit is destroyed, send a replacement unit) or changing the h-plan completely.

Execution of Level-0 Plans. The execution procedure works as follows: each action t_i in the level-0 plan is considered for execution in the order that it appears in the plan. Before executing an action the system checks if the action requires resources from previous actions. If so it will only execute that action if those previous actions' execution is completed. For example, for the level-0 plan in Fig. 1, the plan will begin executing both *Produce Units* but not *Move Units* since they share the same resource: the units that the former produced are used by the latter. The other levels of the h-plan are taken into account when executing the level-0 plan. For example, the action *Move Units* in the portion B of the plan will not be executed until all actions in the portion A are completed because the compound task *Attack Units Direct* occurs after the compound task *Attack Ground Direct.* As a result of this simple mechanism, some actions will be executed in parallel while still committing to the total order of the h-plan.

5   A Hierarchical Case-Based GDA System

Our *LUiGi-H* system combines CBR episodic and hierarchical task network (HTN) representation to attain a GDA system that reasons with expectations, discrepancies and explanations at varied level of abstraction.

Figure 3 shows an overview of *LUiGi-H*. It consists of two main components: the Controller and the Bot. The Controller is the main component of the system and it is responsible for performing the GDA cycle (shown under the box "GDA Cycle"), planning, and reasoning with the ontology.

The Bot is in charge of executing and monitoring the agent's actions. In our experiments *LUiGi-H* plays Starcraft games. Communication between the Controller and the Bot are made with TCP/IP sockets and file share systems.

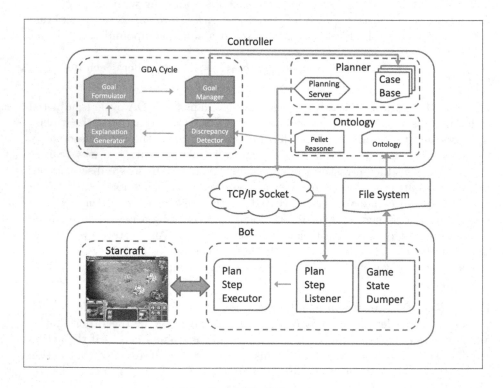

Fig. 3. *LUiGi-H* Overview

5.1   Basic Overview of *LUiGi-H*'s Components

Here we give a brief overview of each component before going into more detail for the Planner in Sect. 5.2 and the Ontology in Sect. 5.3.

The Planner. As explained in Sect. 4, an expert-authored case base is composed of h-plans that encode high-level strategies. Actions are parametrized, for example, the action Produce Units takes a list of pairs of the form (unit-type, count) and the bot will begin to produce that number of units of each type given. All expectations for tasks in the current h-plan are inferred using the ontology Ω, which include all facts in the current state and new facts inferred from the rules in the ontology.

Ontology: The ontology contains entities, relations, and axioms used to infer new facts about the current state of game. It is refreshed every n frames of the Starcraft match, and contains facts such as regions, unit data (health, location, etc.), player data (scores, resources, etc.). The ontology is represented as a semantic web ontology.

Bot: Component that directly interfaces with Starcraft to issue game commands. This component dumps game state data that is loaded into the ontology and listens for actions from the Goal Reasoner.

Game State Dumper: Component within the Starcraft Plan Executor that outputs all of the current game state data to a file which is then used to populate the ontology of the State Model of the controller.

Plan Action Listener: The bot listens for actions from the controller, and as soon as it receives an action it begins executing it independently of other actions. It is the job of the controller to ensure the correct actions are sent to the bot. The bot only has knowledge of how to execute individual actions.

5.2   Planner

While actions in a level-0 plan are the most basic tasks in the context of the h-plans, these actions encode complex behavior in the context of the Starcraft games. For example, Produce Units takes multiple in-game commands to create the desired number and type of units (i.e. 5 Terran Marines). These include commands to to harvest the necessary resources, build required buildings, and issue commands to build each unit. Each action is parametrized to take different arguments. This allows general actions to be used in different situations. For example, Produce Units is used to produce any kind of units, while Move Units is used to move units to any region.

5.3   Ontology

One of the main benefits of using an ontology with GDA is the ability to provide formal definitions of the different elements of a GDA system. The ontology uses

Table 3. Plans

Name	Description	Actions
Attack Ground Direct	Produce ground units and attack the enemy base directly	Produce Units (marine, x)
		Move Units (enemy base)
		Attack Units (enemy base)
Attack Air Direct	Produce air units and attack the enemy base directly	Produce Units (marine, x)
		Move Units (enemy base)
		Attack Units (enemy base)
Attack Both Direct	Produce air and ground units and attack the enemy base directly	Produce Units (marine, x)
		Produce Units (wraith, x)
		Move Units (enemy base)
		Attack Units (enemy base)
Attack Air Sneak	Fly units directly to nearest corner of the map in regards to the enemy base before sending to enemy base	Produce Units (wraith, x)
		Move Units (nearest corner)
		Move Units (enemy base)
		Attack Worker Units (enemy base)
Rush Defend Region	Take all units from a previous plan and defend the home base	Acquire Units (unit ids list)
		Move Units (home base)
		Attack Region (home base)
Attack Ground Surround	Calculates the location of each region surrounding the enemy base, send units to that location, and then attacks the enemy	Attack Ground Direct (adj. regions)
		Attack Ground Direct (enemy base)
Attack Air Surround	Calculates the location of each region surrounding the enemy base, send units to that location, and then attacks the enemy	Attack Air Direct (adj. regions)
		Attack Air Direct (enemy base)
Attack And Distract	Attack directly with ground units while at the same time attacking from behind with air units which focus specifically on killing worker units	Attack Ground Direct (enemy base)
		Attack Air Sneak (enemy base)

facts as its representation of atoms. Facts are ⟨subject, predicate, object⟩ triples. A fact can be an initial fact (e.g. ⟨unit5, hasPosition, (5,6)⟩ which is directly observable) or an inferred fact (e.g. ⟨player1, hasPresenceIn, region3⟩). We use an ontology to represent high-level concepts such as controlling a region. By using a semantic web ontology, that abides by the open-world assumption, it is technically not possible to infer that a region is controlled by a player, unless full knowledge of the game is available. Starcraft is one such domain that intuitively seems natural to abide by the open world assumption because of the fog of war. That is, a player can vie only the portion of the map where it has units deployed. As a result, we can assume local closed world for the areas that are within visual range of our own units. For example, if a region is under visibility of our units and there are no enemy units in that region, we can infer the region is not *contested*, and therefore we can label the region as *controlled*. Similarly, if none of our units are in a region, then we can infer the label of *unknown* for that region.

The following are formal definitions for a GDA agent using a semantic web ontology:

- **State** S: collection of facts
- **Inferred State** S^i: $S \cup \{$ facts inferred from reasoning over the state with the ontology$\}$
- **Goal** g: a desired fact $g \in S^i$
- **Expectation** x: one or more facts contained within the S^i associated with some action. We distinguish between primitive expectations, x_p, and compound expectations, x_c. x_p is a primitive expectation if $x_p \in S$ and x_c is a compound expectation if $x_c \in (S^i - S)$. $(S^i - S)$ denotes the set difference of S^i and S, which is the collection of facts that are strictly inferred.
- **Discrepancy** d: Given an inferred state S^i and an expectation, x, a discrepancy d is defined as:
    1. $d = x$ if $x \notin S^i$, or
    2. $d = \{x\} \cup S^i$ if $\{x\} \cup S^i$ is inconsistent with the ontology
- **Explanation** e: Explanations are directly linked to an expectation. For primitive expectations, such as $x_p = (player1, hasUnit, unit5)$ the explanation is simply the negation of the expectation when that expectation is not met: $\neg x_p$. For compound expectations, x_c (e.g. expectations that are the consequences of rules or facts that are inferred from description logic axioms), the explanation is the trace of the facts that lead up to the relevant rules and/or axioms that cause the inconsistency.

5.4   Discussion

LUiGi-H is composed of two major components, the controller and the bot. The controller handles the goal reasoning processes while the bot interfaces with the game directly. The controller and bot operate separately from each other, and communicate via a socket and file system. There are two methods of data transfer between the controller and the bot. First, every n frames the bot dumps all visible gamestate data to the controller via a file (visible refers to the knowledge that a human player would have access to; the bot does not have global knowledge). The controller then uses this data to populate a semantic web ontology, in which to reason about the game to infer more abstract conclusions (these are used in discrepancy detection). The other method of data transfer is the controller sending messages to the bot which happens via a socket. Both the bot and controller run as completely different processes, use their own memory, and are written in different languages (bot is c++ and controller is java).

The controller's perspective of the game is different than the bot's in a few ways. The controller's game state data is only updated when the Pellet reasoner finishes. The Pellet reasoner is one of a few easily available reasoners for semantic web ontologies. However, the controller's game state data includes more abstract notions such as "I control region x right now". The controller also knows all current actions being executed. As a result, the controller has a overall view of the match but at the loss of some minute details, such as the exact movements of every unit at every frame of the game. This level of detailed information is perceived by the bot but at cost of only having a narrow, instant view of the game. The bot receives actions from the controller, it only receives a single action per plan at a time (when that action finishes, successfully or not, the bot requests the next action of the plan). The bot can execute multiple actions together independently, without knowing which action is going to come next. If the controller decides an action should be aborted while the bot is executing it, it sends a special message to the bot instructing it to stop executing that action.

6   Empirical Evaluation

In order to demonstrate the benefit of h-plans, we ran *LUiGi-H* against the baseline *LUiGi*. Matches occurred on three different maps: Heartbreak Ridge, Challenger, and Volcanis. Heartbreak Ridge is one of the most commonly used maps for Starcraft (it is one of the maps used in AIIDE's annual tournament), while Challenger and Volcanis are commonly played maps. Data was collected every second, and the Starcraft match was run at approximately 20 frames per second (BWAPI[1] function call of setLocalSpeed(20)). The performance metrics are:

- **kill score.** Starcraft assigns a weight to each type of unit, representing the resources needed to create it. For example, a marine is assigned 100 points whereas a siege tank is assigned 700 points.[2] The kill score is the difference between the weighted summation of units that *LUiGi-H* killed minus the weighted summation of units that *LUiGi* killed.
- **razing score.** Starcraft assigns a weight to each type of structure, representing the resources needed to create it. For example, a refinery[3] is assigned 150 points whereas a factory[4] is assigned 600 points. The razing score is the difference between the weighted summation of structures that *LUiGi-H* destroyed minus the weighted summation of structures that *LUiGi* destroyed.
- **total score.** The total score is the summation of the kill score plus the razing score for *LUiGi-H* minus the summation of the kill score plus the razing score for *LUiGi*.

In addition to these performance metrics, the unit score is computed. The unit score is the difference between the total number of units that *LUiGi-H* created minus the total number of units that *LUiGi* created. This is used to assess if one opponent had an advantage because it created more units. This provides a check to ensure that a match wasn't won because one agent produced more units than another.

---

[1] Brood War API: github.com/bwapi/bwapi.

[2] These values come from the Starcraft game engine.

[3] A refinery is a building that allows to harvest gas, a resource needed to produce certain kinds of units. For instance, 100 gas units are needed to produce a single siege tank.

[4] A factory is building that allows the production of certain kinds of units such as siege tanks provided that the required resources have been harvested.

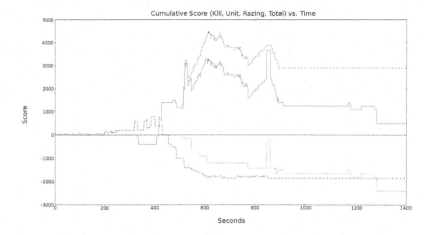

Fig. 4. *LUiGi-H* vs. *LUiGi* on Heartbreak Ridge

We show our results in Fig. 4 below.[5]

The red dashed line shows the kill score, the blue dot-dashed line shows the unit score and the green dotted line is the razing score. The total score, which is the sum of the kill score and razing score is shown as the unbroken cyan line. All lines show the difference in cumulative score of *LUiGi-H* vs. *LUiGi*. A positive value indicates *LUiGi-H* has a higher score than *LUiGi*.

From Fig. 4 we see that *LUiGi-H* ended with a higher total score than *LUiGi*, starting around the 400 second mark. In Fig. 4, the difference in the blue dot-dashed line (unit score) shows that in this match the *LUiGi* system produced far many more units than the *LUiGi-H* system. Despite producing significantly fewer units *LUiGi-H* system outperformed *LUiGi* as can be seen by the total score line (cyan unbroken). *LUiGi-H* scored much higher on the kill score, but less on the razing score. A qualitative analysis revealed that *LUiGi* had slightly more units end game, shown in the graph by the much higher unit score (blue dot dashed), which caused its razing score to be higher than *LUiGi-H*. We expect that as the unit score approaches zero, *LUiGi-H* will exhibit higher kill and especially razing scores. *LUiGi-H* won both this match and the match shown in Fig. 5, on the map Challenger.

Figure 5 shows *LUiGi-H* vs. *LUiGi* on the Challenger map. *LUiGi-H* produces slightly more units in the beginning but towards the end falls behind *LUiGi*. This graph shows a fairer matchup in unit strength. Both the razing score and kill score show *LUiGi-H* outperforming the ablation: *LUiGi*.

LUiGi-H used h-plans with multiple levels of expectations which allowed a more coordinated effort of the primitive actions of a plan. In the situation where *LUiGi-H* and *LUiGi* were executing plans composed of the same primitive actions, in the event of a discrepancy, *LUiGi-H* would trigger discrepancy

---

[5] We plot results for a single run because difference in scores between different runs were small.

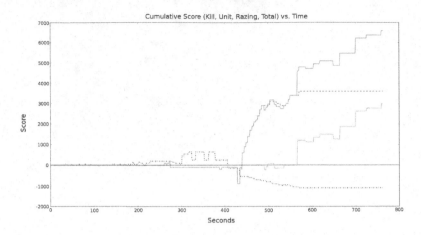

Fig. 5. *LUiGi-H* vs. *LUiGi* on Challenger

detection that would reconsider the broader strategy (the entire h-plan of which the primitive actions were composed from) while *LUiGi* would only change plans related to the single level-0 plan that was affected by the discrepancy. This allows *LUiGi-H* more control in executing high level strategies, such as that depicted in the example in Fig. 1.

A non-trivial task in running this experiment was ensuring that each bot produced roughly the equivalent strength of units (shown in the graph as unit score). While we were unable to meet this ideal in our experiments precisely, including the unit score in the graphs helps identify the chances that a win was more likely because of sheer strength vs. strategy.

We leave out the result from Volcanis due to a loss from a delay due to the reasoning over the ontology. The average time taken by each agent to reason over the ontology is about 1–2 seconds. This is the crucial part of the discrepancy detection step of the GDA cycle. A delay in the reasoning means that discrepancy detection will be delayed. During the match on Volcanis, at the first attack by *LUiGi* on *LUiGi-H* the reasoning hangs and causes discrepancy detection to respond late and fail to change goals before a building is destroyed. This causes *LUiGi-H* a big setback in the beginning of the match and results in a loss of the game. This issue is due to the fact that at any given point in time there are a few hundred atoms in the state (and thus ontology), with greater numbers of atoms during attacks (because the agent now has all the atoms of its enemy units which it can now see). Optimizing the ontology for both reasoning and state space is one possibility for future improvement: an improvement in reasoning time would increase the rate of discrepancy detection. This also demonstrates that even though the GDA cycle is being performed every few seconds while the bot is issuing a few hundred actions per minute, GDA is still beneficial due to the ability to generate and reason about high level strategies.

7   Related Work

To the best of our knowledge, *LUiGi-H* is the first agent to use episodic hierarchical plan representation in the context of goal-driven autonomy where the agent reasons with GDA elements at different levels of abstraction. Nevertheless there are a number of related works which we will now discuss.

Other GDA systems include *LUiGi* [1], GRL [2] and EISBot [3]. As with all GDA systems, their main motivation is for the agents to autonomously react to unexpected situations in the environment. From these, the most closely related is *LUiGi* as it uses ontologies to infer the expectations. However, none of these GDA systems, including *LUiGi*, uses h-plan representations.

The most closely related works are the one from [5] and [6–8], which describe the HTNbots and the ARTUE system respectively. Both HTNBots and ARTUE uses the HTN planner SHOP [9]. SHOP is used because it can generate plans using the provided HTN domain knowledge. This HTN domain knowledge describes how and when to decompose tasks into subtasks. Once the HTN plan is generated, HTNBots and ARTUE discard the k-level plans ($k \geq 1$) and focus their GDA process on the level-0 plans (i.e. the sequence of actions or primitive tasks). That is expectations, discrepancies, explanations, all reason at the level of the actions. There are two main difference versus our work. First, in our work we don't require HTN planning knowledge. Instead, *LUiGi-H* uses episodic knowledge in the form of HTN plans. Second, *LUiGi-H* reasons about the expectations, discrepancies and explanations at all levels of the HTN plan; not just at the level 0. As our empirical evaluation demonstrates, reasoning about all levels of the HTN plans results in better performance of the GDA process compared to a GDA process that reasons only on the level-0 plans.

Other works have proposed combining HTN plan representations and CBR. Included in this group are the PRIAR [10] Caplan/CbC system [11], Process manufacturing case-based HTN planners [12] and the SiN system [13]. None of these systems perform GDA. They use CBR as a meta-level search control to adapt HTN plans as in PRIAR or to use episodic knowledge to enhance partial HTN planning knowledge as in SiN.

Compared to other Starcraft game playing agents, to the best of our knowledge, *LUiGi-H* is the first that plays Starcraft using general plan actions. This enables the use of different groups of units to carry out individual attacks as part of a more complex macro-level strategy at the regional level (e.g. the Attack Ground Surround plan from Table 3). Previous Starcraft bots[6] attack or defend using all units which generally results in giant armies battling one another. Our work is a step towards improving the strategic macro level combat of Starcraft game playing agents.

---

[6] from AIIDE and CIG competitions.

8   Conclusion

In this paper, we presented *LUiGi-H*, a GDA agent that combine CBR episodic knowledge, h-plan knowledge and ontological information enabling it to reason about the plans, expectations, discrepancies, explanations and new goals at different levels of abstraction.

We compared *LUiGi-H* against an ablated version, *LUiGi*. Both agents use the same case base for goal formulation and have access to the same level-0 plans. In our experiments, *LUiGi-H* outperforms *LUiGi* demonstrating the advantage of using episodic hierarchical plan representations over non-hierarchical ones for GDA tasks. We noted one match where *LUiGi-H* lost because of a delay in ontology reasoning time that caused discrepancy detection to respond too slowly to an attack on *LUiGi-H*'s base.

For future work, we will explore using case-based learning techniques to acquire the h-plans automatically from previous problem-solving experiences. Specifically, we envision a situation in which *LUiGi-H* starts with no h-plans and learns these plans from multiple starcraft matches against different opponents. This will in turn allows us to test *LUiGi-H* versus the highly optimized (and hard-coded) entries in the Starcraft competition.

Acknowledgement. This work is funded in part by NSF grant 1217888.

References

1. Dannenhauer, D., Muñoz-Avila, H.: LUIGi: a goal-driven autonomy agent reasoning with ontologies. In: Advances in Cognitive Systems (ACS 2013) (2013)
2. Jaidee, U., Muñoz-Avila, H.: Modeling unit classes as agents in real-time strategy games. In: Ninth AAAI Conference on Artificial Intelligence and Interactive Digital Entertainment (2013)
3. Weber, B.: Integrating learning in a multi-scale agent. Ph.D thesis, University of California, Santa Cruz, June 2012
4. Erol, K., Hendler, J., Nau, D.S.: HTN planning: Complexity and expressivity. In: AAAI **94**, pp. 1123–1128 (1994)
5. Muñoz-Avila, H., Aha, D.W., Jaidee, U., Klenk, M., Molineaux, M.: Applying goal driven autonomy to a team shooter game. In: FLAIRS Conference (2010)
6. Molineaux, M., Klenk, M., Aha, D.W.: Goal-driven autonomy in a navy strategy simulation. In: AAAI (2010)
7. Molineaux, M., Aha, D.W.:P Learning models for predicting surprising events. In: Advances in Cognitive Systems Workshop on Goal Reasoning (2013)
8. Shivashankar, V., Alford, R., Kuter, U., Nau, D.: Hierarchical goal networks and goal-driven autonomy: going where AI planning meets goal reasoning. In: Goal Reasoning: Papers from the ACS Workshop, pp. 95 (2013)
9. Nau, D., Cao, Y., Lotem, A., Muñoz-Avila, H.: SHOP: Simple hierarchical ordered planner. In: Proceedings of the 16th International Joint Conference on Artificial intelligence, vol. 2, pp. 968–973. Morgan Kaufmann Publishers Inc. (1999)
10. Kambhampati, S., Hendler, J.A.: A validation-structure-based theory of plan modification and reuse. Artif. Intell. **55**(2), 193–258 (1992)

11. Muñoz, H., Paulokat, J., Wess, S.: Controlling a nonlinear hierarchical planner using case replay. In: Haton, J.-P., Manago, M., Keane, M.A. (eds.) EWCBR 1994. LNCS, vol. 984. Springer, Heidelberg (1995)
12. Chang, H.-C., Dong, L., Liu, F.X., Lu, W.F.: Indexing and retrieval in machining process planning using case-based reasoning. Artif. Intell. Eng. **14**(1), 1–13 (2000)
13. Muñoz-Avila, H., Aha, D.W., Nau, D.S., Weber, R., Breslow, L., Yaman, F.: Integrating case-based reasoning with task decomposition. Technical report, DTIC Document, Sin (2001)

Evaluating a Textual Adaptation System

Valmi Dufour-Lussier[1,2](✉) and Jean Lieber[2,3,4]

[1] Université de Moncton, Campus de Shippagan, New-Brunswick, Canada
vdl@umcs.ca, lieber@loria.fr
[2] Université de Lorraine, LORIA, 54506 Vandœuvre-lès-Nancy, France
[3] CNRS, 54506 Vandœuvre-lès-Nancy, France
[4] Inria, 54602 Villers-lès-Nancy, France

Abstract. This paper presents a CBR method to retrieve and adapt processes represented as instruction texts, as well as the evaluation methodology that we developed to evaluate it. The evaluation process is user-based, blind and comparative. It is less labour intensive than most existing approaches and is more open to a variety of possible solutions to the same query, among other benefits. It also makes it possible to evaluate separately the textual adaptation process and the underlying formal adaptation process. CRAQPOT, a CBR system that adapts recipe texts, using a case-based process to extract domain knowledge on the fly, is presented and evaluated. We show that it generates recipes of good quality and texts of acceptable quality.

Keywords: Adaptation · Evaluation · Textual case-based reasoning · Process-oriented case-based reasoning

1 Introduction

Textual [26] and process-oriented [18] case-based reasoning are two fields of case-based reasoning (CBR) that tend to use unconventional case structures. They have therefore required the development of specific retrieval techniques, which are rather well established nowadays. Adaptation, on the other hand, has been more problematic. In textual CBR, it has mostly been limited to selecting and aggregating parts of textual cases. Different techniques have been proposed recently to make a deeper level of adaptation possible in process-oriented CBR.

We propose CRAQPOT, a CBR system that retrieves and adapts processes represented as instruction texts. In this system, cases are recipe texts associated with a formal case structure. The structure consists in a network of temporal constraints on events, represented using a qualitative algebra based on Allen's interval calculus [3]. When a user makes a query, a case is retrieved, and both the text and the formal structure are adapted. In order to evaluate the quality of those adaptations, we have made CRAQPOT available as a Facebook application that offers a helpful service to users while encouraging them to provide evaluations.

© Springer International Publishing Switzerland 2015
E. Hüllermeier and M. Minor (Eds.): ICCBR 2015, LNAI 9343, pp. 104–118, 2015.
DOI: 10.1007/978-3-319-24586-7_8

In Sects. 2 and 3, a short introduction to adaptation in textual and process-oriented case-based reasoning is given, and existing evaluation frameworks are discussed. Section 4 presents CRAQPOT, and Sect. 5 details how the domain knowledge that is needed for the adaptation is "simulated". The evaluation methodology is presented in Sects. 6 and 7 presents the results. Concluding remarks and future work are shown in Sect. 8.

2   Adapting Textual Cases

The CBR community recognises that significant knowledge is available in a textual format and, consequently, that being able to exploit this text can be a great help in deploying CBR applications. There has been a significant interest since the very beginning of CBR in systems which use texts as cases. This interest has been expressed, among other things, by a series of workshops on textual CBR at the 1998 AAAI conference, as well as at International and European CBR Conference from 2005 to 2007. Most work in textual CBR has focused on retrieval, but a few have taken an interest in trying to reuse texts.

2.1   Principles

The problem of text reuse in textual case-based reasoning has been addressed in different manners. In [1], textual solutions are reused by identifying small chunks of text to be reused from different solutions and aggregating them, which can be seen as a type of compositional adaptation. An inverse approach is also possible, in which a text is reused in whole but parts that should be modified are identified [14].

Another way is to use a natural language generation system following the adaptation of the underlying formal representation of the textual solution [13]. The approach we propose is based on regeneration, that is starting from existing text or text fragments and making linguistic changes therein. We are aware of only one other system that uses a similar approach, which is CookIIS [21], which performs string substitutions based on ingredient substitutions in recipe texts.

2.2   Evaluation

The part of a textual adaptation system responsible for selecting parts of a case text to be reused as part of a solution text can be evaluated. The typical way to evaluate such a system is to annotate manually the sentences that are expected to be reused in the answer to test queries, then compare the actual result of the system with the expected result, computing a precision and a recall score.

On the other hand, other aspects of textual adaptation have not really been evaluated before. In particular, we are aware of no prior work that aimed at evaluating actual text quality, nor at evaluating the quality of the text adaptation separately from the quality of the underlying adaptation mechanism.

3 Adapting Procedural Cases

More often, in CBR, the temporal aspect is taken into account by considering sequences of events, sometimes integrating relative or absolute time stamps [8,15,25]. The most advanced work in this respect is that of process-oriented CBR (PO-CBR), in which cases are often made of activities structured using workflows. In CBR, workflows are usually expressed in a graph-based formal language, such as the one described in [19], to make retrieval and adaptation possible. Again, retrieval has received substantial research interest, but little work has been done on the adaptation of procedures.

3.1 Principles

Arguably the first approach to workflow adaptation was case-based adaptation: adaptation cases, which are combinations of a source case, a change request and the resulting case, are used as a source of adaptation knowledge [16]. A somewhat similar approach identifies small workflow parts from the case base that attain specific goals, and uses this to make substitutions of parts of the retrieved workflow [20].

3.2 Evaluation

Most evaluation work in procedural adaptation is manual: either test queries are provided along with the expected result, or users are asked to evaluate the quality of the result. This is labour intensive and requires the intervention of both domain experts and of people familiar with the formalism used.

An alternative, automatic approach is to compare the various parts of the solution with the case base, the expectation being that if a generated workflow part describes a feasible activity, it is likely to occur naturally in a large enough case base [17].

4 CRAQPOT

In this section, we present our own approach to textual and procedural adaptation and its software implementation, named CRAQPOT— the Case-based Reasoning Adaptor of Qualitative Procedures Over Texts— which provides an interface to obtain recipes in response to any query. As its case base, CRAQPOT uses the recipe database published for the 2nd Computer Cooking Contest.[1] The processes for case acquisition, for retrieval, for formal case adaptation and for textual adaptation are all implemented as separate modules, and will therefore be presented individually in the following subsections.

---

[1] http://www.wi2.uni-trier.de/shared/eccbr/ccc09/.

4.1 Case Acquisition

The cases are provided as unannotated text, and so their formal counterpart must be extracted to make adaptation possible. This is the case acquisition step, which is detailed in [11]. Our approach is based on natural language processing, and therefore goes through much of the same main steps as any other natural language understanding system:

- Identifying word, clause and sentence boundaries. This task is performed using hand-crafted regular expressions.
- Identifying the part-of-speech of each word, i.e. finding verbs, nouns, etc. This task is performed by a Brill tagger [5], a semi-supervised machine learning tool trained on a small set of annotated recipes.
- Performing syntactic analysis. This is done using a chunker, which is a parser using a regular grammar, implemented using regular expressions, which is not able to compute a complete parse tree, but can find noun and prepositional phrases. Those are sufficient to identify verb complements, which correspond to action parameters.

This is not the only possible approach: for instance, in [23], satisfying results are obtained over the same type of texts, using information extraction. Both approaches are efficient with instruction texts but would require adjustments to give good results with different types of text. Another important and difficult step is resolving anaphoras i.e. associating words from the text with the objects they are referring to. To this end, we implemented certain ideas from dynamic semantics, wherein actions expressed in the text are considered as creating, transforming and removing objects.

Once all the relevant linguistic information has been identified in a text, annotation rules are used to translate it into workflow patterns or, more interestingly in our case, in qualitative constraints between events—cooking actions and states— expressed using the qualitative algebra $\mathcal{INDU}$ [22], an extension of Allen interval calculus [3]. The 9 annotation rules used in this implementation are detailed in [9].

As an example, the following $\mathcal{INDU}$ constraints would be part of a simplified formal case representation of the recipe shown in Fig. 1:

$$
\begin{aligned}
&\text{cook rice ?}^= \text{18 min} && \textit{Rice cooks for 18 min.} \\
&\text{cook mushrooms ?}^= \text{2 min} && \textit{Mushrooms cook for 2 min.} \\
&\text{cook mushrooms } \{f^>\} \text{ cook rice} && \textit{Mushrooms start cooking after rice and} \\
& && \textit{finish at the same time.}
\end{aligned}
\tag{1}
$$

Because case acquisition from text is not perfect, it is essential to evaluate it separately to interpret the overall evaluation results of the system, because any error at this stage will correspond to a decrease in the solution quality further down the road.

Mushroom risotto

Heat the oil and butter. Add the onion and cook until soft, about one minute. Add the rice and cook for two minutes, then add a glass of wine. Once the wine is evaporated, start adding broth, one ladleful at a time. Meanwhile, slice the mushrooms. Add them two minutes before the end.

Fig. 1. A simple mushroom risotto recipe.

4.2   Retrieval

While it could in theory rely on an approach inspired by adaptation-guided retrieval [24], in practice CRAQPOT relies on a reimplementation of TUUUR-BINE [12], a generic, ontology-guided case-based inference engine, using WIKI-TAAABLE[2] [4,7] as its knowledge base. Our evaluation framework is based on the comparison of different adaptation approaches all using the same retrieval engine, so retrieval should not have a strong influence on the evaluation results.

4.3   Case Adaptation

When a solution to a user query cannot be retrieved from the case base, adaptation is required, which in CRAQPOT begins with a substitution. If, for instance, the user wants a recipe for a carrot risotto and the case base does not contain one, the mushroom risotto recipe of Fig. 1 may be retrieved. The system will then adapt the retrieved recipe by replacing mushrooms with carrots, and making whichever modifications are necessary to the instructions to obtain a satisfactory result— for instance, adding the carrots earlier during the cooking because otherwise they would be too crunchy— as described into more details in [10].

Intuitively, this is done by finding the conjunction of the retrieved recipe modified by the necessary ingredient substitutions and of the domain knowledge that is available about the new ingredients— for instance, their required cooking time, which may be represented as

$$\text{cook carrots } ?^= 20 \text{ min} \quad \textit{Carrots cook for 20 min.} \tag{2}$$

Because a *qualitative* algebra is used, metric information must be specified with additional knowledge:

$$\begin{aligned} 2 \text{ min } ?^< 18 \text{ min} \quad &\textit{2 min are shorter than 18.} \\ 18 \text{ min } ?^< 20 \text{ min} \quad &\textit{18 min are shorter than 20.} \end{aligned} \tag{3}$$

Whenever adaptation is actually necessary, though, this will be because there is a contradiction between the retrieved recipe and the domain knowledge, and

---

[2] http://wikitaaable.loria.fr.

so there will be no conjunction. In the example given, replacing mushrooms with carrots will expose a contradiction between "cook ~~mushrooms~~ carrots ?= 2 min" from (1) and "cook carrots ?= 20 min" from (2).

The workaround is to use belief revision theory [2] to make minimal modifications to the recipe in such a way that it becomes consistent with the domain knowledge, an approach that has already been used successfully for adaptation in CBR [6]. In the example this would, among other things, replace the last constraints of (1) with

$$\text{cook } \sout{\text{mushrooms}}\text{ carrots } \{fi^{>}\} \text{ cook rice} \quad \textit{Carrots start cooking before}$$
$$\textit{rice and finish at the same time.}$$
$$(4)$$

The implementation of a belief revision operator is a search algorithm that looks through the possible interpretations of the set of qualitative constraints that come from the domain knowledge to find those closest to the constraints of the source case. A set of constraints has an exponential amount of possible interpretations with respect to the number of intervals used in the case representation, therefore the search takes exponential time. We were able, though, to implement an approximation algorithm that reuses modified constraint satisfaction problem algorithms algorithms to obtain satisfactory results in polynomial time.

4.4   Text Adaptation

Once the formal constraints have been adapted, the text must be modified to reflect the changes. The easiest solution, given that annotation rules exist that associate linguistic features to algebraic constraints, would be to use the inverse of those rules: given a constraint change, find the set of linguistic features that would have generated this constraint, and change the actual linguistic features of the text to reflect those. If the set of annotation rules were a bijection between the set of sets of possible linguistic features and the possible algebraic relations, this would be straightforward. But it is not, and therefore specific strategies are used to make approximate changes in text, with the objective always being to make the smallest possible changes, to limit the risk of introducing mistakes or diminishing the quality of the text.

Additionally, the implementation favours moving events such that they appear in the text in the order in which they begin, which minimises changes inside the sentences at the expense of maximising the movement of whole sentences.

With respect to the change described in (4), CRAQPOT makes the following modifications:

Add the onion and cook until soft, about one minute. Meanwhile, slice the ~~mushrooms~~ carrots. Add them ~~two minutes before the end~~. Add the rice and cook for two minutes, then add a glass of wine. Once the wine is evaporated, start adding broth, one ladleful at a time.

Observe that, because a qualitative algebra is used, it is not possible for the system to know, and therefore indicate, that the rice should be added two minutes after the carrots. This is a tradeoff for the algorithmic feasibility of the approach.

5   Simulating Domain Knowledge

One benefit of revision-based adaptation is that it can use whichever amount of domain knowledge is available. If no domain knowledge is available at all, the system will still work but give a result equivalent to null-adaptation. If complete domain knowledge is available, the system will give a result equivalent to a classic planning system. Any intermediary level of available domain knowledge will be used to improve the results of the adaptation.

The acquisition of domain knowledge for a case-based reasoning application falls outside the scope of this work. On the other hand, in order to get meaningful adaptation from CRAQPOT that makes it possible to evaluate the system, some quantity of knowledge is needed. We have therefore created a system to simulate domain knowledge on the fly.

While it would have been possible, for instance, to consider that the domain knowledge about the cooking of carrots is the disjunction of all the ways that carrots are cooked in our recipe base, this would have given little constrained knowledge, resulting in limited adaptations. For instance, we may have a recipe for a carrot salad in which the cooking time is 0 min, and one for a soup in which the cooking time is 60 min, which would suggest that any cooking time between 0 and 60 min is acceptable, with the effect that the mushroom risotto recipe would not be modified at all. We considered it would be more relevant for an evaluation of adaptation to use highly constrained knowledge, which requires a high adaptation effort.

Therefore, we have developed a system for on-the-fly extraction of relevant domain knowledge. This method can be seen as an additional retrieval stage, during which more cases are retrieved to be used in guiding adaptation. Given a recipe Source and a substitution $p \rightsquigarrow q$, a new recipe KnowledgeSource containing q is retrieved, such that ingredient q in this recipe is treated as much as possible in a similar way as ingredient p in Source.

For instance, if the user requests a carrot risotto recipe and a mushroom risotto recipe is retrieved, the system will attempt to retrieve some recipe with carrot and obtain carrot knowledge from it. Suppose that three recipes with carrots exist: a soup recipe in which carrots are cooked for one hour until they decompose in the broth, a salad recipe in which carrots are shredded and used raw, and a Asian recipe for sauteed pasta and vegetables. The system will retrieve the recipe in which carrots are cooked in the way most similar to how mushrooms are cooked in the mushroom risotto, which will be the Asian recipe. The way the carrot is used in this recipe, e.g. how it is cut and how long it is cooked, will become the carrot knowledge used to perform this adaptation, which is referred to in (2).

In our implementation, the distance function used is a Hamming distance: the distance between the recipe Source containing ingredient p and a candidate

KnowledgeSource recipe containing ingredient q is the amount of actions applied to p in Source that are not applied to q in the KnowledgeSource candidate, plus the amount of actions applied to q in the KnowledgeSource candidate that are not applied to p in Source. An unweighted Hamming distance was chosen because it makes the retrieval engine simple, but a different distance function may be desirable if not all actions or action substitutions are considered to be of equivalent importance.

Using this type of overly constrained domain knowledge can affect the outcome of the adaptation both positively and negatively. For instance, if the carrot recipe most similar to the mushroom risotto is in fact a carrot soup, it may be that our domain knowledge will demand for carrots to be cooked for one hour, and the adaptation result will suffer from this. On the other hand, the alternative, under-constrained approach could result in accepting a two-minute cooking time for carrots on the basis, for instance, of a carrot salad recipe in which carrots are not cooked at all.

6   Evaluation Framework

This section presents the evaluation methodology we propose for adaptation in textual CBR. It is a comparative, blind, user-based approach: a user makes a query, and is shown a result obtained from one of various different adaptation techniques available. They are then asked to evaluate the result based on a set of criteria.

This type of evaluation is less time-consuming than an evaluation based on combinations of test queries and expected results, and we also think it is more accurate because it does justice to the creativity of the system, which may be able to provide results that did not occur to the designers of the test cases yet fully satisfy the users. With respect to user-based evaluation of workflow adaptation systems, it is also very advantageous in that the users need not be fluent in the formalism underlying the adaptation in order to be able to evaluate the system. As all user-based evaluation methodologies though, this evaluation is by definition a black box type evaluation: while we can know, for a given query–result pair, whether it gave satisfaction to the user, there is no automatic way to determine what went wrong in case it doesn't.

In the next subsections, the evaluation interface, the compared methods, and the evaluation criteria are shown.

6.1   Interface

A new user first needs to create an account, which is automatic if they are accessing the application from Facebook. They can then immediately make a query, as shown in Fig. 2.

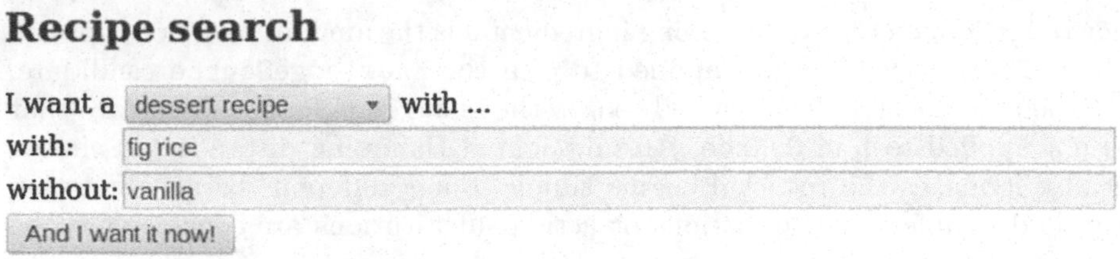

Recipe search

I want a [dessert recipe   ▾] with ...

with: [fig rice   ]

without: [vanilla   ]

[And I want it now!]

Fig. 2. CRAQPOT query interface.

6.2   Presentation of the Adaptation Methods

A method is selected randomly between the two control methods and the experimental method, and the query is processed with this method. If processing fails— which is theoretically possible only with the retrieval method— processing is transferred to another method.

The system has access to two control methods to answer requests:

- A retrieval-only method that performs no adaptation and fails when it is not possible to find a recipe that corresponds exactly to the query. Since CRAQPOT uses the same textual case base, the retrieval system defines the maximum possible scores that could be obtained in text and recipe quality: our proposal is not expected to adapt recipes in such a way that the result is better than the original.
- A method based on a reimplementation of COOKIIS text adaptation, which performs a smart string replacement [21]. This provides a baseline: given the simplicity and efficiency of COOKIIS text adaptation, our proposal would be difficult to justify if it did not offer better quality.

Applying the query from the running example (a carrot risotto) to the retrieval-only method would fail—the query would then be handled by one of the other methods. Applying the same to the COOKIIS method would work, but the instructions would only be modified insofar as the word "mushrooms" would be replaced with the word "carrots", resulting in a two minute cooking time for carrots: "Meanwhile, slice the ~~mushrooms~~ carrots. Add them two minutes before the end."

6.3   Evaluation Questions

The result is then presented to the user, as shown in Fig. 3. The user has no way of knowing which method was used to process their query.

Before the user can make a new query, they are asked to tell how much they agree with the following statements:

- "This recipe seems tasty." We postulate that this provides the most relevant indicator to evaluate the quality of the *content* adaptation of the recipe— independantly of the way it is written.

CRAQPOT
The neverending recipe book

Glutinous rice with figs

Your query: a #dessert# recipe with fig rice without vanilla

- 1 ts Salt
- 1/2 c Sugar
- 3 c Glutinous rice
- 1 1/2 c Coconut cream
- 6 Ripe figs, well chilled
- 2 tb Sesame seeds, toasted

Soak the rice in cold water for 2 hours. Drain. Line a steamer with cheesecloth, heat steamer and lay rice on the cheesecloth. Steam for 30 minutes or until cooked through. The rice will become glossy. Mix the SEASONINGS ingredients in a large bowl and gently mix in the hot steamed rice. Cover tightly and let soak for 30 minutes to absorb the coconut flavour. Blend the SAUCE ingredients in a pot and heat until it just reaches the boiling point. Let cool. Peel the figs, slice lengthwise and remove the pits. Divide the rice among 6 plates. Place fig slices on top and cover with the sauce. Sprinkle with the sesame seeds and serve.

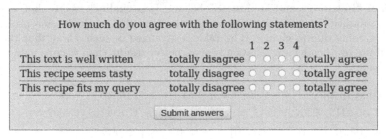

Fig. 3. CRAQPOT response and evaluation interface. This screenshot shows a recipe that was adapted by the COOKIIS method.

- "This text is well written." We postulate that this provides the most relevant indicator to evaluate the quality of the *textual* adaptation of the recipe.
- "This recipe fits my query." We postulate that this provides a general indicator as to whether the adaptation approach used was appropriate.

Users rate their degree of agreement on a 4-point Likert scale— where 1 indicates strong disagreement and 4 indicates strong agreement.

The hypotheses we made are:

H₁ CRAQPOT and COOKIIS will output lower text and recipe quality, and lower fitness—i.e. lower score on the third criterion—than simple retrieval.
H₂ CRAQPOT will output higher recipe quality than COOKIIS.
H₃ CRAQPOT and COOKIIS will output a similar text quality.
H₄ CRAQPOT will leave its users with a better impression that the answer fits the query than COOKIIS.

We postulate H_1 because of the inherent risk of automatic adaptation, H_2 because CRAQPOT, unlike COOKIIS, integrates domain knowledge, H_3 because the risk of adaptation is mitigated by finer linguistic processing, and H_4 because the adaptation is less superficial. The null hypothesis H_0 is that all three systems are comparable and any difference in score would be the result of chance.

7   Results

Raw results based on 9 users performing 50 queries are shown in Fig. 4. Wilcoxon signed-rank tests were performed for each criterion to compare methods pairwise and measure the probability that the observed differences in scores are the result of chance. The resulting p-values thresholds are shown in Table 1. It is commonly assumed that p-values between .05 and .1 offer a weak presumption against H_0, whereas p-values below .05 offer a strong presumption against H_0. Values below .01 offer a very strong presumption.

Table 1. Significance of the pairwise method comparisons. For instance, "< .05" at the intersection of "Retrieval" and "COOKIIS" below "Text quality" means that the difference in text quality between the retrieval and the COOKIIS method has a probability $p \leq 5\%$ of being due to chance. The table is arranged in such a way that the system named in the row header systematically gives better results than the one named in the column header.

	Text quality		Recipe quality		Fitness	
	COOKIIS	CRAQPOT	COOKIIS	CRAQPOT	COOKIIS	CRAQPOT
Retrieval	< .05	< .1	< .01	< .05	< .001	> .5
CRAQPOT	> .5		< .5		< .001	

In all three indicators, retrieval ranked first, and CRAQPOT ranked second. As expected, CRAQPOT's and COOKIIS overall performance is worse than simple retrieval, partially validating H_1. This is mitigated, though, by the fact that simple retrieval was able to process only 46 % of the queries assigned to it. In recipe quality, retrieval performed significantly better than CRAQPOT and strongly significantly better than COOKIIS, but the score difference between CRAQPOT and COOKIIS, while important, was not significant: the exact p-value is .30. This indicates a 30 % chance that CRAQPOT's better scores with respect to COOKIIS were the effect of chance, and therefore H_2 is not supported by the results. Further evaluations may change this. In text quality, retrieval performed strongly significantly better than CRAQPOT and very strongly significantly better than COOKIIS. Although the evaluations surprisingly show that CRAQPOT did better than COOKIIS, the difference is not statistically significant, confirming H_3. In fitness, CRAQPOT performed just as well as retrieval, and both methods were very strongly significantly better than COOKIIS, confirming H_4.

The few available evaluations make it possible to claim that H_3 and H_4 are verified, and that H_1 is partially verified. H_2 is not verified but more evaluations will be necessary.

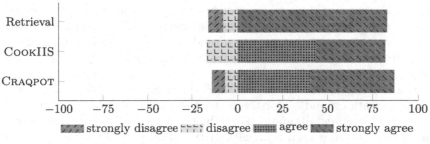

(a) Results for criteria: "This text is well written."

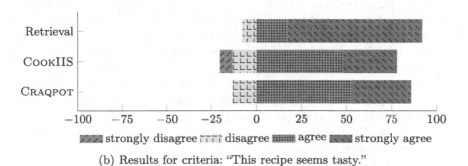

(b) Results for criteria: "This recipe seems tasty."

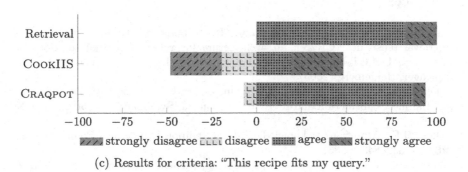

(c) Results for criteria: "This recipe fits my query."

Fig. 4. Detailed user evaluations.

8   Conclusion

We have presented CRAQPOT, a CBR system that retrieves and adapts processes represented as instruction texts and the evaluation methodology we developed to evaluate it.

The evaluation approach we propose has many benefits with respect to existing approaches used in textual and process-oriented CBR. Compared to sets of queries–expected results that are often used, it is much less labour-intensive to put in place, and it gives value to creative solutions proposed by a system. Additionally, because we provide the results as text, we can rely on domain experts that are not fluent in the formalism to provide evaluations. Yet, we are able to obtain a separate evaluation for the textual and for the underlying formal case adaptation. Certain details about the evaluation process are specific to the adaptation of processes, but we believe with further work it would be easy to redefine our methodology in a more generic way for various textual CBR applications.

The evaluation results for our application, CRAQPOT, were mostly satisfactory, although more evaluations would be needed to obtain stronger statistical significance. There are many benefits from developing CRAQPOT as a Facebook application that have been left as future work: for instance, the possibility of using the user's timeline in order to advertise for the application, and even obtain multiple evaluations for the same query, using the different methods, by appealing to their network.

We also proposed a system based on case-based retrieval for on-the-fly extraction of relevant knowledge. This system made it possible to evaluate our application without having to specify complete domain knowledge. We also think that, given further study to make it more generic, this approach could actually be developed into a fully adequate way of integrating the experience of many cases in order to adapt a source case.

References

1. Adeyanju, I., Wiratunga, N., Lothian, R., Sripada, S., Lamontagne, L.: Case retrieval reuse net (CR2N): an architecture for reuse of textual solutions. In: McGinty, L., Wilson, D.C. (eds.) ICCBR 2009. LNCS, vol. 5650, pp. 14–28. Springer, Heidelberg (2009)
2. Alchourrón, C.E., Gärdenfors, P., Makinson, D.: On the logic of theory change: partial meet contraction and revision functions. J. Symbolic Logic **50**(2), 510–530 (1985)
3. Allen, J.F.: Maintaining knowledge about temporal intervals. Commun. ACM **26**(11), 832–843 (1983)
4. Badra, F., Cojan, J., Cordier, A., Lieber, J., Meilender, T., Mille, A., Molli, P., Nauer, E., Napoli, A., Skaf-Molli, H., Toussaint, Y.: Knowledge acquisition and discovery for the textual case-based cooking system WikiTaaable. In: Delany, S.J. (ed.) ICCBR 2009 Workshop Proceedings, pp. 249–258 (2009)
5. Brill, E.: A simple rule-based part of speech tagger. In: Workshop on Speech and Natural Language, Association for Computational Linguistics, pp. 112–116 (1992)
6. Cojan, J., Lieber, J.: Applying belief revision to case-based reasoning. In: Prade, H., Richard, G. (eds.) Computational Approaches to Analogical Reasoning: Current Trends, pp. 133–161. Springer, Heidelberg (2014)

7. Cordier, A., Dufour-Lussier, V., Lieber, J., Nauer, E., Badra, F., Cojan, J., Gaillard, E., Infante-Blanco, L., Molli, P., Napoli, A., Skaf-Molli, H.: Taaable: a case-based system for personalized Cooking. Studies in Computational Intelligence. In: Successful Case-based Reasoning Applications. Springer, Heidelberg (2013, in press)
8. Dojat, M., Ramaux, N., Fontaine, D.: Scenario recognition for temporal reasoning in medical domains. Artif. Intell. Med. **14**(1–2), 139–155 (1998). Selected Papers from AIME 1997
9. Dufour-Lussier, V.: Spatial-temporal qualitative reasoning from textual cases. Ph.D. thesis, Université de Lorraine (2014)
10. Dufour-Lussier, V., Le Ber, F., Lieber, J., Martin, L.: Adapting spatial and temporal cases. In: Agudo, B.D., Watson, I. (eds.) ICCBR 2012. LNCS, vol. 7466, pp. 77–91. Springer, Heidelberg (2012)
11. Dufour-Lussier, V., Le Ber, F., Lieber, J., Nauer, E.: Automatic case acquisition from texts for process-oriented case-based reasoning. Inf. Syst. **40**, 153–167 (2014)
12. Gaillard, E., Infante-Blanco, L., Lieber, J., Nauer, E.: Tuuurbine: a generic CBR engine over RDFS. In: Lamontagne, L., Plaza, E. (eds.) ICCBR 2014. LNCS, vol. 8765, pp. 140–154. Springer, Heidelberg (2014)
13. Gervás, P., Hervás, R., Recio-García, J.A.: The role of natural language generation during adaptation in textual CBR. In: 4th Workshop on Textual Case-Based Reasoning: Beyond Retrieval (ICCBR 2007), pp. 227–235 (2007)
14. Lamontagne, L., Lee, H.-H.: Textual reuse for email response. In: Funk, P., González Calero, P.A. (eds.) ECCBR 2004. LNCS (LNAI), vol. 3155, pp. 242–256. Springer, Heidelberg (2004)
15. Ma, J., Knight, B.: A framework for historical case-based reasoning. In: Ashley, K.D., Bridge, D.G. (eds.) Case-Based Reasoning Research and Development. LNCS, vol. 2689, pp. 246–260. Springer, Heidelberg (2003)
16. Minor, M., Bergmann, R., Görg, S., Walter, K.: Towards case-based adaptation of workflows. In: Bichindaritz, I., Montani, S. (eds.) ICCBR 2010. LNCS, vol. 6176, pp. 421–435. Springer, Heidelberg (2010)
17. Minor, M., Islam, M.S., Schumacher, P.: Confidence in workflow adaptation. In: Agudo, B.D., Watson, I. (eds.) ICCBR 2012. LNCS, vol. 7466, pp. 255–268. Springer, Heidelberg (2012)
18. Recio-García, J.A., Minor, M., Montani, S.: Process-oriented case-based reasoning. Inf. Syst. **40**, 103–105 (2014)
19. Minor, M., Schmalen, D., Bergmann, R.: XML-based representation of agile workflows. In: Bichler, M., Hess, T., Krcmar, H., Lechner, U., Matthes, F., Picot, A., Speitkamp, B., Wolf, P. (eds.) Multikonferenz Wirtschaftsinformatik, pp. 439–440. GITO, Berlin (2008)
20. Müller, G., Bergmann, R.: Workflow streams: a means for compositional adaptation in process-oriented CBR. In: Lamontagne, L., Plaza, E. (eds.) ICCBR 2014. LNCS, vol. 8765, pp. 315–329. Springer, Heidelberg (2014)
21. Newo, R., Bach, K., Hanft, A., Althoff, K.D.: On-demand recipe processing based on CBR. In: ICCBR 2010 Workshop Proceedings, pp.209–218 (2010)
22. Pujari, A.K., Kumari, G.V., Sattar, A.: INDU: an interval & duration network. In: Foo, N. (ed.) Advanced Topics in Artificial Intelligence. Lecture Notes in Computer Science, vol. 1747, pp. 291–303. Springer, Heidelberg (1999)
23. Schumacher, P., Minor, M., Schulte-Zurhausen, E.: On the use of anaphora resolution for workflow extraction. In: Bouabana-Tebibel, T., Rubin, S.H. (eds.) Integration of Reusable Systems. Advances in Intelligent Systems and Computing, vol. 263, pp. 151–170. Springer, Cham (2014)

118   V. Dufour-Lussier and J. Lieber

24. Smyth, B., Keane, M.T.: Adaptation-guided retrieval: questioning the similarity assumption in reasoning. Artif. Intell. **102**(2), 249–293 (1998)
25. Sánchez-Marré, M., Cortés, U., Martínez, M., Comas, J., Rodríguez-Roda, I.: An approach for temporal case-based reasoning: episode-based reasoning. In: Muñoz-Ávila, H., Ricci, F. (eds.) ICCBR 2005. LNCS (LNAI), vol. 3620, pp. 465–476. Springer, Heidelberg (2005)
26. Weber, R.O., Ashley, K.D., Brüninghaus, S.: Textual case-based reasoning. Knowl. Eng. Rev. **20**(3), 255–260 (2005)

Visual Case Retrieval for Interpreting Skill Demonstrations

Tesca Fitzgerald[(✉)], Keith McGreggor, Baris Akgun,
Andrea Thomaz, and Ashok Goel

School of Interactive Computing, Georgia Institute of Technology,
30332 Atlanta, Georgia
{tesca.fitzgerald,bakgun3,athomaz,goel}@cc.gatech.edu,
keith.mcgreggor@venturelab.gatech.edu

Abstract. Imitation is a well known method for learning. Case-based reasoning is an important paradigm for imitation learning; thus, case retrieval is a necessary step in case-based interpretation of skill demonstrations. In the context of a case-based robot that learns by imitation, each case may represent a demonstration of a skill that a robot has previously observed. Before it may reuse a familiar, *source* skill demonstration to address a new, *target* problem, the robot must first retrieve from its case memory the most relevant source skill demonstration. We describe three techniques for visual case retrieval in this context: feature matching, feature transformation matching, and feature transformation matching using fractal representations. We found that each method enables visual case retrieval under a different set of conditions pertaining to the nature of the skill demonstration.

Keywords: Visual case retrieval · Case-based agents · Imitation learning

1 Introduction

Learning by imitation is a well-researched methodology, both in human cognition and in cognitive robotics [2,18,26]. Robot learning by demonstration is an approach which aims to enable imitation by having the robot receive a demonstration of a skill from a human teacher. The robot perceives the workspace and objects involved in completing the skill during the demonstration, while also recording the actions required to complete the skill. At a later time, the robot may be asked to repeat the learned skill in the same or in a new workspace.

Case-based reasoning is an important paradigm for learning by imitation (e.g. [6,7]). In the case-based approach to imitation, the robot would (i) store the observed skill demonstrations as cases in a case memory, (ii) given a new, related problem, retrieve the most similar case from the case memory, (iii) adapt the demonstrated actions from the retrieved case to the new problem, and (iv) execute the adapted actions to address the new problem. We refer to the first two steps

E. Hüllermeier and M. Minor (Eds.): ICCBR 2015, LNAI 9343, pp. 119–133, 2015.
DOI: 10.1007/978-3-319-24586-7_9

of this approach as *skill demonstration interpretation*. Note that a necessary step in skill demonstration interpretation is for the robot to recall the skill demonstration most similar to the current configuration of objects. Thus, in this paper, we focus *solely on this task of case retrieval* to enable case-based interpretation of skill demonstrations in the context of interactive robot learning by imitation. The goal of case retrieval in this context is to return a source case demonstrating the same skill as shown in a new, uncategorized skill demonstration.

A critical question in case-based interpretation is that of case representation. A case of a previously observed skill should be represented such that, given a new skill demonstration, it is feasible for the robot to recognize the similarity between the two. In the rest of this paper, we make the following contributions:

1. Propose three visual representations for skill demonstration cases, with corresponding source case retrieval algorithms.
2. Present experiments testing each representation on skill demonstrations provided in a table-top environment.
3. Test the effectiveness of Fractal reasoning on real-world images perceived during skill demonstrations.
4. Compare the efficacy of the three case retrieval methods by providing an analysis of situations in which each method performs better than the others.

2   Background

Case-based reasoning is a cognitively inspired paradigm for reasoning and learning [1,11–13,22,23]; Thagard [25] views case-based reasoning as a paradigm for modeling human cognition. In case-based reasoning, new problems are addressed by retrieving and adapting solutions to similar problems stored as cases in a case memory. In case-based reasoning, (a) learning is incremental, (b) learning is problem-specific in that the robot adapts the most similar case to address the current problem, and (c) learning is lazy, meaning that the robot learns the abstraction only when needed.

Ontanon et al. [19] studied case-based learning from demonstration in the context of online case-based planning in real-time strategy games. While an important domain for case-based reasoning, games do not offer the low-level challenges of perception and action to the same degree that interactive robots immediately pose. Floyd, Esfandiari and Lam [7] describe a case-based method for learning soccer team skills by observing spatially distributed soccer team plays. Ros et al. [24] present a case-based approach to action selection in robot soccer. More recently, Floyd and Esfandiari [6] describe a preliminary scheme for separating domain-independent case-based learning by observation from domain-dependent sensors and effectors on a physical robot.

We seek to use visual case-based reasoning to recognize that a new target demonstration, such as the overhead view of a box-closing skill shown in the top row of Fig. 1, is similar to skill demonstrations previously stored in the robot's memory, such as the related box-closing demonstrations shown in the bottom two

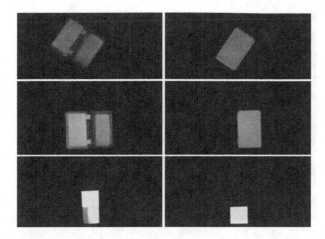

Fig. 1. Similar box-closing skill demonstrations

rows of Fig. 1. Visual case-based reasoning has been previously studied in tasks ranging from interpreting line drawings [5,28] to image interpretation [8,20] in domains ranging from molecular biology [4,10] to design [3,9,28]. Perner, Hold and Richter [21] provide a review of some of these applications. Techniques for visual case retrieval in these applications range from heuristic [9] to graph matching [5] to constraint satisfaction [28]. Images in these applications typically are static and often discrete (e.g., in the form of line drawings). In contrast, images in case-based interpretation of skill demonstrations are dynamic and continuous, requiring the development of new techniques for visual case retrieval.

The first column of Fig. 1 depicts the observed initial states of three demonstrations of the same skill, and the second column depicts the corresponding final states. As Fig. 1 illustrates, our current focus is on case-based interpretation of skill demonstrations in a table-top learning environment. Our aim is to first develop approaches for case-based interpretation, leaving the task of perception in cluttered, occluded, messy, or poorly-lit environments to future work.

We first approach the problem of case-based skill interpretation using a Fractal representation [17]. Instead of encoding the features detected within visual scenes, the Fractal method encodes the visual transformations between initial and final states of a skill demonstration. We wanted to use the Fractal method because it allows automatic adjustment of the level of spatial resolution for evaluating similarity between two sets of images. While the Fractal method has been applied to geometric analogies on intelligence tests, it has not yet been applied to real-world images such as those a robot would perceive. To fully evaluate the Fractal method for case-based interpretation, we chose to compare it to a baseline method which uses the Scale-Invariant Feature Transform (SIFT) algorithm to select image features. The SIFT algorithm identifies features regardless of the image's scale, translation, or rotation [14,15] and is widely used for computer vision tasks in robotics research.

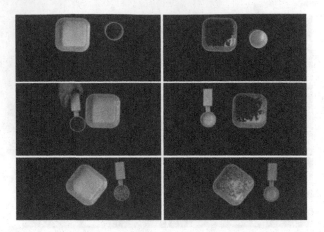

Fig. 2. Analogical pouring skill demonstrations

3   Problem Characterization

We refer to a *source case* as a skill demonstration that has been provided to the robot and is stored in the robot's case memory. Thus, we use the terms *demonstration*, *skill demonstration*, and *case* interchangeably. Each demonstration is defined as $d = <p, a>$, where p encodes the *problem* the demonstration seeks to address and a encodes the demonstrated action. We focus on representing demonstrations that illustrate only one *action label* (e.g. "pouring", "opening", "stacking"). The list of observed objects, o, and the list of observed features of the objects (color, size, etc.), f, are also elements of the demonstration representation. A skill demonstration then consists of the following elements:

- The problem description $p = <o, f, v>$, where o and f are as described above, and v is a set of parameters (e.g. initial object locations, initial end-effector position)
- The action model $a = \{j_0, j_1, \ldots, j_i\}$ encoding the robot's end-effector position at each time interval i.

The case-based interpretation process uses the problem descriptions of sources cases in memory as input, such as the demonstrations shown in the first two rows in Fig. 1, and a target problem, such as the third row in Fig. 1, and maps the target problem to the most similar case in memory. Case-based interpretation is completed by evaluating the similarity between the visual representations of the target problem and the source cases, i.e., on o and f, and the visual transformations in them, and does not require semantic information that specifies the demonstrated action label.

We define a visual transformation as the tuple $<S_i, S_f, T>$, where S_i is an overhead view of the initial state (the first column of images in Fig. 1), S_f is the observed goal state that is reached following the skill completion (the second column of images in Fig. 1), and T is the visual relation, or transformation, between the two images S_i and S_f.

4   Algorithms

4.1   Fractal Method

Our first approach uses fractal representations to encode the visual transformation function T between two images [16], and is expressed as the set of operations that occur to transform the initial state image S_i into the final state image S_f. Thus, the transformation function T encodes a set of sub-transformations between S_i and S_f. The Fractal method evaluates similarity at several levels of abstraction, allowing automatic adjustment of the level of spatial resolution. The similarity between two image transformations can be determined using the *ratio model*:

$$sim(T, T') = f(T \cap T')/f(T \cup T')$$

In this model, T encodes the first set of image transformations, T' encodes the second set of image transformations, and $f(x)$ returns the number of features in the set x [16,27]. Thus, $f(T \cap T')$ returns the number of transformations common to both transformation sets, and $f(T \cup T')$ returns the number of transformations in either set. The following process encodes a visual transformation as a fractal [16]:

1. The initial state image is segmented into a grid containing a specified number of partitions, $S = \{s_0, s_1, \ldots, s_p\}$, where p is determined by the abstraction level n.
2. For each sub-image $s \in S$, the destination image is searched for a sub-image d such that for some transformation $k \in K$, $k(s)$ is most similar to d.
3. The transformation k and shift c, the mean color-shift between d and $k(s)$, are used to create a fractal code f_s.
4. The resulting fractal is defined by $F = \{f_0, f_1, \ldots, f_p\}$

This encoding process is repeated for multiple values of n, resulting in an encoding of the transformation at n levels of abstraction, where n is derived from the images' pixel dimensions. Here, we partition each 300 px by 180 px image at $n = 7$ levels of abstraction. A code is defined by the tuple

$$<<s_x, s_y>, <d_x, d_y>, k, c>$$

where:

- s_x and s_y are the coordinates of the source sub-image
- d_x and d_y are the coordinates of the destination sub-image
- $k \in K$ represents the affine transformation between the source and destination sub-images where $K = \{$ 90° clockwise rotation, 180° rotation, 270° clockwise rotation, horizontal reflection (HR), vertical reflection (VR), identity (I) $\}$. k is the transformation that converts sub-image s into sub-image d minimally, while requiring minimal color changes.

– c is the mean color-shift between the two sub-images

A set of fractal features is derived as combinations of different aspects of each fractal code. While the fractal code does describe the transformation from a section of a source image into that of a target image, the analogical matching occurs on a much more robust set of features than merely the transformation taken by itself. The illustrations which visualize the fractal representation therefore demonstrate only those transformations, and not the features.

4.2 SIFT Feature-Matching

The SIFT algorithm selects keypoint features using the following steps [14]. First, candidate keypoints are chosen. These candidates are selected as interest points with high visual variation. Candidate keypoints are tested to determine their robustness to visual changes (i.e., illumination, rotation, scale, and noise). Keypoints deemed "unstable" are removed from the candidate set. Each keypoint is then assigned an orientation invariant to the image's orientation. Once each keypoint has been assigned a location, scale, and orientation, a descriptor is allocated to each keypoint, representing it in the context of the local image.

Our second approach to source demonstration retrieval using SIFT features is based on feature-matching. The target skill demonstration is represented by the image pair (S_i, S_f). Using the SIFT algorithm, features are extracted from each image and matched to features from the initial and final states of source skill demonstrations. Each feature consists of the 16×16 pixel area surrounding the feature keypoint. A brute-force method is used to determine that two features match if they have the most similar 16×16 surrounding area. The source demonstration sharing the most features with the target demonstration is retrieved using the following process:

```
 1: Let D be a set of source skill demonstration images
 2: c ← null; m ← 0
 3: Uᵢ ← SIFT features extracted from Sᵢ
 4: U_f ← SIFT features extracted from S_f
 5: for each demonstration d ∈ D do
 6:     Cᵢ ← SIFT features extracted from dᵢ
 7:     C_f ← SIFT features extracted from d_f
 8:     T ← (Uᵢ ∩ Cᵢ) ∪ (U_f ∩ C_f)
 9:     If size(T) > m, then: m ← size(T), c ← d
10: end for
11: return c
```

Figure 3(e) illustrates a retrieval result, where the left-side image is S_i and the right-side image is the d_i selected with the highest number of matching SIFT features.

4.3   SIFT Feature-Transformation

Our final approach to source demonstration retrieval via the SIFT algorithm serves as an intermediate method which incorporates aspects of the Fractal method's emphasis on visual transformations, while adopting the same feature selection strategy as the previous SIFT feature-matching method. This approach focuses on the transformation of SIFT features between a demonstration's initial and final states. Rather than retrieve a source demonstration based on the explicit features it shares with the target demonstration, this approach retrieves a source demonstration according to the similarities between its feature transformations and those of the transformations observed in the target demonstration.

Each feature of the demonstration's S_i is matched to its corresponding feature in S_f, as shown in Fig. 3(b). This method uses the same features and feature-matching method as in the feature-matching approach described previously. We define each SIFT feature transformation as the tuple

$$<<s_x, s_y>, \theta, l>$$

where s_x and s_y are the coordinates of the feature in the initial state, θ is the angular difference between the feature location in the initial and final states, and l is the distance between the feature location in the initial and end state images. Each feature transformation occurring between S_i and S_f in the target demonstration is compared to each transformation occurring between S_i and S_f in each source skill demonstration. The difference between two SIFT feature transformations is calculated by weighting the transformations' source location change, angular difference, and distance.

Each comparison is performed over seven blurring levels, which serves to reduce the number of irrelevant or noisy features comparably to the Fractal method's usage of multiple abstraction levels. At each blur level, a normalized box filter kernel is used to blur the target and source demonstrations' visual states, with the kernel size increasing by a factor of two at each level. The SIFT feature-transformation method retrieves a source demonstration as follows:

1: Let D be a set of source skill demonstration images
2: $c \leftarrow null$; $m \leftarrow 0$; $x \leftarrow 0$
3: **for** each demonstration $d \in D$ **do**
4:      $n \leftarrow 0$
5:      **while** $n <$ maximum abstraction level **do**
6:          Blur S_i, S_f, d_i, and d_f by a factor of 2^n
7:          $U_i \leftarrow$ SIFT features extracted from S_i
8:          $U_f \leftarrow$ SIFT features extracted from S_f
9:          $T_u \leftarrow getTransformations(U_i \cap U_f)$
10:         $C_i \leftarrow$ SIFT features extracted from d_i
11:         $C_f \leftarrow$ SIFT features extracted from d_f
12:         $T_c \leftarrow getTransformations(C_i \cap C_f)$
13:         **for** each transformation $t_u \in T_u$ **do**
14:             Find $t_c \in T_c$ that minimizes $diff(t_u, t_c)$

15:         **end for**
16:         $x \leftarrow 0$
17:         **for** each transformation $t_u \in T_u$ **do**
18:             $x \leftarrow x + \mathit{diff}(t_u, t_c)$
19:         **end for**
20:         If c is *null* or $x < m$, then: $c \leftarrow d$, $m \leftarrow x$
21:         $n \leftarrow n + 1$
22:     **end while**
23: **end for**
24: **return** c

5 Experiment

Each approach was used to retrieve a source skill demonstration for three test sets of target demonstrations. Each skill demonstration is a pair of two recorded keyframe images depicting the initial state and end state of a box-closing or cup-pouring skill performed by a human participant, as shown in Figs. 1 and 2. Nine participants demonstrated the two skills, and were recorded using an overhead camera above the tabletop workspace. Participants indicated the initial and final states verbally, and were asked to remove their hands from view when the initial and final states were recorded. Each participant's demonstration set consisted of nine demonstrations per skill, each skill being performed at the orientations shown in Figs. 1 and 2.

We evaluated the algorithms on three test sets, each representing retrieval problems of a different difficulty level. In the *aggregate* set, a source demonstration is retrieved for two participants' demonstrations (two skills each performed with two objects at three configurations, resulting in a total of 12 target demonstrations) from a library of 48 source demonstrations, which included 24 demonstrations of each skill. All box-closing and pouring demonstrations used the same two boxes and two pouring objects, respectively, shown in the first two rows of Figs. 1 and 2. In the *individual* set, a source skill demonstration was retrieved for each of 54 target demonstrations (27 per skill). Within each participant's demonstration set, the target demonstration was compared to the other demonstrations by the same participant. As a result, a source was retrieved for each target demonstration from a library containing two source demonstrations of the same skill and three of the opposite skill. As in the aggregate test set, demonstrations used the same two boxes and two pouring objects.

In the *analogical* set, a source demonstration was retrieved for each of 161 target demonstrations (80 box-closing, 81 pouring). Within each participant's demonstration set, the target demonstration was compared only to other demonstrations performed by the same participant. Unlike the previous test sets, target demonstrations were compared to source demonstrations involving different objects, as in Figs. 1 and 2. As a result, demonstrations involving a third kind of box and pouring object were introduced, shown in the last row of Figs. 1 and 2. A source demonstration was retrieved for each target demonstration from a library

Table 1. Source Case Retrieval Results

Test set	Fractal	SIFT feature-matching	SIFT feature-transformations
Aggregate	100 %	100 %	91.7 %
Individual	87 %	100 %	35.2 %
Analogical	65.3 %	93.8 %	84.5 %

containing six source demonstrations of the same skill and nine of the opposite skill. One box-closing demonstration was incomplete and could not be included in the test set; as a result, 17 target demonstrations were compared to one fewer box-closing demonstration. The purpose of the analogical test set was to test each retrieval method's ability to retrieve a source skill demonstration, despite containing a different set of objects than the target demonstration.

6 Experimental Results

Table 1 lists the overall accuracy of each method when applied to each test set. Since the aggregate test contained a large set of source demonstrations and was most likely to contain a demonstration similar to the target problem, we expected that this test set would be the easiest test set for any of the three methods to address.

6.1 Detailed Analysis

While the experimental results provide useful information about the accuracy of the three methods, it is useful to also analyze the strengths of each method.

Case Study: Fractal Method Success. First, we analyze an example in which only the Fractal method retrieved an appropriate source demonstration. Figure 3(a) depicts the target problem demonstration, which the Fractal method correctly matched to the source demonstration shown in Fig. 3(d). The Fractal method offers both a decreased susceptibility to noise as well as a plethora of fractal features over which to calculate a potential match (beyond the transformation itself).

The SIFT feature-matching method incorrectly classified Fig. 3(a) as a pouring skill demonstration, due to the many features matched between the target demonstration and pouring demonstration's final states. Features of the demonstrator's hand were incorrectly matched to features of the pouring instrument, as shown in Fig. 3(e). The SIFT feature-transformation method also incorrectly classified the demonstration as a pouring skill demonstration. Figure 3(b) illustrates the feature transformations used to represent the target problem. Each feature in the initial state was matched to the single feature identified in the final state. Thus, the resulting feature transformations did not properly represent the skill being performed, which led to the retrieval of an incorrect source demonstration (see Fig. 3(c)).

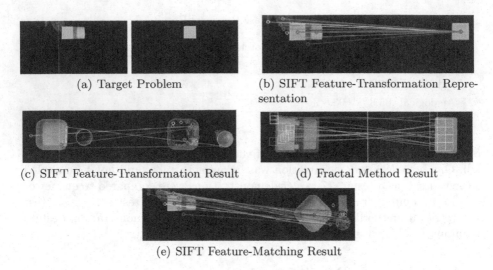

(a) Target Problem

(b) SIFT Feature-Transformation Representation

(c) SIFT Feature-Transformation Result

(d) Fractal Method Result

(e) SIFT Feature-Matching Result

Fig. 3. Case study 1: retrieval method results

We conclude that the Fractal method can be applied to source retrieval problems in which the visual transformation, rather than keypoint features, are indicative of the skill being performed. The Fractal method is also applicable to demonstrations that include some clutter, such as the demonstrator's hand or other objects unrelated to the skill being performed. This case study also demonstrates that the feature-matching method is sensitive to clutter. Additionally, the feature-transformation method is less effective in classifying demonstrations in which there are few features in the initial or final state, or in which there is a poor correspondence between features matched between the initial and final state images. As an example, the feature-transformation method would perform poorly given a demonstration of a book-closing skill, where initial-state SIFT features detected on the inside pages of the book cannot be matched to final-state SIFT features on the cover of the book.

Case Study: SIFT Transformation Success. In the next case, only the SIFT feature-transformation method retrieved an appropriate source demonstration for the target problem shown in Fig. 4(a). The SIFT feature transformation method retrieves visually analogical source demonstrations by identifying visual transformations at multiple abstraction levels. The transformations in Fig. 4(c) were deemed similar to those in the target problem. Features in the initial and final states were matched correctly, which is why this method was able to succeed.

The Fractal method incorrectly retrieved the source demonstration shown in Fig. 4(d) due to its emphasis on visual transformations independent of features, and thus is less effective in distinguishing between skills that have similar visual transformations. The more similar the visual transformations, the more common and therefore the less salient are the Fractal method's generated features derived

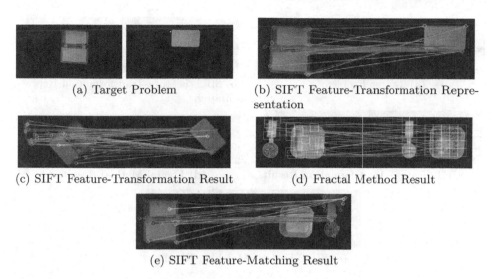

(a) Target Problem

(b) SIFT Feature-Transformation Representation

(c) SIFT Feature-Transformation Result

(d) Fractal Method Result

(e) SIFT Feature-Matching Result

Fig. 4. Case study 2: retrieval method results

from those transformations. The Fractal method chose this source demonstration due to the similarity between the movement of the box lid from one part of the target demonstration image to another, and the movement of coffee beans from one part of the source demonstration image to another. The SIFT feature-matching method also returned an incorrect source demonstration in this case, as it erroneously matched features of the target demonstration's initial state to features of a pouring instrument (see Fig. 4(e)).

This case study teaches us that the feature-transformation method is best applied to situations in which there are a large number of features in both the initial and final state images, and the two sets of features have been mapped correctly. Additionally, we find that the Fractal method is less effective in distinguishing between skills that have similar visual transformations. Finally, this case study demonstrates how the feature-matching method relies on having a correct mapping between features of the target demonstration and features extracted from a potential source demonstration.

Case Study: SIFT Feature-Matching Success. In the final case study, only the feature-matching method retrieved the correct source demonstration to address the target problem shown in Fig. 5(a). This method correctly corresponds features between the target problem and source demonstration's initial and final state features. The initial state feature mapping is shown in Fig. 5(e).

Just as in the first case study, the feature-transformation method does not retrieve the correct source demonstration because there are not enough features in the final state image. All features in the source demonstration's initial state are mapped to the single feature in the final state image, causing the feature transformations to poorly reflect the skill being performed. The Fractal method

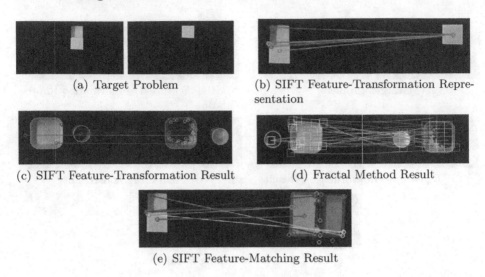

(a) Target Problem

(b) SIFT Feature-Transformation Representation

(c) SIFT Feature-Transformation Result

(d) Fractal Method Result

(e) SIFT Feature-Matching Result

Fig. 5. Case study 3: retrieval method results

retrieves an incorrect source demonstration due to its emphasis on the visual transformation between the two states, without any weight to the objects being moved. In this example, the Fractal method determined the movement of the box lid to be analogical to the movement of coffee beans from the left side of the image to the right side, as shown in Fig. 5(d).

Thus, the feature-matching method is most effective when there is a correct correspondence between features of the target problem and matching features in the potential source demonstration, and there are enough features in both demonstrations to represent the objects being used. As it turns out, even our analogical test set used objects that were similar enough for feature-matching to achieve the highest success rate (e.g., even after switching from pouring coffee beans to white beans, black flecks made them look enough like coffee beans to match). We expect that for analogical images with less object feature correspondence, this result would dramatically change.

The feature-matching method performed best on each test set. However, we anticipate that this method would not perform well on skill demonstrations in which irrelevant features are present, such as clutter or the demonstrator's hand. Additionally, this method would mistake skill demonstrations with the same feature set; block-sorting and block-stacking demonstrations could be performed using the same objects, and thus the two demonstrations would be matched as a result of having the same set of features.

6.2   Discussion

Several variables may affect the accuracy of each skill interpretation method. The Fractal method is affected by the heuristic used to select the abstraction level at which two demonstrations should be compared. We currently use the heuristic of summing the similarity scores that are calculated at multiple abstraction levels. However, this heuristic may negatively impact the Fractal method's overall accuracy if skill types are most accurately classified at a certain abstraction level. Additionally, the SIFT feature-transformation method is affected by the scoring function used to determine the similarity of two transformations. The weight values applied to the angular difference, change in transformation distance, and change in start location between two feature transformations will impact how accurately the method can determine the similarity between visual feature transformations. These two variables, the abstraction-level selection heuristic and the transformation similarity metric, may become the focus of future work.

7 Conclusion

We have explored visual case retrieval for case-based interpretation of skill demonstrations as a precursor to case-based robot learning by imitation. We have presented three methods for this task: SIFT feature-matching, SIFT feature-transformation, and Fractal feature-transformation. Although the general SIFT algorithm is widely used for computer vision tasks, the use of fractal and SIFT features in case-based skill interpretation is new insofar as we know.

No single method works best for all case-based skill interpretation problems. Rather, each method discussed in this paper is best suited for a particular type of problem. The feature-matching method is best suited for interpretation problems in which enough visual features can be extracted to identify the skill and no clutter is present. The SIFT feature-transformation method is most effective in problems where many features can be extracted from the demonstrations, and correspondences between features can be identified correctly. Finally, the Fractal method is most effective in identifying skills in which the visual transformation should be emphasized, rather than features of the demonstration images themselves. This suggests the use of a multi-strategy technique for visual case retrieval in the domain of interpreting skill demonstrations.

Acknowledgments. This material is based upon work supported by the United States' National Science Foundation through Graduate Research Fellowship Grant #DGE-1148903 and Robust Intelligence Grant #1116541. Any opinion, findings, and conclusions or recommendations expressed in this material are those of the authors and do not necessarily reflect the views of NSF.

References

1. Aamodt, A., Plaza, E.: Case-based reasoning: foundational issues, methodological variations, and system approaches. AI commun. **7**(1), 39–59 (1994)
2. Argall, B.D., Chernova, S., Veloso, M., Browning, B.: A survey of robot learning from demonstration. Robot. Auton. Syst. **57**(5), 469–483 (2009)

3. Cheetham, W., Graf, J.: Case-based reasoning in color matching. In: Leake, D.B., Plaza, E. (eds.) ICCBR 1997. LNCS, vol. 1266, pp. 1–12. Springer, Heidelberg (1997)
4. Davies, J., Goel, A.K., Nersessian, N.J.: A computational model of visual analogies in design. Cogn. Syst. Res. **10**(3), 204–215 (2009)
5. Ferguson, R.W., Forbus, K.D.: GeoRep: a flexible tool for spatial representation of line drawings. In: AAAI/IAAI, pp. 510–516 (2000)
6. Floyd, M.W., Esfandiari, B.: A case-based reasoning framework for developing agents using learning by observation. In: 2011 23rd IEEE International Conference on Tools with Artificial Intelligence (ICTAI), pp. 531–538. IEEE (2011)
7. Floyd, M.W., Esfandiari, B., Lam, K.: A case-based reasoning approach to imitating robocup players. In: FLAIRS Conference, pp. 251–256 (2008)
8. Grimnes, M., Aamodt, A.: A two layer case-based reasoning architecture for medical image understanding. In: Smith, I., Faltings, B.V. (eds.) EWCBR 1996. LNCS, vol. 1168, pp. 164–178. Springer, Heidelberg (1996)
9. Gross, M.D., Do, E.Y.L.: Drawing on the back of an envelope: a framework for interacting with application programs by freehand drawing. Comput. Graph. **24**(6), 835–849 (2000)
10. Jurisica, I., Glasgow, J.: Applications of case-based reasoning in molecular biology. AI Mag. **25**(1), 85 (2004)
11. Kolodner, J.: Case-Based Reasoning. Morgan Kaufmann, San Mateo (1993)
12. Leake, D.B.: Case-Based Reasoning: Experiences, lessons and future directions. MIT press, Menlo Park (1996)
13. Lopez De Mantaras, R., McSherry, D., Bridge, D., Leake, D., Smyth, B., Craw, S., Faltings, B., Maher, M.L., Cox, M.T., Forbus, K., et al.: Retrieval, reuse, revision and retention in case-based reasoning. Knowl. Eng. Rev. **20**(03), 215–240 (2005)
14. Lowe, D.G.: Object recognition from local scale-invariant features. In: Proceedings of the Seventh IEEE International Conference on Computer Vision, vol. 2, pp. 1150–1157. IEEE (1999)
15. Lowe, D.G.: Distinctive image features from scale-invariant keypoints. Int. J. Comput. Vision **60**(2), 91–110 (2004)
16. McGreggor, K., Goel, A.: Fractal analogies for general intelligence. In: Bach, J., Goertzel, B., Iklé, M. (eds.) AGI 2012. LNCS, vol. 7716, pp. 177–188. Springer, Heidelberg (2012)
17. McGreggor, K., Kunda, M., Goel, A.: Fractals and ravens. Artif. Intell. **215**, 1–23 (2014)
18. Meltzoff, A.N.: Imitation and other minds: the "like me" hypothesis. Perspectives on Imitation: From Neuroscience to Social Science, vol. 2, pp. 55–77 (2005)
19. Ontañón, S., Mishra, K., Sugandh, N., Ram, A.: Case-based planning and execution for real-time strategy games. In: Weber, R.O., Richter, M.M. (eds.) ICCBR 2007. LNCS (LNAI), vol. 4626, pp. 164–178. Springer, Heidelberg (2007)
20. Perner, P.: An architecture for a CBR image segmentation system. Eng. Appl. Artif. Intell. **12**(6), 749–759 (1999)
21. Perner, P., Holt, A., Richter, M.: Image processing in case-based reasoning. Know. Eng. Rev. **20**(3), 311–314 (2005)
22. Richter, M.M., Weber, R.: Case-Based Reasoning. Springer, Heidelberg (2013)
23. Riesbeck, C., Schank, R.: Inside Case-Based Reasoning. Lawrence Erlbaum Associates, Hillsdale (1989)
24. Ros, R., Arcos, J.L., De Mantaras, R.L., Veloso, M.: A case-based approach for coordinated action selection in robot soccer. Artif. Intell. **173**(9), 1014–1039 (2009)

25. Thagard, P.: Mind: Introduction to Cognitive Science. MIT press, Cambridge (2005)
26. Tomasello, M., Kruger, A.C., Ratner, H.H.: Cultural learning. Behav. Brain Sci. **16**(03), 495–511 (1993)
27. Tversky, A.: Features of similarity. Psychol. Rev. **84**(4), 327 (1977)
28. Yaner, P.W., Goel, A.K.: Analogical recognition of shape and structure in design drawings. Artif. Intell. Eng. Des. Anal. Manuf. **22**(02), 117–128 (2008)

Improving Trust-Guided Behavior Adaptation Using Operator Feedback

Michael W. Floyd[1]([✉]), Michael Drinkwater[1], and David W. Aha[2]

[1] Knexus Research Corporation, Springfield, VA, USA
{michael.floyd,michael.drinkwater}@knexusresearch.com
[2] Navy Center for Applied Research in Artificial Intelligence,
Naval Research Laboratory (Code 5514), Washington, DC, USA
david.aha@nrl.navy.mil

Abstract. It is important for robots to be trusted by their human teammates so that they are used to their full potential. This paper focuses on robots that can estimate their own trustworthiness based on their performance and adapt their behavior to engender trust. Ideally, a robot can receive feedback about its performance from teammates. However, that feedback can be sporadic or non-existent (e.g., if teammates are busy with their own duties), or come in a variety of forms (e.g., different teammates using different vocabularies). We describe a case-based algorithm that allows a robot to learn a model of feedback and use that model to adapt its behavior. We evaluate our system in a simulated robotics domain by showing that a robot can learn a model of operator feedback and use that model to improve behavior adaptation.

Keywords: Inverse trust · Behavior adaptation · Adaptable autonomy

1 Introduction

Robots can be valuable members of teams if they provide the team with additional skills, reduce task load, or minimize potential risks to humans. In some scenarios, the robots' contributions could be vital to achieving team goals or successfully completing missions. We focus on semi-autonomous robots that can be issued high-level commands by an operator (e.g., *"move to the river"*, *"patrol for threats"*). However, for the operator to use the robot to its full potential it needs to trust the robot. A lack of trust could result in unnecessarily monitoring the robot's actions, underutilizing the robot, or not using it at all [1].

We have previously examined how a robot can evaluate its trustworthiness using an *inverse trust metric* and adapt its behavior in an effort to engender trust [2]. Unlike traditional trust metrics that can be used to measure a robot's trust in its operator, an inverse trust metric estimates how much trust the operator has in the robot. Our inverse trust metric uses the robot's performance to estimate trust and measures general trends in trustworthiness (i.e., increasing, decreasing, remaining constant). Behavior adaptation is performed using case-based reasoning (CBR) and allows the robot to adapt to changes in operators,

© Springer International Publishing Switzerland 2015
E. Hüllermeier and M. Minor (Eds.): ICCBR 2015, LNAI 9343, pp. 134–148, 2015.
DOI: 10.1007/978-3-319-24586-7_10

missions, or contexts. However, our earlier approach assumes that no explicit feedback is provided by the operator so only observable indicators of the robot's performance are used. Although such an assumption is beneficial in scenarios where the operator does not have time to provide explicit feedback, it limits the robot's ability to use this information when it is available.

In this paper we describe an extension of our previous work, which only used implied feedback (i.e., the operator allowed the robot to complete a task or interrupted it), to allow the robot to utilize *explicit* operator feedback. No assumptions are made about the format of feedback so it is applicable when feedback comes in different modes (e.g., text, gestures, interface commands) or expressions (e.g., synonymous phrases, different languages). Similarly, no assumptions are made about the frequency of feedback. The robot uses case-based reasoning to learn a model for what the various pieces of feedback mean and uses that model to assist in behavior adaptation (e.g., if the operator says *"go faster"* the robot should increase its speed).

The remainder of this paper describes how a robot can learn a model of operator feedback and use that feedback to adapt its behavior. In Sect. 2, we provide a summary of our previous work on how a robot can adapt its behavior in an attempt to be a more trustworthy member of a team. Section 3 provides a discussion of related work. Section 4 describes the type of feedback that an operator can give to the robot, and Sect. 5 presents an approach for learning the meaning of feedback and using feedback to improve behavior adaptation. In Sect. 6, we evaluate our approach using scenarios from a simulated robotics domain. They indicate that the robot can use a learned feedback model to improve its behavior adaptation performance. Concluding remarks are discussed in Sect. 7.

2   Trust-Based Behavior Adaptation

The robot receives high-level commands from the operator (e.g., *"move to the building"*, *"patrol for threats"*, *"transport supplies to the hospital"*) and autonomously performs the tasks it is assigned. As the robot performs these tasks, it can control and modify its behavior by changing *modifiable components*. Each modifiable component i represents a single aspect of the robot's behavior and the robot is responsible for selecting a value m_i for that component from the set $\mathcal{M}_i$ of possible values ($m_i \in \mathcal{M}_i$). For example, the robot could select a parameter value from a set of possible values, an algorithm to use from a set of path planning algorithms, or data to use from a set of alternative data sources. Without loss of generality, we assume that the possible values are totally ordered, with an ordering relation between each pair of values ($m_i^j \prec m_i^k$, where $m_i^j, m_i^k \in \mathcal{M}_i$).

If the robot has n modifiable components, its current behavior B is a tuple containing the currently selected value for each modifiable component ($B = \langle m_1, m_2, \ldots, m_n \rangle$). In our work, the robot modifies its behavior in an attempt to increase its trustworthiness. Unlike traditional trust metrics [3] where the robot measures its trust in another agent, an *inverse trust metric* [2] is used to estimate how much trust an agent has in the robot. Since inverse trust is measured from the robot's perspective, only observable indicators of human-robot trust can be used (i.e., none of the human's internal reasoning information). The robot estimates its trustworthiness based on when it successfully completes assigned tasks, when it fails to complete assigned tasks, and when it is interrupted while performing a task.

The inverse trust metric evaluates the trustworthiness of the current behavior B and tracks trends in trust over time. If the robot has received c commands, the metric will include information from each of those commands. Successfully completed commands will increase the trust estimate, and failed commands or interruptions will decrease the estimate ($cmd_i \in \{-1, 1\}$, with a weight w_i based on the command's relative importance):

$$Trust_B = \sum_{i=1}^{c} w_i \times cmd_i$$

The robot updates this value as more commands are issued and compares it to two thresholds: the trustworthy threshold (τ_T) and the untrustworthy threshold (τ_U). If the estimate reaches the trustworthy threshold ($Trust_B \geq \tau_T$), the robot concludes it has found a sufficiently trustworthy behavior but continues to estimate trust in case there is a change in operator, mission, or goals. If the estimate reaches the untrustworthy threshold ($Trust_B \leq \tau_U$), the robot concludes its behavior is untrustworthy and should be changed. Otherwise ($\tau_U < Trust_B < \tau_T$), the robot continues to monitor the trust estimate until it is more confident about its trustworthiness.

When the untrustworthy threshold is reached, the robot changes its behavior from B to a new behavior B' and begins measuring the trustworthiness of that behavior (i.e., $Trust_{B'}$). The behavior that was found to be untrustworthy, along with the time t it took to reach the untrustworthy threshold, are stored as an *evaluated pair* E ($E = \langle B, t \rangle$).

As the robot evaluates more behaviors, it maintains a set of previously evaluated behaviors $\mathcal{E}_{past}$ that contains all of the behaviors it has found to be untrustworthy ($\mathcal{E}_{past} = \{E_1, E_2, \dots\}$). These previously evaluated behaviors represent the search path the robot has taken while searching for a trustworthy behavior B_{final}. In a CBR context, the previously evaluated behaviors are the *problem* and the final behavior is the *solution*. We use the following case representation:

$$C = \langle \mathcal{E}_{past}, B_{final} \rangle$$

When the robot successfully finds a trustworthy behavior, it creates a new case and stores it in its case base. The motivation for using this case representation is that two operators who find similar behaviors untrustworthy in a similar amount of time may find similar behaviors to be trustworthy. When the robot needs to adapt its behavior (i.e., the trust metric reaches the untrustworthy threshold), it compares the behaviors it has previously evaluated ($\mathcal{E}_{past}$) to each of the cases. If a case is sufficiently similar and its final behavior has not already been evaluated as untrustworthy, the robot switches to use that case's final behavior.

If no cases are sufficiently similar, the robot performs a random walk. This form of behavior adaptation finds the evaluated behavior that took the longest to be labeled as untrustworthy and modifies that behavior. This works under the assumption that the behavior that took the longest to reach the untrustworthy threshold was the closest to being trustworthy, so a slight modification could make it more trustworthy. A more detailed description of case creation, case retrieval, case-based behavior adaptation, and random walk behavior adaptation can be found in [2].

Our previous work only accounted for *implied* feedback (i.e., successes, failures, and interruptions). The primary contribution of this paper is extending our previous work to allow for *explicit* feedback from the operator.

3   Related Work

Kaniarasu et al. [4] have also examined an online, performance-based estimate of operator trust. Their measure tracks only negative factors (e.g., how often the robot is warned of poor performance, or the operator manually controls the robot), so it can identify only decreases in operator trust. To also track increases in trust, they extended their measure to incorporate performance information from the operator at regular intervals [5] (e.g., the operator provides the robot feedback about its performance every 30 s). However, this approach is unable to track trust for any periods where explicit feedback in unavailable. Saleh et al. [6] estimate operator trust using a set of expert-authored rules. Since the rules could be different for each operator, mission, or context, this measure requires an expert redefine the rules whenever a change occurs.

In case-based reasoning, research has focused on traditional trust rather than inverse trust, and has generally been examined in the context of agent collaboration [7] or recommendation systems [8]. Case provenance [9] also deals with trust but focuses on the trustworthiness of a case's source rather than the trustworthiness of an agent or system.

Our work on inverse trust is related to learning a user's preferences. The ability to incorporate a user's preferences has been examined in areas such as learning interface agents and preference-based planning [10]. Learning interface agents build a model of a user by watching the user perform a task (e.g., e-mail sorting [11] or schedule management [12]) and later assisting the user in performing that task. Similarly, preference-based planners can learn a user's planning preferences by observing the user perform a planning task [13]. In our work, these demonstration-based approaches would be equivalent to the operator manually controlling the robot and performing the task. This would not be practical in time-sensitive situations or when the operator does not have a fully constructed plan for how a task should be performed.

Both user preferences and feedback play an important role in human-in-the-loop CBR systems, such as conversational case-based reasoning systems [14]. These interactions tend to be in the form of dialogs between the user and the system, whereas in our work interactions are one-sided (i.e., information is passed only from the operator to the robot) and sporadic. Conversational recommender systems [15] use feedback to refine a model of the user and iteratively improve the recommendations that are provided. Similarly, feedback can also be used to tailor what questions to ask a user [16], thereby influencing what feedback will be provided in the future. Whereas our system is designed to work in a variety of tasks and missions, these approaches focus on a single task (i.e., recommendation).

The meaning of explicit operator feedback is learned by the robot by determining a relationship between its behavior when the feedback was received and its final trustworthy behavior. Relationships are similar to compound critiques [17] in recommender systems in that they represent the changes that should be made to a set of features. More generally, learning behavior relationships is similar to rule-induction [18]. The primary difference between these approaches and our own is that behavior relationships are generated using two data points (e.g., the behavior when feedback was received and the trustworthy behavior), rather than a full or partial subset of observations.

4   Operator Feedback

The primary methods used by the operator to interact with the robot are issuing commands and interrupting (i.e., implied feedback). However, the operator can also provide additional information to the robot in the form of *explicit feedback*. For the remainder of this paper, when referring to feedback we mean explicit feedback rather than implied feedback.

Feedback is provided at the operator's discretion, so no assumptions are made about when it will occur or how often it will be provided. The frequency of feedback is operator-specific (i.e., some operators prefer to provide more feedback) but is also influenced by the operator's workload. For example, an operator would likely have less time to provide feedback to the robot during an emergency situation. In the extreme case, the robot would not receive any feedback from the operator.

For a robot that supports multimodal interaction, feedback can be provided by any of the available modes (e.g., written text, speech, gestures, user interface commands). This allows the operator to provide the same feedback in a variety of ways. For example, the operator can tell the robot to stop by saying the word *"stop"*, making a hand gesture, or pressing a keyboard button. Similarly, the operator can provide the same feedback in a variety of ways using a single mode of interaction (e.g., *"go faster"*, *"speed up"*, *"get going"*, saying it in other languages). The robot could use computational semantics or machine translation to group similar utterances, but this might not be feasible due to the robot's computational constraints. Similarly, the robot would need a method for grouping similar pieces of feedback from different modalities.

In some domains, the format of feedback can be formally defined such that the robot has a prior model of what feedback it can receive and what each piece of feedback means. However, this requires the operator to be aware of the format and structure its feedback accordingly. It would be difficult to enforce this requirement if the robot is expected to interact with a variety of operators with minimal training (e.g., a robot that is part of an ad-hoc team or a mass-produced consumer robot). Even if the operator is fully aware of how to correctly provide feedback, the format might limit how expressive the feedback can be. This would make it difficult to provide feedback if the team encounters new environments, new tasks, or unforeseen events. Instead, we will examine how the robot can learn a model of operator feedback without any prior knowledge about the frequency, format, or modality of feedback.

Each time feedback is provided by the operator, the robot stores a pair F containing the feedback f and the behavior B that was being used by the robot when the feedback was received ($F = \langle f, B \rangle$). This representation encodes the circumstances under which the operator decided to provide feedback (i.e., how the robot was behaving) as well as the information the operator was trying to convey to the robot (i.e., the feedback). This makes the assumption that the operator's feedback is a direct response to how the robot is currently behaving. If the operator provides feedback about a previous behavior (e.g., *"You were driving too slowly five minutes ago."*), this encoding will erroneously attribute that feedback to the current behavior. However, we anticipate that such delayed feedback will be relatively rare compared to feedback about the current behavior or delayed feedback that is still valid for the current behavior (e.g., the robot was driving slowly five minutes ago and is still driving slowly).

Over the course of operation, the robot will maintain a set $\mathcal{F}_{received}$ of received feedback ($\mathcal{F}_{received} \subseteq \mathcal{F}$, where $\mathcal{F}$ is the set of all possible feedback items). This set, which will be empty initially, will be extended when the robot receives a new feedback item ($\mathcal{F}_{received} = \bigcup_{i=1}^{n} F_i$, where n is the number of feedback items received).

5   Feedback Model

We have described how the robot can record feedback but not how it can leverage that information. This section will present methods that allow the robot to learn from feedback and use that feedback in an attempt to improve its behavior adaptation performance.

5.1   Learning the Feedback Model

Since the meaning of feedback is initially unknown to the robot, it needs to learn a feedback model. The feedback items themselves do not provide enough information to build the model because they capture only what the robot's behavior was at the time the feedback was received. The robot also needs to know what it should have done in response to the feedback. Since feedback is received while searching for a trustworthy behavior, when the robot finds a trustworthy behavior it can use that behavior to build its feedback model.

The feedback model is structured as a case base that contains guidelines for how the robot should adapt its behavior in response to feedback. We refer to this case base as the *feedback base* to differentiate it from the case base used for case-based behavior adaptation. A case FR is defined as:

$$FR = \langle f, R, cnt \rangle$$

Each case contains a piece of feedback f, a relationship R, and a frequency count cnt. The relationship represents guidelines for how the robot should adapt its behavior in response to the feedback. For any pair of behaviors (e.g., the behavior when feedback was received and a final trustworthy behavior), the relationship encodes how the two behaviors differ ($relation : \mathcal{B} \times \mathcal{B} \rightarrow \mathcal{R}$, where $\mathcal{B}$ is the set of all behaviors and $\mathcal{R}$ is the set of all relationships). More specifically, the relationship encodes how the modifiable components of each behavior differ. A relationship can be determined for each pair of modifiable components ($rel : \mathcal{M}_i \times \mathcal{M}_i \rightarrow \mathcal{O}, \mathcal{O} = \{\prec, \succ, =\}$). The relationship R_{ij} between two behaviors B_i and B_j contains the relationship between each of their modifiable components ($|B_i| = |B_j| = |R_{ij}|, R_{ij} = \langle rel(B_i.m_1, B_j.m_1), rel(B_i.m_2, B_j.m_2), \dots \rangle$).

Consider an example where the robot has two modifiable components: its speed and its object padding (how far it attempts to stay away from obstacles when planning its movement). If the robot receives the feedback *"go faster"* while using a speed of 1 meter/second and a padding of 0.5 meters ($B_1 = \langle 1, 0.5 \rangle$), and eventually finds a behavior with a speed of 5 meters/second and a padding of 0.5 meters ($B_2 = \langle 5, 0.5 \rangle$) to be trustworthy, the relationship will show the speed increased and the padding remained constant ($R_{12} = \langle \prec, = \rangle$).

The cnt value stores the number of times that feedback f resulted in the relationship R being observed. The motivation for storing this value is that it is possible to observe unnecessary relationships or erroneous relationships. An unnecessary relationship would occur if the robot changed one or more modifiable components when it did not need to (e.g., in an attempt to go faster the robot changed both its speed and its padding), whereas an erroneous relationship would occur when the operator gives incorrect feedback (e.g., telling the robot to go faster when it is already driving fast enough). We make the assumption that correct relationships, even if they contain unnecessary modifications, will occur more frequently than erroneous relationships. Using this case definition, the feedback can be thought of as the *problem*, the relationship as the *solution*, and the frequency count a measurement of the quality of a relationship.

Algorithm 1 shows the process the robot uses to refine its feedback model. The algorithm is used at the end of a search when the robot has found a trustworthy behavior (this is also when it creates and stores behavior adaptation cases). It receives as input the set of received feedback items $\mathcal{F}_{received}$, the trustworthy behavior B_{final} that was found, and its feedback base $FeedbackBase$. The algorithm iterates through each of the feedback items (line 1) and checks to see if the behavior when feedback was received differs from the final behavior (line 2). This check is performed to ensure the robot stores only cases when feedback required it to adapt its behavior (i.e., it would never store the relationship

$\langle =, =, =, \ldots \rangle$). If the behaviors differ, the relationship between the behaviors is computed (line 3). If the feedback base already contains a case with that feedback and relationship, the frequency count for that case is increased (line 5–8). Otherwise, a new case is created and added to the feedback base (lines 9–11). Once all feedback items have been processed, the set is emptied (line 12) and can again be extended as new feedback is received.

Algorithm 1. Process the feedback items received during a search

---

Function: *processFeedback($\mathcal{F}_{received}$, B_{final}, FeedbackBase)*;

```
1  foreach Fᵢ ∈ ℱ_received do
2      if Fᵢ.B ≠ B_final then
3          Rᵢ ← relation(Fᵢ.B, B_final);
4          exists ← false ;
5          foreach FRⱼ ∈ FeedbackBase do
6              if FRⱼ.f = Fᵢ.f and FRⱼ.R = Rᵢ then
7                  FRⱼ.cnt ← FRⱼ.cnt + 1;
8                  exists ←true ;
9          if !exists then
10             FR_new ← ⟨Fᵢ.f, Rᵢ, 1⟩;
11             FeedbackBase ← FeedbackBase ∪ FR_new;
12  ℱ_received ← ∅;
```

5.2 Using the Feedback Model

We have previously described how the robot stores the feedback it receives (Sect. 4) and will now describe how the robot uses the feedback model it has learned to adapt its behavior. Algorithm 2 is called when the operator provides the robot with feedback. A new feedback item is created from the received feedback and current behavior (line 1), and is stored in the set of feedback items (lines 2). The algorithm iterates through all feedback relationships in the feedback base (line 5) and stores the most frequent feedback relationship for the given feedback (lines 6–8). This is because there can be multiple feedback relationships for each type of feedback, so only the best relationship (i.e., the one with the highest frequency value) is used. If no feedback relationship is found (i.e., the feedback base is empty or no relationship has been found for that feedback yet), the robot does not change its behavior (lines 9–10). However, if a feedback relationship is found, then the robot uses the *applyRelationship(. . .)* function to modify its behavior.

The *applyRelationship(. . .)* function does the following:

1. The current behavior B_{curr} is stored along with its current trust estimate $Trust_{B_{curr}}$ and evaluation time t_{curr}. These are stored because behavior

---

Algorithm 2. Receive feedback from the operator

Function: *receiveFeedback(f, B_{curr}, $\mathcal{F}_{received}$, FeedbackBase) **returns** B_{new};*

1 $F_{new} \leftarrow \langle f, B_{curr} \rangle$;
2 $\mathcal{F}_{received} \leftarrow \mathcal{F}_{received} \cup F_{new}$;
3 $bestFrequency \leftarrow 0$;
4 $R_{best} \leftarrow \emptyset$;
5 **foreach** $FR_i \in FeedbackBase$ **do**
6   **if** $FR_i.f = f$ **and** $FR_i.cnt > bestFrequency$ **then**
7          $bestFrequency \leftarrow FR_i.cnt$;
8          $R_{best} \leftarrow FR_i.R$;

9 **if** $R_{best} = \emptyset$ **then**
10   **return** B_{curr};
11 **return** $applyRelationship(B_{curr}, R_{best})$;

---

adaptation is triggered by feedback, not by the behavior being labelled as trustworthy or untrustworthy. Since the feedback can result in unnecessary behavior changes (e.g., erroneous feedback or incorrect feedback relationships), this allows the robot to continue evaluating the behavior at a later time.

2. A new behavior B_{new} is selected under the conditions that it has not already been found to be untrustworthy ($\forall E_i \in \mathcal{E}_{past}, E_i.B \neq B_{new}$) and it satisfies the relationship R_{best}. Ideally, the new behavior will satisfy the entire relationship ($relation(B_{curr}, B_{new}) = R_{best}$). However, if no behaviors meet the entire relationship (e.g., the relationship requires decreasing the robot's speed but the speed is already at its minimum value), B_{new} will be a behavior that partially satisfies the relationship.

3. If the new behavior has already been partially evaluated (i.e., its trust estimate and time were previously stored in Step 1), the trust estimate and evaluation time are loaded. This allows the robot to continue its previous evaluation of the behavior and avoids spending longer than necessary evaluating behaviors.

The feedback process works under the assumption that errors, either in the feedback provided by the operator or in the feedback model learning, are unavoidable. However, the relationships' frequency counts are used to reinforce correct relationships while ignoring poor relationships. For example, consider the situation where feedback is received, a relationship is selected, and applying the relationship results in a trustworthy behavior being found. Since the feedback is stored (lines 1–2 of Algorithm 2), the robot will generate the relationship again when it processes the feedback items (using Algorithm 1). This increases the relationship's frequency count and can increase the chance that it is used again in the future (i.e., that it will have the highest frequency count for that feedback). Similarly, if applying a relationship does not result in a trustworthy behavior being found, the robot will continue to adapt its behavior until a trustworthy behavior is found (e.g., using further feedback, case-based adaptation, or random walk adaptation). When feedback items are eventually processed, a different relationship will likely be generated and have its frequency count increased. Since the unsuccessful relationship does not increase its frequency count it may be less likely to be used in the future (i.e., it may no longer have the highest frequency for that feedback).

6 Evaluation

In this section, we evaluate our claim that *learning a feedback model and using operator feedback can improve the performance of behavior adaptation.* Case-based behavior adaptation has previously been found to allow a robot to efficiently locate a trustworthy behavior [2]. However, it requires using the significantly less efficient random walk behavior adaptation to acquire cases. Since the robot starts with an empty case base and learns cases during deployment, random walk behavior adaptation serves as a bottleneck. We focus on evaluating the improvements the feedback model provides compared to random walk adaptation. Our evaluation tests the following hypotheses:

H1: Learning and using a feedback model will demonstrate improved performance compared to random walk behavior adaptation.

H2: The performance improvement will increase as the model learns from feedback.

6.1 eBotworks Simulator

We use the eBotworks simulator [19] for our evaluation. eBotworks allows autonomous agents to control simulated robotic vehicles while interacting with human operators using a variety of command modalities (e.g., speech, text, user interface commands). This simulator was selected based on its built-in agent design framework, autonomy modules (e.g., natural language command interpretation and path planning), and experimentation and data collection capabilities. Additionally, eBotworks allows for non-deterministic environments and noisy sensory inputs.

In our evaluation, the robot is a wheeled unmanned ground vehicle (UGV) operating in an urban environment that is composed of ground features (e.g., paved roads, grass), objects (e.g., houses, vehicles, road barriers, traffic cones), and other agents (e.g., humans, other robots). The scenario we use in the evaluation involves the robot receiving commands from an operator to patrol between an initial location and a goal location. While patrolling, the robot continuously scans for suspicious objects. If a suspicious object is found, the robot moves toward it and uses its sensor for detecting explosives to determine if the object is a threat or harmless. After classifying each suspicious object, the robot continues patrolling.

In this scenario, the robot has four modifiable components of its behavior: speed, padding, scan time, and scan distance. Speed, measured in meters per second, controls how quickly the robot moves through the environment, while padding, measured in meters, controls how far the robot attempts to stay away from obstacles when planning its path (i.e., lower padding makes it more likely to bump into objects). Scan time, measured in seconds, is how much time the robot spends scanning each suspicious object, and scan distance, measured in meters, is how close the robot gets to suspicious objects while scanning. Longer scan times and smaller scan distances increase the probability that the robot will successfully classify objects as threats or harmless. The possible values for each modifiable component are: $\mathcal{M}_{speed} = \{0.5, 1.0, \ldots, 10.0\}$, $\mathcal{M}_{padding} = \{0.1, 0.2, \ldots, 2.0\}$, $\mathcal{M}_{scantime} = \{0.5, 1.0, \ldots, 5.0\}$, $\mathcal{M}_{scandistance} = \{0.25, 0.5, \ldots, 1.0\}$.

6.2   Experimental Conditions

Our study uses simulated operators that issue natural language commands to the robot and monitor its performance. The simulated operators were selected to represent a subset of the control strategies of human operators, and each operator's preferences influence when the robot is able to complete a task and when it is interrupted. The operators evaluate the robot based on how quickly the task is completed, how safely it is completed, and how well it identifies and correctly classifies suspicious objects. Two simulated operators are used: *speed-focused* and *detection-focused*. The speed-focused operator prefers the task to be completed quickly (i.e., 95 % probability of interrupting if the robot exceeds 120 s) and correctly (i.e., 100 % probability of interrupting if the robot misses a suspicious object or incorrectly classifies it), with less focus on safety (i.e., 5 % probability of interrupting if the robot hits an obstacle). The detection-focused operator prefers the task be completed correctly, but is less concerned with speed (i.e., 5 % probability of interrupting if the robot exceeds 120 s) or safety.

The operators can give four types of natural language feedback in the following categories: speed feedback, safety feedback, false positive feedback (i.e., classifying a harmless object as a threat), and false negative feedback (i.e., missing a suspicious object or classifying a threat as harmless). Each category of feedback has three synonymous pieces of feedback that the operators can use interchangeably and with equal probability (e.g., *"go faster"*, *"speed up"*, *"get going"*). Although we use a simulated operator, this is done to represent that human operators may not use a fixed vocabulary for feedback. Every time an operator interrupts the robot it can, with probability p_f, give the robot feedback.

For each feedback probability $p_f \in \{0.00, 0.05, 0.10, \ldots, 1.00\}$, we perform 50 experimental *trials* and start from an initially empty feedback base (i.e., the robot has no feedback model at the start of the first trial with each feedback probability). At the start of each trial the robot is assigned a random initial behavior and a random operator (both with uniform distribution). A trial concludes when the robot successfully finds a trustworthy behavior or has evaluated all possible behaviors. Each trial is composed of numerous experimental *runs*. At the start of each run the environment is reset, the robot is placed at the start position, and

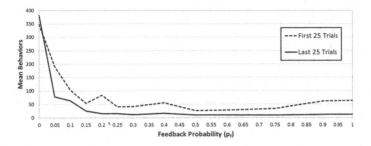

Fig. 1. Mean number of behaviors evaluated before a trustworthy behavior is found using a variety of feedback probabilities.

six suspicious objects are placed in the environment (their appearance and location are randomly selected each run). Between 0 and 3 of the objects (inclusive) are selected randomly to be threats while the remaining objects are harmless. A run concludes when the robot successfully completes the assigned tasks, fails, or is interrupted. At the end of a trial the robot updates its trust estimate and may adapt its behavior (either using random walk behavior adaptation or based on feedback). The robot stores and uses feedback (Algorithm 2) at the end of any run where feedback is provided, and updates the feedback base (Algorithm 1) at the end of each trial where a trustworthy behavior is found.

Since we are assessing how using feedback improves random walk behavior adaptation, which is used by case-based behavior adaptation to acquire cases, the robot uses only random walk adaptation. The robot uses a trustworthy threshold of $\tau_T = 5.0$ and an untrustworthy threshold of $\tau_U = -5.0$. These thresholds were selected to allow some fluctuation between increasing and decreasing trust while still identifying trustworthy and untrustworthy behaviors quickly.

6.3 Results

The mean number of behaviors that were evaluated before a trustworthy behavior was found is shown in Fig. 1. The results are further divided into the mean for the first 25 trials and last 25 trials. When comparing the results when no feedback model is learned or used (i.e., $p_f = 0.0$) to when feedback is used (i.e., $p_f > 0.0$), using feedback results in a statistically significant improvement (using a paired t-test with $p < 0.001$). This provides evidence that hypothesis **H1** is supported.

Figure 1 also shows evidence that when feedback is used the performance increases in later trials. When $p_f > 0.0$, the performance in the last 25 trials (i.e., when the robot has had time to build a feedback model) is an improvement over the first 25 trials (i.e., when the model is empty or still being refined). Figure 2 examines this further by displaying the running mean (i.e., the value for trial N is the mean of the first N trials) using four feedback probabilities ($p_f \in \{0.00, 0.05, 0.50, 1.00\}$). In early trials, performance is poor because the feedback model is still being learned. The differences in performance in the first

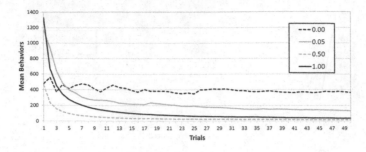

Fig. 2. Running mean number of behaviors evaluated over 50 trials.

trials is because each of those trials starts at a random behavior, some of which are further from a trustworthy behavior than others. However, regardless of their early performance, all evaluations that used feedback (i.e., all but $p_f = 0.0$) had a mean that decreased as the number of trials increased. The improvement occurs because the robot refines its feedback model over time and improves its ability to adapt in response to feedback. This shows support for hypothesis **H2**.

6.4 Discussion

Even when feedback is relatively rare (e.g., $p_f = 0.05$), the robot can still improve its performance significantly. Additionally, there is no statistically significant difference in performance when p_f values between 0.15 and 1.0 are used. This indicates that this approach does not require near-constant feedback, but can perform well using moderate amounts of feedback. Similarly, since feedback is most important when the robot needs to do random walk behavior adaptation, the robot could request additional feedback when case-based behavior adaptation fails. This would be beneficial because it would not only improve the robot's ability to acquire additional behavior adaptation cases but would also inform the operator that a period of sub-optimal behavior should be expected (i.e., using random walk behavior adaptation to acquire cases rather than the more efficient case-based behavior adaptation).

At the end of the evaluation, the feedback bases contained between 81 and 309 feedback cases (mean of 175.25), with the majority of cases having low frequency counts (i.e., their relations were rarely found for their feedback item). The cases with the highest frequency counts tended to contain the relationships we would expect given the feedback. However, some cases with high frequency counts displayed unexpected relationships. For example, with speed-related feedback the relationships often indicated that speed should be increased and padding decreased. This relationship arises because lower padding allows the robot to navigate through narrow pathways and make tighter turns, ultimately increasing its speed.

7   Conclusions

In this paper, we presented an extension of our work on trust-guided behavior adaptation to allow for the incorporation of explicit operator feedback. Since the robot learns the feedback model, it does not require that the operator limits feedback to a fixed vocabulary (e.g., the operator can use synonyms for feedback). Similarly, behavior adaptation is not dependent on feedback so feedback is used only when it is available. Our approach is beneficial because it does not require a predefined feedback model but learns one over time. This model is continuously refined and updated as more information becomes available, improving the robot's response to feedback over time. However, a limitation of our approach is that new feedback is incorporated into the feedback model only after a trustworthy behavior is found. Until that point, the robot can use feedback to adapt but cannot refine the feedback model.

We evaluated our approach in a simulated robotics environment where the robot was responsible for patrolling an urban environment, identifying suspicious objects, and classifying them as threats or harmless. Our results indicate that by learning a feedback model and using it to assist in behavior adaptation the robot can significantly improve its behavior adaptation performance. Although the robot did not initially have a feedback model, it quickly learned one and used it to improve future performance.

One area of future work we plan to address is using the feedback base to allow the robot to explain its reasoning behind behavior adaptation. In this sense, the robot would search for similar solutions to its proposed solution (i.e., the relationship between the current behavior and the new behavior) and retrieve their associated problems (i.e., what feedback the operator might have been considering). This adds transparency between the robot and operator by providing information about the robot's reasoning process and can further increase trust [20]. We also plan to investigate how the robot can reason about its goals and the team's goals to ensure they compliment each other, and to detect any unexpected goal changes. Additionally, we plan to evaluate our trust-guided behavior adaptation approach in a series of user studies.

Acknowledgments. Thanks to the Naval Research Laboratory and the Office of Naval Research for supporting this research.

References

1. Oleson, K.E., Billings, D.R., Kocsis, V., Chen, J.Y., Hancock, P.A.: Antecedents of trust in human-robot collaborations. In: Proceedings of the 1st International Multi-disciplinary Conference on Cognitive Methods in Situation Awareness and Decision Support, pp. 175–178 (2011)
2. Floyd, M.W., Drinkwater, M., Aha, D.W.: How much do you trust me? Learning a case-based model of inverse trust. In: Lamontagne, L., Plaza, E. (eds.) ICCBR 2014. LNCS, vol. 8765, pp. 125–139. Springer, Heidelberg (2014)

3. Sabater, J., Sierra, C.: Review on computational trust and reputation models. Artif. Intell. Rev. **24**(1), 33–60 (2005)
4. Kaniarasu, P., Steinfeld, A., Desai, M., Yanco, H.A.: Potential measures for detecting trust changes. In: 7th International Conference on Human-Robot Interaction, pp. 241–242 (2012)
5. Kaniarasu, P., Steinfeld, A., Desai, M., Yanco, H.A.: Robot confidence and trust alignment. In: 8th International Conference on Human-Robot Interaction, pp. 155–156 (2013)
6. Saleh, J.A., Karray, F., Morckos, M.: Modelling of robot attention demand in human-robot interaction using finite fuzzy state automata. In: International Conference on Fuzzy Systems, pp. 1–8 (2012)
7. Briggs, P., Smyth, B.: Provenance, trust, and sharing in peer-to-peer case-based web search. In: Althoff, K.-D., Bergmann, R., Minor, M., Hanft, A. (eds.) ECCBR 2008. LNCS (LNAI), vol. 5239, pp. 89–103. Springer, Heidelberg (2008)
8. Tavakolifard, M., Herrmann, P., Öztürk, P.: Analogical trust reasoning. In: Ferrari, E., Li, N., Bertino, E., Karabulut, Y. (eds.) IFIPTM 2009. IFIP AICT, vol. 300, pp. 149–163. Springer, Heidelberg (2009)
9. Leake, D.B., Whitehead, M.: Case provenance: the value of remembering case sources. In: Weber, R.O., Richter, M.M. (eds.) ICCBR 2007. LNCS (LNAI), vol. 4626, pp. 194–208. Springer, Heidelberg (2007)
10. Baier, J.A., McIlraith, S.A.: Planning with preferences. AI Mag. **29**(4), 25–36 (2008)
11. Maes, P., Kozierok, R.: Learning interface agents. In: 11th National Conference on Artificial Intelligence, pp. 459–465 (1993)
12. Horvitz, E.: Principles of mixed-initiative user interfaces. In: 18th Conference on Human Factors in Computing Systems, pp. 159–166 (1999)
13. Li, N., Kambhampati, S., Yoon, S.W.: Learning probabilistic hierarchical task networks to capture user preferences. In: 21st International Joint Conference on Artificial Intelligence, pp. 1754–1759 (2009)
14. Aha, D.W., McSherry, D., Yang, Q.: Advances in conversational case-based reasoning. Knowl. Eng. Rev. **20**(3), 247–254 (2005)
15. McGinty, L., Smyth, B.: On the role of diversity in conversational recommender systems. In: Ashley, K.D., Bridge, D.G. (eds.) ICCBR 2003. LNCS, vol. 2689, pp. 276–290. Springer, Heidelberg (2003)
16. Mahmood, T., Ricci, F.: Improving recommender systems with adaptive conversational strategies. In: 20th ACM Conference on Hypertext and Hypermedia, pp. 73–82 (2009)
17. McCarthy, K., Reilly, J., McGinty, L., Smyth, B.: On the dynamic generation of compound critiques in conversational recommender systems. In: De Bra, P.M.E., Nejdl, W. (eds.) AH 2004. LNCS, vol. 3137, pp. 176–184. Springer, Heidelberg (2004)
18. Quinlan, J.R.: Generating production rules from decision trees. In: 10th International Joint Conference on Artificial Intelligence, pp. 304–307 (1987)
19. Knexus Research Corporation: eBotworks (2015). http://www.knexusresearch.com/products/ebotworks.php. Accessed 6 May 2015
20. Kim, T., Hinds, P.: Who should I blame? Effects of autonomy and transparency on attributions in human-robot interaction. In: 15th IEEE International Symposium on Robot and Human Interactive Communication, pp. 80–85 (2006)

Top-Down Induction of Similarity Measures Using Similarity Clouds

Thomas Gabel$^{(\boxtimes)}$ and Eicke Godehardt

Faculty of Computer Science and Engineering,
Frankfurt University of Applied Sciences, 60318 Frankfurt am Main, Germany
{tgabel,godehardt}@fb2.fra-uas.de

Abstract. The automatic acquisition of a similarity measure for a CBR system is appealing as it frees the system designer from the tedious task of defining it manually. However, acquiring similarity measures with some machine learning approach typically results in some black box representation of similarity whose magic-like combination of high precision and low explainability may decrease a human user's trust in the system. In this paper, we target this problem by suggesting a method to induce a human-readable and easily understandable – and thus potentially trustworthy – representation of similarity from a previously learned black box-like representation of similarity measures. Our experimental evaluations support the claim that, given some highly precise learned similarity measure, we can induce a less powerful, but human-understandable representation of it while its corresponding level of accuracy is only marginally impaired.

1 Introduction

Similarity measures represent an integral part of a CBR system, but providing an accurate and suitable definition of these functions represents a difficult task for any designer of a case-based application. A natural way out of this problem is to cast the task as a function learning problem and, instead of defining similarity measures by hand, to apply some machine learning algorithm to generate useful similarity measures. Different such learning techniques exist, each of which comes up with its own model for representing the learned function. A common characteristic of various machine learning methods is, however, that the models learned represent black boxes from the user's perspective.

A substantial portion of the success and the acceptance of CBR systems by users can be tributed to the fact that case-based systems do inherently generate trust. The solutions suggested by a CBR system always relate to previous experience and the reasons why a specific solution is proposed to a user are, in general, easily explainable and, hence, perceived to be trustworthy.

If machine learning techniques are employed for the acquisition of highly accurate similarity measures and if, in doing so, these measures become represented by some black box machine learning model, then understandability and traceability of the CBR inference process are reduced and, as a consequence, the trustworthiness of the system is impaired. This is exactly the point we want to

© Springer International Publishing Switzerland 2015
E. Hüllermeier and M. Minor (Eds.): ICCBR 2015, LNAI 9343, pp. 149–164, 2015.
DOI: 10.1007/978-3-319-24586-7_11

address in the context of this paper. We first propose a powerful machine learning approach based on neural networks for learning high-quality, but black box-like similarity measures (Sect. 2). In a second step, we capture the essence of these neural network-based similarity measures in so-called similarity clouds and utilize these to induce easily interpretable similarity measures that are represented using some established human-readable formalism (Sect. 3). Finally (Sect. 4), we evaluate our approach in the context of a large number of benchmark application domains and, in so doing, analyze the trade-off made between a highly accurate black-box representation and a mapping to a human-readable, but probably less powerful representation of similarity measures.

2   Neural Similarity Measures

Artificial neural networks are known for their excellent performance in different areas of machine learning. Specifically, multi-layer perceptron neural networks have been shown to be universal function approximators [12]. Recent advances in the training of so-called "deep architectures" have once more boosted the attention to neural network-based learning architectures for high-dimensional input data [11]. Thus, employing neural networks for the representation of similarity measures seems to be a natural choice. While neighboring research communities have frequently addressed the topic of representing distance or similarity measures with neural networks (cf. Sect. 2.4) the core CBR community and conferences have paid comparatively little attention to that topic so far.

2.1   Multi-layer Perceptron Neural Networks

A multi-layer perceptron is an artificial neural network whose units (perceptrons) are connected in an acyclic graph. All of its neurons are arranged in layers that are disjoint from one another in that there are no connections among units within the same layer and that two successive layers are fully connected with one another. Data is propagated through the network (forward propagation) by providing inputs to the network's first layer (input layer) and, subsequently, calculating the activations of all neurons in all successive layers (hidden layers) till the final, so-called output layer. For a given training set

$$\mathbb{P} = \{(x^p, t^p) | p \in \{1, \ldots, |\mathbb{P}|\}\} \tag{1}$$

of training patterns (x^p, t^p) with input vectors $x^p = (x_1^p, \ldots, x_m^p) \in \mathbb{R}^m$ and target values $t^p = (t_1^p, \ldots, t_n^p) \in \mathbb{R}^n$, a multi-layer perceptron can be trained using the back-propagation algorithm which essentially performs a gradient descent-based adaptation of the net's connection weights such that the error

$$E = \sum_{p=1}^{|\mathbb{P}|} \sum_{i=1}^{n} (t_i^p - o_i^p)^2$$

is minimized where o^p denotes the net's output under input of pattern x^p [19].

When training neural networks, the resilient propagation update rule (Rprop [17]) has frequently been shown to provide robust and convincing results. Rprop is a batch method which calculates the gradient of the error as in standard back-propagation, but which does not use the magnitude of the gradient, but its direction for determining the weight change. To this end, the step length of a weight change is calculated by a simple heuristic that is stored separately for each connection weight. If the direction of the gradient has been the same in successive update steps, then the step width is incremented, if the sign of the gradient changes, however, the step width is decremented. An appealing feature of Rprop is that it introduces relatively few parameters and that the setting of these parameters has been shown to be quite robust with respect to the results obtained. When training neural networks within this work, we stick to the use of Rprop with its default parameter setting published in [17].

2.2   Supervised Training of Neural Net-Based Similarity Functions

In the remainder of this paper we focus on case characterizations with m describing attributes $A_1, \ldots, A_m$ and an additional solution attribute A_s. While the methods we present are generic enough to accommodate complex solutions (e.g. object-oriented or multi-dimensional ones), we will specifically focus on the case where the solution can be represented by a single value, e.g. by a class label from a finite set $D_{A_s} = \{g_1, \ldots, g_k\}$ in classification tasks or by a numeric value in regression tasks. More importantly, in what follows we will use the notation $c = (c_p, c_s)$ when speaking about case c and, in doing so, emphasize the distinction between the case's problem and solution part.

2.2.1   Utility Feedback and Case Order Feedback

In [21], a framework for learning similarity measures is described which we have used for our research repeatedly in the previous years. Its core ideas are that

1. some "similarity teacher" provides information (utility feedback) about the desired order of cases as it should result from a retrieval for a given query
2. and some machine learning module employs that information in order to learn an improved similarity measure – ideally one which matches perfectly the feedback the similarity teacher has specified.

Concerning 2, we have worked with gradient-based feature weighting techniques [21] as well as with evolutionary algorithms [20]. Within this paper, we focus on neural methods to learn and represent the similarity measure.

2.2.2   Solution Similarity

During the retrieval phase of a CBR application, the similarity between a query q and several cases c must be determined in order to find the most similar cases. To this end, the similarity is calculated between q and the problem part c_p of cases $c = (c_p, c_s)$ using some function $Sim : \mathcal{M} \times \mathcal{M} \to [0, 1]$ where $\mathcal{M}$ denotes the set of all problem descriptions (in our case $\mathcal{M} \subset \mathbb{R}^m$). One established way for obtaining utility feedback in the form of the desired case retrieval order for some query is to employ a solution similarity measure [22]. The idea here is to define an additional similarity measures $sim_s : A_s \times A_s \to [0, 1]$ for the cases' solution parts, i.e. for the solution attribute A_s. Then, the similarity assessments received from sim_s can be used to generate the training data for optimizing the similarity measure $sim_p := Sim$ for the cases' problem parts. If, for example, the similarity between the solutions of two cases is very different from the similarity between their problem parts, then this may indicate that sim_p is poorly defined.

The solution similarity measure may either be a rather simple, distance-based syntactical similarity measure or a more sophisticated one defined by an expert. The fundamental assumption, however, is that it is in general by far easier to settle on a similarity measure for the cases' solution part than to define an appropriate similarity measure for the problem part on the basis of which the retrieval will be carried out. Instead, the former one ought to be used to learn the latter one. Based on these assumptions, we can rewrite our definition of a training set (cf. Eq. 1) such that it contains pairs of cases as input values and the solution similarities of those case pairs as targets.

Definition 1 (Case-Based Training Pattern). *Given two cases c and d each of which consists of a problem and solution part ($c = (c_p, c_s)$ and $d = (d_p, d_s)$), we define a* case-based training pattern *as a triple $(c_p, d_p, sim_s(c_s, d_s))$ where sim_s is the solution similarity measure and $sim_s(c_s, d_s)$ the target value.*

Matching this definition with the notion of Eq. 1, where the training set is $\mathbb{P} = \{(x^p, t^p) | p = 1, \ldots, |\mathbb{P}|\}$ we can say that each x^p corresponds to a pair of case problem parts (c_p, d_p) and the target value t^p corresponds to a solution similarity value $sim_s(c_s, d_s)$. On top of this, we define the full training pattern set $\mathbb{P}_{CB}$ for a given case base CB as

$$\mathbb{P}_{CB} = \{((c_p, d_p), sim_s(c_s, d_s)) | \forall c, d \in CB, c \neq d\}. \tag{2}$$

If we assume case problem parts to be made up of m (numerically represented) attributes, then the space of the supervised learning problem is $2m$-dimensional.

2.2.3 Classification Tasks

A special case arises in classification domains. Here, the utility of a case c for a given query q is either zero or one, depending on the real class membership of q, i.e. on whether it matches c's class or not. The solution similarity is then

$$sim_s(c_s, d_s) = \begin{cases} 1 & if \ c_s = d_s \\ 0 & else. \end{cases} \tag{3}$$

Therefore, the resulting utility feedback cannot be said to be very "substantial", since for any case $c \in CB$ two sets of training examples can be generated – one with maximal utility, the rest with zero. In previous work [7], we found that learning similarity measures given such "knowledge-poor" training information is only of limited success, when using evolutionary optimization techniques. In this paper, we demonstrate, however, that this kind of feedback is suitable when learning and representing similarity measures with artificial neural networks.

2.3   Exemplary Results

Throughout this paper, we employ the data set on the evaluation of car values, taken from the UCI Machine Learning Repository [14], as explanatory example. In Sect. 4 we will present results for a larger selection of application domains.

The car data set consists of 1728 cases describing six features of cars like their maintenance costs, number of doors, and others (see Fig. 1, right). The solution attribute denotes the index for one out of four classes denoting how good (or acceptable) the car is. We split the data set into a train and independent test set using 5-fold cross-validation. Each time we trained a multi-layer perceptron neural network on the basis of a training set which was made up of $|\mathbb{P}_{CB}| \approx$ 1.9 million training examples according to Eq. 2 where we employed a solution similarity as given by Eq. 3. The network was made up of an input layer wit 12 inputs, two hidden layers with 13 neurons each using sigmoidal activation functions and a single output neuron with linear activation. We trained the network for 2000 epochs of Rprop and tested its performance on the independent test data set). The results shown in Fig. 1 correspond to the progress of the average classification error that results from a nearest neighbor classification on the basis of some knowledge-poor default similarity measure Sim_{def} (its accurate definition will be given in Sect. 3.1) as well as the neural network-based similarity measure (Sim_{NN}). Using Sim_{def} an average error of $24.1 \pm 2.4\%$ is obtained, whereas Sim_{NN} yields an average error of 7.2% with a standard deviation of 1.0% within a 5-fold cross-validation.

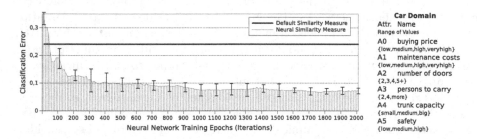

Fig. 1. Classification error of the neural similarity measure Sim_{NN} compared to the knowledge-poor default similarity measure for the car evaluation domain.

2.4   Related Work on and Discussion of Neural Similarities

Neural networks are an established tool within machine learning. In this section, we point to related work that is of highest relevance to the first part of this paper insofar as it focuses on the combination of neural net-based and case-based approaches and on the use of neural networks for assessing similarity.

In [6], Dieterle and Bergmann target the regression task of internet domain appraisal. They employ knowledge-intensive similarity measures in conjunction with a neural network for a form of feature weighting. Interestingly, their trained networks receive (feature-specific) local similarity values as inputs and produce an estimate of the target value (numeric solution attribute) as output. While their approach to obtaining target values from some kind of solution similarity measure is similar to what we described in Sect. 2.2, a core difference is that in our approach the network represents a similarity measure in itself, i.e. a function that produces a similarity value from $[0, 1]$, whereas in [6] the network generates a prediction of the solution.

The approach described in Sect. 2.2 is highly related to the work of Maggini et al. on similarity neural networks [15]. These authors train a multi-layer perceptron that is used as a similarity measure for a k-nearest neighbor retrieval. Input to the net are the two full case representations of query and case, output is a single scalar similarity value from $[0, 1]$ which is similar to our approach. This work differs from ours in that they do not use the notion of utility feedback and solution similarity, but instead employ so-called pairwise constraints to generate the training data set. This is also related to the work of Hüllermeier et al. [13] who, however, do not focus on neural network-based architectures when learning or representing similarity measures. Additionally, Maggini et al. make use of a specialized, sophisticated network topology which, for example, also ensures the resulting similarity measure to be symmetric (a restriction we do not desire). By contrast, in this work we understand the utilization of neural networks as a useful and easy-to-use standard tool and therefore stick with established standards and defaults to the largest degree possible.

While the papers mentioned so far are of high relevance to our work, there exists also a number of further pieces of work on hybrid approaches using CBR methods and neural networks. These include applications for case adaptation [9], sequential case-based decision-making [25], as well as for case retrieval [16].

Discussion: Given the well-known generalization and approximation capabilities of neural networks, their usage for representing similarity measures in CBR appears highly attractive, but it misses two important facts. First, neural networks are black boxes and as those are completely untransparent to the user. While this issue may often seem acceptable to researchers, it quite as often represents a no-go for industrial applications where decision-makers and stakeholders critically scrutinize the decisions or recommendations of any AI-based system, specifically those involving subsymbolic approaches like neural networks. Second, the solutions and outputs of a neural network are not at all self-explaining. So, while a similar case or a set of cases represent actual episodical experience and, in doing so, generate trust by a potential (re-)user, neural networks completely fail to do so. These issues are, however, not new and have been acknowledged by other authors as well, e.g. [6] complain about the lack of transparency of neural similarity measures.

3   Similarity Measure Induction with Similarity Clouds

The discussion at the end of the preceding section represents the point of departure for the remainder of this paper. We proceed on the assumption that some neural similarity Sim_{NN} has been trained on the basis of a training data set $\mathbb{P}_{CB}$ with the method described in Sect. 2. While Sim_{NN} may yield excellent performance (in terms, for example, of classification or regression accuracy), it is a black box. To this end, it is our goal to extract the essence of the knowledge encoded in the trained network Sim_{NN} into a human-readable and understandable similarity measure Sim_{HR}. In so doing, of course, we want to lose as little of Sim_{NN}'s capabilities as possible.

3.1   The Local-Global Principle

In what follows, we are going to model similarity measures using the so-called *local-global principle* [4] which disassembles the overall similarity calculation into

1. local similarity measures $sim_i : D_{A_i} \times D_{A_i} \rightarrow [0,1]$ used to compute similarities between values of individual attributes A_i with domain D_{A_i},
2. feature weights w_i used to express the importance of individual attributes,
3. an amalgamation function used to combine local similarities and feature weights. Here, we stick to a weighted average calculation of similarity

$$Sim(q,c) = \frac{\sum_{i=1}^{n} w_i \cdot sim_i(q_i, c_i)}{\sum_{i=1}^{n} w_i} \tag{4}$$

With respect to local similarity measures, we focus on the following two commonly used representation formalisms for numeric and symbolic data types.

Definition 2 (Similarity Table and Similarity Function). *Let S be a symbolic attribute with a defined list of allowed values $D_S = \{v_1, \ldots, v_d\}$. A $d \times d$-matrix with entries $x_{i,j} \in [0,1]$ representing the similarity between the query value $q_S = v_i$ and the case value $c_S = v_j$ is called* similarity table *for D_S.*
Let N be a numeric attribute with a value range of $D_N = [D_N^{min}, D_N^{max}]$. A difference-based similarity function $sim_N : D_N \times D_N \rightarrow [0,1]$ is defined as to compute a similarity value based on the difference between the case value $c_N = d_x$ and query value $q_N = d_y$ with $d_x, d_y \in D_N$, i.e. it calculates $sim_N(q_N, c_N) = f(q_N - c_N)$ for some function $f : [D_N^{min} - D_N^{max}, D_N^{max} - D_N^{min}] \rightarrow [0,1]$.

So, a knowledge-poor default similarity measure Sim_{def} is defined to be made up of identical weights for all features ($w_i = 1$ for $1 \leq i \leq n$) as well as *default* local similarity measures. These are defined as

$$sim_{N,def} : D_N \times D_N \rightarrow [0,1]$$
$$(q,c) \mapsto f(q-c) \text{ with } f(x) = 1 - \frac{|x|}{D_N^{max} - D_N^{min}}$$

for a numeric attribute N with domain range $[D_N^{min}, D_N^{max}]$, and as

$$sim_{S,def} : D_S \times D_S \to [0,1] \; with \; (q,c) \mapsto \begin{cases} 1 & if \; q = c \\ 0 & else \end{cases}$$

for a symbolic attribute S with domain D_S.

3.2 Similarity Clouds

The central concept on the basis of which our similarity induction procedure is based is the *similarity cloud*.

Definition 3 (Similarity Cloud). *Let a case c be described by m attributes A_i ($1 \leq i \leq m$) such that $c = (c_1, \ldots, c_m)$. For a given case base CB and neural similarity measure Sim_{NN} a similarity cloud SC_i for attribute A_i is defined as*

$$SC_i : CB \times CB \times D_{A_i} \to [0,1] \; with \; (q,c,v) \mapsto sim_{NN}(q,c^v)$$

where case $c^v = (c_1^v, \ldots, c_n^v)$ is defined such that $c_j^v = \begin{cases} v & if \; j = i \\ c_j & else \end{cases}$.

So, a similarity cloud is defined for a given set of cases and a specific attribute. It provides access to the neural net-based similarity for any combination of a query (from CB) and a full range of cases (also from CB) where the latter, however, have been "modified" such they take all possible values from domain D_{A_i}. The similarity cloud thus captures not just the similarities between any pair of cases from CB, but also the variations in similarity along the domain of attribute A_i (i.e. fluctuations, if that attribute value is altered).

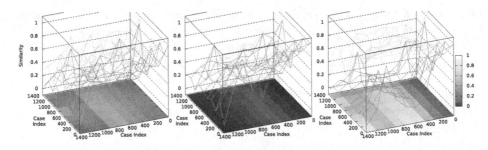

Fig. 2. Visualization of an excerpt of the similarity clouds for three of the attributes of the car domain (left A_0, middle A_2, and right A_5) which show average, little and high variability and yield corresponding feature weights. See the text for a full explanation.

In Fig. 2, we attempt to visualize parts of similarity clouds for the car evaluation domain introduced in Sect. 2.3. While the x and y axes correspond to queries and cases taken from the case base, respectively, we use the z values to plot the minimal (solid, μ_i) as well as maximal (dashed, ν_i) similarity values from the cloud for three exemplary attributes, where

$$\mu_i(q,c) = \min_{v \in D_{A_i}} SC_i(q,c,v) \; and \; \nu_i(q,c) = \max_{v \in D_{A_i}} SC_i(q,c,v).$$

The volume between μ_i and ν_i corresponds to the variability in neural similarity, when varying the value of attribute A_i. This volume is actually filled with a large number of similarity data points which, when watched in 3D, conveys the impression of a cloud (hence, the name "similarity cloud"). For clarity, however, the visualization shows only min and max values for each query-case combination.

3.3 Feature Weighting Using Similarity Clouds

Within this section, our goal is to derive feature weights w_i for Sim_{HR}. We employ a fixed amalgamation scheme (according to Eq. 4) as well as fixed local *default* similarity measures as specified in Sect. 3.1. Thus, the method introduced subsequently relies solely on the modification of the feature weights.

The variability of the similarity within the similarity cloud mentioned and visualized above represents changes in the *overall* (global) similarity between queries and cases. Therefore, it is an indicator of the respective feature's importance for the overall similarity assessment. Based on this observation we derive feature weights from the variability in the similarity cloud as follows.

Definition 4 (Similarity Cloud-Based Feature Weight). *Let SC_i be the similarity cloud for case base CB and attribute A_i. If A_i is a discrete attribute, we let $D := D_{A_i}$, otherwise (i.e. if A_i is numeric) we discretize D_{A_i} equidistantly according to $D := \{\frac{j}{S}(D_{max} - D_{min}) + D_{min} | j = 0, \ldots, S\}$. The similarity cloud-based feature weight w_i is then defined as*

$$w_i = \frac{1}{|CB|^2} \sum_{c \in CB} \sum_{q \in CB} \sqrt{\sum_{v \in D} \left(\left(\sum_{u \in D} \frac{SC_i(q,c,u)}{|D|} \right) - SC_i(q,c,v) \right)^2}$$

The induced feature weight of an attribute essentially corresponds to the average standard deviation of neural similarity when altering the value of the attribute considered. In Fig. 2, samples of the standard deviation of neural similarity are visualized at the bottom of the plots using colored map views. As can be seen, for attribute A_2 there is very little variability (almost all samples are black) corresponding to a low feature weight, whereas for A_5 there are much higher variations which correspond to a significantly higher value of w_5.

3.4 Induction of Local Similarity Measures Using Similarity Clouds

In the following, we aim at the induction of local similarity measures from similarity clouds. We cover both types of local measures specified in Definition 2.

Symbolic Attributes: For a symbolic attribute with $D_{A_i} = \{v_1, \ldots, v_s\}$ we fill the similarity table sim_i which contains an entry for each combination (v_q, v_c) of the query's and case's ith attribute value. For a specific combination of $v, w \in D_{A_i}$ we assume that $sim_i(v, w)$ is low, if the neural similarity $Sim_{NN}(q, c)$ for some query q and case c with $c_i = v$ changes "a lot", if we modify c_i's value and set it from v to w. Clearly, the notion "a lot" needs to be formalized. To this end, we utilize the concept of a similarity cloud, which already captures this information, and define the following scoring function.

Definition 5 (Cloud-Based Local Distance Scoring). *Given a symbolic attribute A_i and two values $v, w \in D_{A_i}$ as well as the attribute's similarity cloud SC_i, the cloud-based local distance scoring δ is defined as*

$$\delta(v, w) = \sum_{q \in CB} \sum_{c \in CB} (SC_i(q, c, v) - SC_i(q, c, w))^2.$$

For given $v, w \in D_{A_i}$, the scoring function from Definition 5 considers all pairs of cases from CB. In so doing, the value of attribute i is, however, altered to be v and w, respectively, and the squared differences in the corresponding neural network-based similarities are summed up.

While the scores we obtain from Definition 5 represent a local distance function and may be utilized for determining a query's nearest neighbor, our emphasis is on inducing an easily human-interpretable similarity measure representation from the similarity cloud. Therefore, we apply the following normalization and transformation from a distance to a compatible similarity measure.

Definition 6 (Cloud-Based Local Similarity Table). *Given a cloud-based local distance scoring function $\delta : D_{A_i} \times D_{A_i} \to \mathbb{R}$ we induce a local similarity table for attribute A_i according to*

$$sim_i : D_{A_i} \times D_{A_i} \to [0, 1] \; with \; (v, w) \mapsto 1.0 - \frac{\delta(v, w) - \delta_{min}(v)}{\delta_{max}(v) - \delta_{min}(v)}$$

where $\delta_{min}(v) = \min_{w \in D_{A_i}} \delta(v, w)$ and $\delta_{max}(v) = \max_{w \in D_{A_i}} \delta(v, w)$.

When normalizing the induced local similarity measure according to Definition 6 we gain better interpretability, but lose information about the relevance of individual attributes. So, the induction of local similarity measures should be used in conjunction with the feature weighting method presented in Sect. 3.3.

Numeric Attributes: For attributes A_i with a numeric domain $D_{A_i} = [D_{A_i}^{min}, D_{A_i}^{max}]$ and, thus, for the induction of a difference-based similarity function we proceed in a similar manner. Essentially, we iterate over all query-case combinations from the case base, i.e. over all $q, c \in CB$, and measure the variability of the neural similarity, if we alter the value of the ith attribute in q and c, respectively, such that the difference $q_i - c_i$ takes some specific difference value s.

Of course, handling that infinite number of possible real-valued differences is numerically infeasible, which is why our goal is to come up with a similarity function that is represented by a set $\mathcal{S}$ of sampling points (i.e. a finite set of possible real-valued differences) that are distributed equidistantly over $[D^{min}_{A_i} - D^{max}_{A_i}, D^{max}_{A_i} - D^{min}_{A_i}]$. This approach is not new and has been successfully applied in [20]. When calculating the local similarity with such a sampled similarity function, the similarity is interpolated linearly between the two neighboring sampling points of $q_i - c_i$. For the purpose of representing sim_i it is therefore sufficient to provide a mapping from $\mathcal{S}$ to $[0,1]$, i.e. it is sufficient to specify sim_i as $sim_i : \mathcal{S} \to [0,1]$. Based on this observation we define:

Definition 7 (Cloud-Based Sampled Distance Scoring). *Let A_i denote a numeric attribute, SC_i the attribute's similarity cloud, and $\mathcal{S}$ denote the set of sampling points (with $|\mathcal{S}|$ assumed to be odd). Then,*

$$D'_{A_i} := \left\{ D^{min}_{A_i} + 2k \frac{D^{max}_{A_i} - D^{min}_{A_i}}{|\mathcal{S}| - 1} \mid 0 \le k \le \frac{|\mathcal{S}| - 1}{2} \right\}$$

represents a discretization of D_{A_i} on the basis of which we compute the cloud-based sampled distance scoring ε according to

$$\varepsilon : \mathcal{S} \to \mathbb{R} \text{ with } s \mapsto \sum_{v \in D'_{A_i}} \sum_{w \in D'_{A_i}} \begin{cases} \delta(v,w) & \text{if } v - w = s \\ 0 & \text{else} \end{cases}$$

where $\delta(v,w)$ is calculated according to Definition 5.

Similarly to the case of symbolic attributes we induce the sampled similarity function by applying a normalization and transformation from a distance to a compatible similarity measure.

Definition 8 (Cloud-Based Sampled Similarity Function). *Given a cloud-based sampled distance scoring function $\varepsilon : \mathcal{S} \to \mathbb{R}$ we induce a local sampled similarity function for an attribute A_i with numeric domain according to*

$$sim_i : \mathcal{S} \to [0,1] \text{ with } s \mapsto 1.0 - \frac{\varepsilon(s) - \min_{x \in \mathcal{S}} \varepsilon(x)}{\max_{x \in \mathcal{S}} \varepsilon(x) - \min_{x \in \mathcal{S}} \varepsilon(x)}.$$

3.5   Exemplary Results

We return to our explanatory example introduced in Sect. 2.3 and apply the similarity measure induction procedure presented above to induce both, feature weights and local similarity measures for the car evaluation domain. The left chart of Fig. 3 compares the classification errors of Sim_{def}, Sim_{NN} as well as induced similarity measures Sim_{HR} (with only weights and only local measures induced from Sim_{NN} as well as the combination of both). The numbers indicate the general effectiveness of the approach, since the high accuracy of the neural similarity can be preserved when inducing a similarity measure that is represented according to the local-global principle.

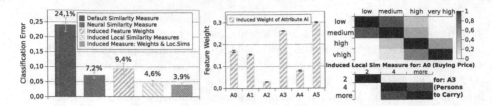

Fig. 3. Results of the similarity measure induction procedure applied to the car domain.

Another interesting question is whether the resulting measure Sim_{HR} is easy to read and understand. To this end, the central chart shows the normalized feature weights induced for the car evaluation domain. Please note how the weight of w_0, w_2 and w_5 (buying price, number of doors, car safety) correspond to the visualization of the attribute's corresponding similarity clouds in Fig. 2. Additionally, the right part of Fig. 3 visualizes the local similarity tables for attributes A_0 and A_3 (buying price, persons to carry).

3.6 Related Work on Feature Weighting and Similarity Learning

Feature weighting and similarity measure learning have a long history in case-based reasoning. Wettschereck and Aha [24] provide an overview on various early approaches. Recent approaches that go beyond pure feature weighting and instead aim at learning full similarity measure representations include neighborhood component analysis [8], large margin nearest neighbor classifiers [23] and various similarity-based classifiers reviewed by Chen et al. [5]. Another line of research on similarity learning focuses on exploiting binary pairwise information about similarity or dissimilarity, so-called pairwise constraints or preferences. This stream of research includes work on kernel-based learning methods [2], the boosting variant DistBoost [10], solution similarity learning in preference-based CBR [1], relevant component analysis [3] as well as similarity learning from case-order feedback [21]. A contrasting feature of the work presented in this paper compared to the pieces of related work mentioned here as well as in Sect. 2.4 is that we aim at learning a human-understandable and easily interpretable similarity measure which may increase trust by the users of the system. To this end, our work bears also some relatedness to the research field on explanations and case-based reasoning [18].

4   Experimental Results

In order to empirically evaluate the approach proposed in this paper we selected a batch of data sets from the UCI Machine Learning Repository. These included both, 14 classification and 5 regression tasks. For each domain, we split the data randomly into a training set S_{train} and a test set S_{test} within a 5-fold cross validation. In each learning run we successively applied the following steps:

1. Create a case base CB from S_{train}.
2. Evaluate the performance of the default similarity measure Sim_{def} (cf. Sect. 3.1) for all queries from S_{test}.
3. Train a neural network-based similarity measure Sim_{NN} as described in Sect. 2 and evaluate it on all queries from S_{test}.
4. Induce feature weights w_i from Sim_{NN} according to the procedure presented in Sect. 3.3, create a similarity measure $Sim_{HR,w}$ which employs the induced weights in combination with default local similarity measures.
5. Induce local similarity measures from Sim_{NN} according to the techniques described in Sect. 3.4, create a similarity measure $Sim_{HR,l}$ which uses those induced local measures in combination with default weights ($w_i = 1 \forall i$).
6. Create a similarity measure Sim_{HR} which uses both, induced feature weights as well as local measures ($4 + 5$) and evaluate it on all queries from S_{test}.

Speaking about the evaluation of similarity measures, we perform k-nearest neighbor classification/regression and report the average classification or regression accuracy on S_{test} and the corresponding standard deviation. Throughout all experiments we set $k = 1$ for better comparability. This value might be optimized for each application domain and thus yield superior overall results.

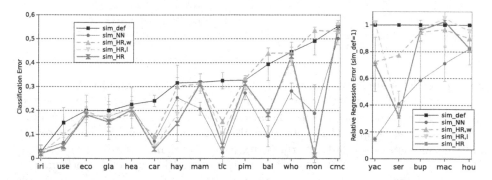

Fig. 4. Visualization of the performance of the considered similarity measures when used for classification/regression in various domains.

Results: As can be seen in Fig. 4, the quality of the learned neural similarity measure varies substantially over the domains considered. Averaging over all those domains the classification/regression error of the neural similarity measure is 39.2 % lower than the error made by default similarity measures (Table 1).

The neural measure, Sim_{NN}, as well as the knowledge-poor default similarity measure Sim_{def} represent, informally speaking, the upper and lower limit of what we can expect from the similarity induction technique proposed in this paper. On the one hand, we can observe that each induced measure is at least as performant as the default measure – which is a minimal goal from an evaluation perspective, because otherwise the entire approach would be rendered pointless. On the other hand, we can observe that in almost all cases the induced

Table 1. Overview of experimental results per domain. The task column distinguishes classification (C, with number of classes in brackets) from regression (R) tasks.

Domain	$|CB|$	#Attributes		Task	Classification/Regression error on S_{test} using				
		discr	num		Sim_{def}	Sim_{NN}	$Sim_{HR,w}$	$Sim_{HR,l}$	Sim_{HR}
balancescale	625	4	0	C(3)	.394 ± .068	.093 ± .043	.440 ± .022	.184 ± .024	.182 ± .024
car	1728	6	0	C(4)	.241 ± .024	.072 ± .010	094. ± .019	.046 ± .009	.039 ± .010
cmc	1473	9	0	C(3)	.553 ± .024	.501 ± .026	.531 ± .015	.557 ± .033	.546 ± .022
ecoli	336	0	7	C(8)	.200 ± .036	.179 ± .085	.197 ± .022	.182 ± .029	.182 ± .032
glass	214	0	9	C(2)	.200 ± .069	.162 ± .049	.176 ± .082	.152 ± .073	.176 ± .062
hayes-roth	132	4	0	C(3)	.315 ± .074	.254 ± .084	.300 ± .063	.185 ± .079	.146 ± .074
heart	270	8	5	C(2)	.226 ± .033	.211 ± .084	.185 ± .029	.207 ± .033	.204 ± .037
iris	150	0	4	C(3)	.027 ± .028	.033 ± .024	.027 ± .028	.020 ± .018	.020 ± .018
mammograph	830	4	0	C(2)	.318 ± .036	.207 ± .027	.312 ± .039	.312 ± .041	.312 ± .041
monks2	432	6	0	C(3)	.491 ± .058	.188 ± .119	.535 ± .039	.007 ± .016	.014 ± .031
pima	768	0	8	C(2)	.327 ± .020	.273 ± .027	.333 ± .024	.299 ± .033	.312 ± .026
tictactoe	958	9	0	C(2)	.325 ± .034	.024 ± .016	.154 ± .034	.101 ± .078	.054 ± .080
userknowl	258	0	5	C(4)	.149 ± .063	.067 ± .033	.047 ± .022	.098 ± .024	.051 ± .018
wholesale	440	0	6	C(3)	.445 ± .040	.282 ± .034	.441 ± .029	.411 ± .044	.427 ± .038
bupa	345	0	5	R	2.26 ± .27	1.32 ± .66	2.13 ± .26	2.18 ± .18	2.17 ± .17
housing	506	2	11	R	2.58 ± .11	2.13 ± .31	2.32 ± .16	2.47 ± .20	2.13 ± .04
machines	209	0	6	R	49.5 ± 10.1	35.2 ± 6.5	47.7 ± 10.6	51.9 ± 10.9	50.7 ± 9.5.
servo	167	4	0	R	.661 ± .139	.271 ± .063	.511 ± .199	.205 ± .046	.207 ± .048
yacht	308	0	6	R	3.78 ± .310	0.55 ± .059	2.73 ± .367	3.83 ± .421	2.66 ± .765
Average of Error Relative to Sim_{def}					100.00 %	60.84 %	82.44 %	69.98 %	63.86 %

human-readable similarity measure brings about almost as good results as the neural ones with only minor impairments, which are actually to be expected since Sim_{NN} is the input to the similarity clouds and, thus, to the entire induction procedure. Despite this, there are also some notable exceptions where the induced measures even outperform Sim_{NN}. Summarizing, from the 19 domains considered, there are 11 domains in which the resulting system's performance using Sim_{HR} is approximately equivalent or even superior to Sim_{NN}, 4 domains where the performance of the induced measures is somewhere in between Sim_{NN} and Sim_{def}, and finally 4 domains where the induced measures' performance is rather close the performance of the default measure. Averaging over all domains, the average classification/regression error of the induced measures, when compared to the error made by Sim_{def}, is reduced by 17.6 % if only weights are induced, by 30.0 % if only local measures are induced, and by 36.1 % if both approaches are combined, whereas Sim_{NN} lowered that error by 39.2 % as stated above. This, in conjunction with the fact, that the induced similarity measures are easily understandable for humans and domain experts, clearly stresses the usability of our approach.

5   Conclusion

In this paper, we have presented an approach that induces a similarity measure according to the local-global principle from a previously learned neural network-based representation of similarity. In so doing, we aimed at generating a human-readable representation of similarity which is likely to be accepted and trusted more by human users of the CBR system. We evaluated this approach on a large set of classification and regression domains from the UCI Repository and found that the induction procedure generates similarity measures that yield levels of accuracy that are only slightly inferior to the accuracy of the neural network-based templates.

References

1. Abdel-Aziz, A., Strickert, M., Hüllermeier, E.: Learning solution similarity in preference-based CBR. In: Lamontagne, L., Plaza, E. (eds.) ICCBR 2014. LNCS, vol. 8765, pp. 17–31. Springer, Heidelberg (2014)
2. Baghshah, M., Shouraki, S.: Semi-supervised metric learning using pairwise constraints. In: Proceedings of the 21st International Joint Conference on Artificial Intelligence (IJCAI), San Francisco, USA, pp. 1217–1222 (2009)
3. Bar-Hillel, A., Hertz, T.: Shental, weinshall: learning a mahalanobis metric from equivalence constraints. J. Mach. Learn. Res. **6**, 937–965 (2005)
4. Bergmann, R., Richter, M., Schmitt, S., Stahl, A., Vollrath, I.: Utility-oriented matching: a new research direction for case-based reasoning. In: Proceedings of the 9th German Workshop on Case-Based Reasoning (GWCBR) (2001)
5. Chen, Y., Garcia, E., Gupta, M., Rahimi, A., Cazzanti, L.: Similarity-based classification: concepts & algorithms. J. Mach. Learn. Res. **10**, 747–776 (2009)
6. Dieterle, S., Bergmann, R.: A hybrid CBR-ANN approach to the appraisal of internet domain names. In: Lamontagne, L., Plaza, E. (eds.) ICCBR 2014. LNCS, vol. 8765, pp. 95–109. Springer, Heidelberg (2014)
7. Gabel, T., Stahl, A.: Exploiting background knowledge when learning similarity measures. In: Funk, P., González Calero, P.A. (eds.) ECCBR 2004. LNCS (LNAI), vol. 3155, pp. 169–183. Springer, Heidelberg (2004)
8. Goldberger, J., Roweis, S., Hinton, G.: Salakhutdinov: Neighborhood Component Analysis. In: Neural Information Processing Systems 18 (NIPS), pp. 513–520 (2005)
9. Henriet, J., Leni, P.-E., Laurent, R., Roxin, A., Chebel-Morello, B., Salomon, M., Farah, J., Broggio, D., Franck, D., Makovicka, L.: Adapting numerical representations of lung contours using case-based reasoning and artificial neural networks. In: Agudo, B.D., Watson, I. (eds.) ICCBR 2012. LNCS, vol. 7466, pp. 137–151. Springer, Heidelberg (2012)
10. Hertz, T., Bar-Hillel, A., Weinshall, D.: Boosting margin-based distance functions for clustering. In: Proceedings of the International Conference on Machine Learning (ICML), New York, USA, pp. 393–400 (2004)
11. Hinton, G., Salakhutdinov, R.: Reducing the dimensionality of data with neural networks. Science **313**, 504–507 (2006)
12. Hornick, K., Stinchcombe, M., White, H.: Multilayer feedforward networks are universal approximators. Neural Netw. **2**, 359–366 (1989)

13. Hüllermeier, E., Cheng, W.: Preference-based CBR: general ideas and basic principles. In: Proceedings of the 23rd International Joint Conference on Artificial Intelligence (IJCAI), Beijing, China, pp. 3012–3016 (2013)
14. Lichman, M.: UCI Machine Learning Repository (2013). archive.ics.uci.edu/ml
15. Maggini, M., Melacci, S., Sarti, L.: Learning from pairwise constraints by similarity neural networks. Neural Netw. **26**, 141–158 (2012)
16. Main, J., Dillon, T.S.: A hybrid case-based reasoner for footwear design. In: Althoff, K.-D., Bergmann, R., Branting, L.K. (eds.) ICCBR 1999. LNCS (LNAI), vol. 1650, pp. 497–509. Springer, Heidelberg (1999)
17. Riedmiller, M., Braun, H.: A direct adaptive method for faster backpropagation learning: the RPROP algorithm. In: Proceedings of the IEEE International Conference on Neural Networks (ICNN), San Francisco, USA, pp. 586–591 (1993)
18. Roth-Berghofer, T.R.: Explanations and case-based reasoning: foundational issues. In: Funk, P., González Calero, P.A. (eds.) ECCBR 2004. LNCS (LNAI), vol. 3155, pp. 389–403. Springer, Heidelberg (2004)
19. Rumelhart, D., Hinton, G.: Learning representations by back-propagating errors. Nature **323**, 533–536 (1986)
20. Stahl, A., Gabel, T.: Using evolution programs to learn local similarity measures. In: Ashley, K.D., Bridge, D.G. (eds.) ICCBR 2003. LNCS, vol. 2689, pp. 537–551. Springer, Heidelberg (2003)
21. Stahl, A., Gabel, T.: Optimizing similarity assessment in case-based reasoning. In: Proceedings of the 21st National Conference on Artificial Intelligence (AAAI 2006). AAAI Press, Boston (2006)
22. Stahl, A., Schmitt, S.: Optimizing retrieval in CBR by introducing solution similarity. In: Proceedings of the International Conference on Artificial Intelligence (IC-AI 2002). CSREA Press, Las Vegas (2002)
23. Weinberger, K., Saul, L.: Distance metric learning for large margin nearest neighbor classification. J. Mach. Learn. Res. **10**, 207–244 (2009)
24. Wettschereck, D., Aha, D.: Weighting features. In: Proceedings of the 1st International on Case-Based Reasoning (ICCBR), London, UK, pp. 347–358 (1995)
25. Zehraoui, F., Kanawati, R., Salotti, S.: CASEP2: hybrid case-based reasoning system for sequence processing. In: Funk, P., González Calero, P.A. (eds.) ECCBR 2004. LNCS (LNAI), vol. 3155, pp. 449–463. Springer, Heidelberg (2004)

Improving Case Retrieval Using Typicality

Emmanuelle Gaillard[1,2,3]([✉]), Jean Lieber[1,2,3],
and Emmanuel Nauer[1,2,3]

[1] Université de Lorraine, LORIA, 54506 Vandœuvre-lès-Nancy, France
{Emmanuelle.Gaillard,Jean.Lieber,Emmanuel.Nauer}@loria.fr
[2] CNRS, 54506 Vandœuvre-lès-Nancy, France
[3] Inria, 54602 Villers-lès-nancy, France

Abstract. This paper shows how typicality can be used to improve the case retrieval of a case-based reasoning (CBR) system, improving at the same time the global results of the CBR system. Typicality discriminates subclasses of a class in the domain ontology depending of how a subclass is a good example for its class. Our approach proposes to partition the subclasses of some classes into atypical, normal and typical subclasses in order to refine the domain ontology. The refined ontology allows a finer-grained generalization of the query during the retrieval process. The benefits of this approach are presented according to an evaluation in the context of TAAABLE, a CBR system designed for the cooking domain.

Keywords: Typicality · Ontology refinement · Case retrieval · Cooking

1 Introduction

This paper shows how typicality can be used to improve the case retrieval of a case-based reasoning (CBR) system, improving at the same time the global results of the system.

Usually, CBR systems store knowledge in four different knowledge containers including the domain knowledge container [1]. In many CBR systems (e.g. [2]), the domain knowledge relates to an ontology composed of a hierarchy of classes facilitating the retrieval process guided by similarity measures. When using such a hierarchical structure, no difference is made between subclasses at the same level (i.e. directly subsumed by the same class). However, in natural categorization, some subclasses of a class are better examples than other ones for their subsuming class. These subclasses are considered as the most typical subclasses of the class. For example, it can be argued that the `Apricot` class is a more typical subclass of the `StoneFruit` class than the `Avocado` class, which is atypical.

In CBR systems using a case retrieval process based on generalization of the query (e.g. TAAABLE [2]) no difference is made between subclasses belonging to the same class. The generalization of a typical class B into a class A will retrieve cases linked to typical subclasses of A as well as cases linked to atypical classes of A. For example, in a CBR system using an ontology of the cooking domain, the generalization of `Peach` into `StoneFruit` will retrieve cases linked to any kind

© Springer International Publishing Switzerland 2015
E. Hüllermeier and M. Minor (Eds.): ICCBR 2015, LNAI 9343, pp. 165–180, 2015.
DOI: 10.1007/978-3-319-24586-7_12

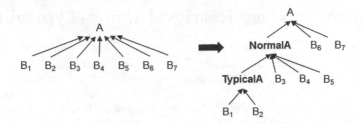

Fig. 1. Ontology reorganization based on typicality where an edge $B \rightarrow A$ means $B \sqsubseteq A$. In this illustration, for A, B_1 and B_2 are typical, B_3, B_4 and B_5 are normal (and not typical), and B_6 and B_7 are atypical.

of `StoneFruit`, including `Apricot` which is considered as a typical stone fruit, as well as cases linked to `Avocado` which is considered as an atypical stone fruit.

In this paper, we propose to exploit the typicality of subclasses with respect to their direct subsuming class. The idea is to divide a set of classes subsumed by the same class into three subsets: the atypical classes, the normal classes and the typical classes, and to use these three sets to reorganize the ontology as shown in Fig. 1. The refined ontology allows a finer-grained generalization of the query during the retrieval process.

The contribution of the approach is shown experimentally by comparing results of two systems, CBR_{std} and CBR_{typ}, using the same inference engine on the same set of queries. CBR_{std} will use a domain ontology $\mathcal{O}_{std}$ which is not refined according to typicality, while CBR_{typ} will use the $\mathcal{O}_{typ}$ domain ontology, which is $\mathcal{O}_{std}$ refined thanks to typicality.

The paper is organized as follows: Section 2 introduces the context of application and gives the motivation of this work. Section 3 presents the state of the art about typicality and its use in knowledge modeling. Section 4 describes the typicality-based approach we propose to refine a domain ontology. Section 5 presents the evaluation methodology. The results are discussed in Sect. 6 and conclusions are presented in Sect. 7.

2   Context and Motivation

2.1   The TAAABLE Use-Case

TAAABLE [2] is a CBR system in the cooking domain which retrieves and creates new recipes by adaptation. TAAABLE, like classical CBR systems, uses a case base (a set of cooking recipes) to retrieve and adapt, using domain knowledge, the cases that are the most similar to the user constraints.

The domain knowledge. The domain knowledge (`DK`) is an ontology composed of a set of atomic classes into several hierarchies (food, dish type, localization, etc.) organized according to the subsumption relation. Given two classes `A` and `B` of this ontology, `A` subsumes `B`, denoted by $B \sqsubseteq A$, if the set of instances of `B` is included in the set of instances of A. For instance, Apricot ⊑ StoneFruit, means that all apricots are stone fruits.

Case base. The case base is a set of recipes. Each recipe R in the case base is represented by its index denoted by $idx(R)$ which is a conjunction of classes of the domain ontology. For example, $idx(R_1) =$ TartDish ∧ PieCrust ∧ Leek ∧ GoatCheese ∧ Cream ∧ Egg ∧ Salt ∧ Pepper is the index of a tart recipe whose ingredients are pie crust, leek, goat cheese, cream, egg, salt and pepper.

Query. In TAAABLE, a query is composed of a set of (user) constraints. Formally, a query, denoted by Q, is also a conjunction of classes. For example, $Q_{ex} =$ TartDish ∧ Zucchini ∧ Parmesan means that the user searches for tart recipes with zucchini and parmesan.

Case retrieval. The retrieval process consists in searching cases that best match the query. If an exact match exists, the corresponding cases are returned. Otherwise, the query is relaxed using a generalization function Γ composed of one-step generalizations, which transforms Q (with a minimal cost) until at least one recipe of the case base matches $\Gamma(Q)$.

A one step-generalization is denoted by $\gamma =$ B ⤳ A, where A and B are classes and B ⊑ A belongs to the domain ontology. Each one-step generalization is associated to a cost denoted by $cost($B ⤳ A$)$. The generalization Γ of Q is a composition of one-step generalizations $\gamma_1, \gamma_2, \ldots \gamma_n$: $\Gamma = \gamma_n \circ \ldots \circ \gamma_2 \circ \gamma_1$, with $cost(\Gamma) = \sum_{i=1}^{n} cost(\gamma_i)$. How this cost is computed is detailed in [2].

In the space of generalization functions Γ, the least costly generalization such that at least one case matches $\Gamma(Q)$ is searched. For Q_{ex}, TAAABLE produces $\Gamma =$ Parmesan ⤳ Cheese ∘ Zucchini ⤳ Vegetable and so $\Gamma(Q_{ex}) =$ TartDish ∧ Cheese ∧ Vegetable, which matches $idx(R_1)$.

TAAABLE returns all the adapted cases that match the generalized query. It may occur that a user of TAAABLE wants to find other cases, less similar than the ones retrieved in the first-step generalization. In this case, the user may trigger the system again to resume the generalization process which will search for the *next* least costly generalization for which new cases are returned. The generalization process is also automatically resumed when a given number of results is requested and the first generalizations do not return enough results.

Adaptation. The result of the case retrieval process is a set of recipes that match the generalized query $\Gamma(Q)$. The adaptation process consists of a specialization of the generalized query produced by the retrieval step. For example, according to $\Gamma(Q_{ex})$, to R_1, and to DK, the system proposes to replace GoatCheese with Parmesan in R_1 because Cheese of $\Gamma(Q_{ex})$ subsumes both GoatCheese and Parmesan. In the same way, Leek has to be replaced with Zucchini in R_1 because Vegetable of $\Gamma(Q_{ex})$ subsumes both Leek and Zucchini. In TAAABLE, the adapted cases matching one generalized query are not ranked. For example, R_2 such that $idx(R_2) =$ TartDish ∧ PieCrust ∧ Carrot ∧ GoatCheese ∧ Egg ∧ Salt ∧ Pepper is also retrieved by $\Gamma(Q_{ex})$ and adapted by replacing Carrot with Zucchini and GoatCheese with Parmesan. R_1 and R_2, with their respective adaptations, are returned with no preference ranking.

2.2 Motivation

Let Olive $\sqsubseteq$ StoneFruit and Apricot $\sqsubseteq$ StoneFruit be elements of the domain ontology. Olive and Apricot are two sibling classes which have different levels of typicality for the subsuming concept StoneFruit; apricot is usually considered as a more typical stone fruit than olive. Let $Q_{ex} =$ Peach $\wedge$ TartDish be a new example query. If no recipe matches exactly Q_{ex}, let $\Gamma =$ Peach $\leadsto$ StoneFruit be the generalized function produced during the retrieval process. Let R_1 and R_2 be two recipes such that $id(R_1) =$ TartDish $\wedge$ PieCrust $\wedge$ Apricot $\wedge$ Almond $\wedge$ Butter and $id(R_2) =$ TartDish $\wedge$ PieCrust $\wedge$ Olive $\wedge$ Tomato $\wedge$ OliveOil. These two recipes match $\Gamma(Q_{ex})$ and the adaptation process proposes to "Replace apricots with peaches" in R_1 and to "Replace olives with peaches" in R_2. The two adaptations involve sibling classes: Apricot and Olive. These two results are presented without ranking preference to the user. The objective of our work is to favor the replacement of a typical stone fruit with another typical stone fruit and more generally to replace food requested in Q by another food according to their typicality closeness, assuming that the more typical two foods are, the more similar they are.

3 State of the Art

This section presents some theories about the use of typicality in knowledge models in general and then in CBR systems, in particular.

3.1 Concept and Classification Models Based on Typicality

A concept is a mental representation of a class whose main function is to classify knowledge of a domain. Smith and Medin [3] present three main types of classification models: the classical view, the exemplar view and the probabilistic view. The classical view taken from Aristotle, argues that all instances of a concept share common properties which are necessary and sufficient to define the concept. The exemplar-based view and the probabilistic view are more flexible and based on "natural" concepts. Using the notion of family resemblance in classes, Beckner [4] defines the notion of polythetic classes, that contrasts with the notion of monothetic classes where members of a class share all the defined necessary and sufficient properties. Members of a polythetic class share a large set of common but not necessary properties, and properties are shared by a majority of members of the class. For example, for the Bird class, the property "can fly" is a property shared by many members, but not shared by all the members, since an ostrich, which is a bird, does not have this property. However, the feature "can fly" is characteristic for Bird. In this way, a member of Bird is typical (i.e. is a good example) of Bird, if this member has this characteristic feature. Such a member is called a *prototype* of Bird. The notion of typicality is related here to the idea that not all the members of a class are "good" (or characteristic) examples. For the psychologists Rosch and Mervis [5], the typicality of an instance of a class A depends on its similarity with other instances of A and on dissimilarity with instances of the other classes. For Rosch and Mervis, a whale is considered to be atypical of the class Mammal, not just because a whale is not very similar to other mammals, but also because a whale shares common properties with types of fish. Barsalou [6], for his part, argues that the typicality of an instance of a class depends on its similarity with other instances and how frequently people have experienced the instance as a member of a class. A degree of typicality may be associated to a pair (i, C) where i is an instance of a class C. The most typical instance(s) of a class is/are its prototype(s).

The probabilistic view, based on prototypicality theory, assumes that a class is represented by a single abstract prototype. The properties shared by the instances of a class are more or less characteristic, and each property is associated with a weight. The degree of typicality of an instance of a class depends on its similarity with the prototype represented by a property vector [7]. For example, for the Bird class, the property vector could be (can_fly, can_sing, has_feathers, is_oviparous) and each property is associated with a weight between 0 and 1. In the probabilistic view, a class has a unique prototype.

The exemplar-based representation [8,9] is more flexible since it may be represented by multiple prototypes which are real examples of the class. For example, Robin and Swallow can be considered to be the most representative subclasses of the Bird class.

Other theories have studied the typicality notion. In [10], fuzzy classes are described with attributes where values are typical or not. For two classes A and B, such that $B \sqsubseteq A$, three types of specialization from A to B are proposed: typical, normal and atypical, depending on the comparison of typical and not typical values of A and B attributes.

3.2 Typicality in CBR

Typicality is a notion that has already been used in CBR systems. Founded on the theory of typicality [11], Weber-Lee et al. [12] use typicality to select the best case in a set of similar cases during the retrieval process. Retrieved cases are gathered into clusters using a geometrical fuzzy clustering algorithm; the *Most Typical Value* measure is used in order to compute the best case.

While [12] uses a measure of typicality in order to retrieve the best case, Protos [9], an exemplar-based learning system, uses typicality knowledge which is a part of the domain knowledge. In Protos, an example is a case. The domain knowledge and cases are represented in a single structure: a category structure. The category structure is a semantic network where nodes consist of categories which are the features and the exemplars (the retained cases), and edges are explanations acquired from an expert triggered by a classification failure. Protos classifies a new problem described with features by searching the closest case in two major steps. The first step consists in retrieving the most similar category by tracking links which associate categories and features. The second step consists in choosing the best case according to edges describing prototype links and difference links. Protos retrieves first the most typical case of the most similar category.

4   Refining the Ontology According to Typicality

Considering the class A as the root of the hierarchy to refine, our approach consists in dividing a set of classes B_i, such that $B_i \sqsubseteq A$, into three subsets: the atypical classes, the normal classes and the typical classes. These subsets will be exploited to refine the domain ontology used by the CBR system in order to take into account typicality during the query generalization.

4.1   Computing the Class Typicality

Typicality acquisition. The degree of typicality of a subclass in a class consists in measuring how a subclass is a good example with respect to its class. Some psychologists propose to acquire the degree of typicality according to cognitive tasks or questionnaires. In [13], a class name is given to participants of the experiment, for example Fruit, and participants have to list subclasses of this class. The most often cited subclasses are the most typical for the class. Rosch [14] has given a class name to participants of her experiment, such as for example, Fruit, in addition to a list of subclasses, such as Apple, Mango or Avocado and subjects have to rate on a 7-point scale, how a subclass is a good example for its class.

   In this paper, the acquisition of the typicality degree is based on Rosch's work. An online questionnaire has been created, where participants must provide a rating about how they consider each B_i as being a good example of A. The statements to evaluate are of the form "Is B_i a good example of A?" A 3 smiley face rating scale is used, where numerical scores are associated on each face: unhappy face $= -1$, neutral face $= 0$, and happy face $= 1$. For each B_i, the average of the numerical scores associated to the user ratings is computed and is considered as the typicality degree of B_i with respect to A. The typicality degree of B_i with respect to A is denoted by $\mathtt{typ}(B_i, A) \in [-1, 1]$. This typicality degree will be used to assign each B_i to one among the three typical, normal and atypical class subsets.

Typicality subset assignment. The assignment of B_i, where $B_i \sqsubseteq A$, to one of the three (typical, normal and atypical) subsets consists first in partitioning the subclasses B_i into two sets: the set of atypical subclasses and the set of non atypical subclasses, according to $\mathtt{typ}(B_i, A)$.

If $\mathtt{typ}(B_i, A) < 0$,

then B_i is atypical for A.

else B_i is non atypical for A.

Table 1. Average of the typicality scores for the subclasses of the StoneFruit, PomeFruit and Berry classes. In each table, S_T is the set of typical subclasses, S_N is the set of normal subclasses and S_A is the set of atypical subclasses, according to the typicality subsets assignment procedure (evaluation with 14 users).

StoneFruit		
S_T	Apricot	1.00
	Mirabelle	1.00
	Peach	1.00
	Cherry	0.92
S_N	Date	0.42
	Litchi	0.17
	Mango	0.08
S_A	Avocado	−0.17
	Olive	−0.25

PomeFruit		
S_T	Apple	0.67
	Pear	0.50
S_N	Grape	0.33
	Lemon	0.17
	Melon	0.08
S_A	Orange	−0.17
	Pumpkin	−0.58
	Tomato	−0.58
	Pepper	−0.67
	Cucumber	−1.00

Berry		
S_T	Blackberry	1.00
	Blueberry	1.00
	Raspberry	1.00
	RedCurrant	0.67
S_N	Strawberry	0.17
S_A	Grape	−0.33
	Kiwi	−0.92

Second, the set of non atypical subclasses is also divided in two sets: the set of normal subclasses and the set of typical subclasses.[1]

If $\mathtt{typ}(B_i, A) \geq 0$ (i.e. B_i is non atypical for A),

then if $\mathtt{typ}(B_i, A) < 0.5$

then B_i is normal for A.

else B_i is typical for A.

Let typicalClasses(A), normalClasses(A), and atypicalClasses(A) be respectively the set of the typical classes of A, the set of the normal classes of A, and the set of the atypical classes of A.

Table 1 presents in three sub-tables, the average of the acquired typicality scores for the subclasses of StoneFruit, PomeFruit (i.e., fruits with pips) and Berry. Each table of Table 1 is divided in three sets: S_T = typicalClasses(A), S_N = normalClasses(A), and S_A = atypicalClasses(A). For example, Avocado is an atypical stone fruit because $\mathtt{typ}(\mathtt{Avocado}, \mathtt{StoneFruit}) = -0.17 < 0$, Mango is a normal stone fruit because $0 \leq \mathtt{typ}(\mathtt{Mango}, \mathtt{StoneFruit}) = 0.08 < 0.5$, and Apricot is a typical stone fruit because $\mathtt{typ}(\mathtt{Apricot}, \mathtt{StoneFruit}) = 1 \geq 0.5$.

4.2   Refinement of the Ontology by Taking into Account Typicality

Refining the ontology consists in adding intermediate classes in the initial hierarchy in order to obtain a finer-grained hierarchy, involving a more accurate

---

[1] Another way to determine the three sets of typicality is to cluster the values of $\mathtt{typ}(B_i, A)$. The clustering method could, for example, use the k-means approach with $k = 3$. Some tests have been run to do so and they show only a little difference with the choice of the thresholds 0 and 0.5. Moreover, for the evaluation we present, this small threshold shifts do not impact the results.

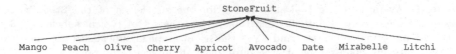

Fig. 2. Initial part of the ontology related to `StoneFruit`.

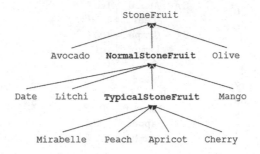

Fig. 3. Refined part of the ontology related to `StoneFruit`.

retrieval. In our approach, the ontology refinement is based on typicality. A class `A` of the ontology is refined like this: `TypicalA` $\sqsubseteq$ `NormalA` $\sqsubseteq$ `A` where `A` directly subsumes $B_i \in$ `atypicalClasses(A)`, `NormalA` directly subsumes $B_i \in$ `normalClasses(A)`, and `TypicalA` directly subsumes $B_i \in$ `typicalClasses(A)`. For example, let us consider the part of the ontology related to `StoneFruit` as the part of the ontology to refine (see Fig. 2). According to Table 1, `Mirabelle`, `Peach`, `Apricot` and `Cherry` are some typical subclasses of `StoneFruit`, `Date`, `Litchi` and `Mango` are some normal subclasses of `StoneFruit`, and `Olive` and `Avocado` are some atypical subclasses of `StoneFruit`. Applying our refinement procedure to `StoneFruit` produces the new ontology presented in Fig. 3. Two new classes (`NormalStoneFruit` and `TypicalStoneFruit`), in bold, have been added.

The underlying hypotheses we make is that the CBR system results are improved when:

- (H1): for a query about a typical ingredient X, the results in which X replaces another typical ingredient are better than the results in which X replaces a normal ingredient, and, in the same way, that results in which X replaces a normal ingredient are better than the results in which X replaces an atypical ingredient.
- (H2): for a query about a normal ingredient X, the results in which X replaces another normal or typical ingredient are better than the results in which X replaces an atypical ingredient.

We do not make any hypothesis for a query about atypical ingredients because we think there is no reason to prefer to replace an atypical ingredient (e.g. `Tomato`) with another atypical ingredient (e.g. `Orange`) rather than by a normal ingredient (e.g. `Lemon`) or a typical ingredient (e.g. `Apple`). Consequently, no additional class is added for structuring the atypical classes (atypical subclasses are directly subsumed by PomeFruit).

According to the refined ontology, when an atypical stone fruit (e.g. Avocado) is first generalized during the case retrieval process into StoneFruit, all cases linked to stone fruits will be retrieved, regardless to their atypical, normal or typical status. When a normal stone fruit (e.g. Mango) is generalized into NormalStoneFruit, all the cases linked to NormalStoneFruit or TypicalStoneFruit will be retrieved. Finally, the first generalization of a typical stone fruit (e.g. Peach) in TypicalStoneFruit which will only retrieve cases linked to typical stone fruits, for example cases linked to Apricot, and only at a next step of generalization, cases linked to normal stone fruits.

4.3   Example of TAAABLE Results with and Without Taking into Account Typicality

The effects of the typicality-based ontology refinement in TAAABLE are now illustrated through an example. This example compares CBR_{std}, a version of TAAABLE where the ontology is not refined according to typicality, with CBR_{typ}, a version of TAAABLE where the ontology is refined according to typicality.

Let $Q_{ex} = $ TartDish $\wedge$ Peach be the example query. Table 2 presents the five first results returned by CBR_{std} and CBR_{typ} according to Q_{ex}. Each result is composed of a retrieved recipe id, its index and the adaptation proposed by TAAABLE.

With the part of the ontology presented in Fig. 2, in which StoneFruit is not refined, CBR_{std} returns for Q_{ex} the five results presented in Table 2 at the first generalization step, which is $\Gamma(Q_{ex}) = $ Peach $\rightsquigarrow$ StoneFruit. In these results, CBR_{std} proposes with no preference the recipe adaptations given in the last column of Table 2. The first adaptation replaces a typical stone fruit, Apricot, with another typical stone fruit, Peach. The last adaptation replaces an atypical stone fruit, Olive, with a typical stone fruit, Peach. Thus, these five results are returned at the same rank without preferring one adaptation from another.

With the refined StoneFruit part of the ontology presented in Fig. 3, CBR_{typ} returns the same results but at different steps of generalization (see Table 3). Peach, a typical stone fruit is first generalized in TypicalStoneFruit, producing the retrieval of recipes linked to typical stone fruits: R_1, linked to Apricot, is adapted by replacing Apricot with Peach, and R_2, linked to Cherry, is adapted by replacing Cherry with Peach. If the user triggers the system again, Peach is then generalized (at the second step of generalization) in NormalStoneFruit, producing the retrieval and adaptation of R_3. Finally, the system will only propose R_4 and R_5 at the third step of generalization, in which Peach is generalized in StoneFruit. So, instead of returning at the same level R_1, R_2, R_3, R_4 and R_5 as CBR_{std} does, CBR_{typ} returns first R_1 and R_2. It is only if the user wants additional results that R_3, and then, R_4 and R_5 will be displayed.

Table 2. The results returned by CBR_{std} and CBR_{typ}, according to $Q_{ex} = \text{TartDish} \wedge$ Peach, with their respective index and adaptation.

id	$idx(\text{R}_i)$	Adaptation
R_1	$\text{TartDish} \wedge \text{PieCrust} \wedge \text{Apricot} \wedge \text{Sugar} \wedge \dots$	$\text{Apricot} \rightsquigarrow \text{Peach}$
R_2	$\text{TartDish} \wedge \text{PieCrust} \wedge \text{Cherry} \wedge \text{Almond} \wedge \dots$	$\text{Cherry} \rightsquigarrow \text{Peach}$
R_3	$\text{TartDish} \wedge \text{PieCrust} \wedge \text{Mango} \wedge \text{Caramel} \wedge \dots$	$\text{Mango} \rightsquigarrow \text{Peach}$
R_4	$\text{TartDish} \wedge \text{PieCrust} \wedge \text{Avocado} \wedge \text{Crab} \wedge \dots$	$\text{Avocado} \rightsquigarrow \text{Peach}$
R_5	$\text{TartDish} \wedge \text{PieCrust} \wedge \text{Olive} \wedge \text{Bacon} \wedge \dots$	$\text{Olive} \rightsquigarrow \text{Peach}$

Table 3. Generalization for which CBR_{std} and CBR_{typ} return the five first results detailed in Table 2 for Q_{ex}. Γ_i refers to the ith generalization step.

id	Generalization for CBR_{std}	Generalization for CBR_{typ}
R_1	Γ_1: $\text{Peach} \rightsquigarrow \text{StoneFruit}$	Γ_1: $\text{Peach} \rightsquigarrow \text{TypicalStoneFruit}$
R_2	Γ_1: $\text{Peach} \rightsquigarrow \text{StoneFruit}$	Γ_1: $\text{Peach} \rightsquigarrow \text{TypicalStoneFruit}$
R_3	Γ_1: $\text{Peach} \rightsquigarrow \text{StoneFruit}$	Γ_2: $\text{Peach} \rightsquigarrow \text{NormalStoneFruit}$
R_4	Γ_1: $\text{Peach} \rightsquigarrow \text{StoneFruit}$	Γ_3: $\text{Peach} \rightsquigarrow \text{StoneFruit}$
R_5	Γ_1: $\text{Peach} \rightsquigarrow \text{StoneFruit}$	Γ_3: $\text{Peach} \rightsquigarrow \text{StoneFruit}$

5   Evaluation Methodology

The objective of the evaluation is to show that refining the ontology according to typicality improves the user satisfaction about the results returned by the CBR system. For that, the two versions of TAAABLE, CBR_{std} and CBR_{typ}, are compared in order to test (H1) and (H2). In addition, a more global hypothesis will also be examined: (H) CBR_{typ} returns results which better satisfy, in average, the user than CBR_{std}.

5.1   Acquiring Evaluations About TAAABLE Results

In order to test (H1), (H2) and (H), a set of queries is submitted to the two systems. Users who participated to the experiment gave their feedback on the results returned by the two systems, through a web interface, as illustrated in Fig. 4. For a given recipe, the possible adaptations are given, but the users do not know which system produces a given adaptation. For a possible adaptation, users have to evaluate how much they are satisfied by the new recipe produced. Each adaptation has been evaluated by at least 3 users. Such an evaluation allows to know how relevant each result is with respect to a query. This knowledge about the results allows to compare the two systems in the evaluation. A similar approach is used for comparing information retrieval systems in the context of TREC (Text Retrieval Conference) [15] or for comparing recommender systems [16].

Apricot almond tart

Ingredients

- Puff pastry
- 1 kg apricots
- 20 cl cream
- 150 g sugar
- 1 sach vanilla sugar
- 5 tsp almond powder
- 2 large eggs

Preparation

Preheat oven to 240°C. Line 9-inch-diameter tart pan with removable bottom with pastry. Using fork, pierce dough all over. Dispose the sliced apricots ...

Adaptation	Satisfied?
Replace Apricot by Peach	
Replace Apricot by Avocado	
Replace Apricot by Mango	

Submit

Fig. 4. Evaluation interface. The user has to evaluate how she is satisfied with some adaptations of the same recipe for different queries. In this screenshot proposing one recipe with three possible adaptations, the user is very satisfied with the substitution `Apricot` ⤳ `Peach`, very unsatisfied with `Apricot` ⤳ `Avocado` and satisfied with `Apricot` ⤳ `Mango`.

Many possibilities exist to collect the satisfaction feedback of the user. One of them is the smiley code system, similar to the Likert scale [17], which is based on a set of grades, allowing the user to qualify her satisfaction degree. We choose to use five grades, so that the user has not too many nor too few options to express her opinion. There are, from left to right, two negative grades (very unsatisfied and unsatisfied), one neutral grade (neither satisfied nor unsatisfied), and two positive grades (satisfied, very satisfied). An advantage of this scale is that it can be easily be turned into a numerical score for a quantitative analysis purpose. The associated scores are, from left to right, -2, -1, 0, 1 and 2.

5.2 Parameters of the Systems

The two systems are evaluated using ATAAABLE, a French instance of TAAABLE, which draws case-based inferences on knowledge built through the collaborative ATAAABLEweb site (http://ataaable.loria.fr).

Data Preparation. To evaluate CBR$_{\text{typ}}$, refinements of the `StoneFruit`, `PomeFruit` and `Berry` parts of the ontology have been performed according to typicality. To obtain typicality ratings, 14 participants have completed the online questionnaire presented in Sect. 4.1: they were asked to rate the typicality degree of subclasses of the `StoneFruit`, `PomeFruit` and `Berry` classes. Table 1 presents the average of the typicality scores collected during this experiment. The results allow to refine the `StoneFruit`, `PomeFruit` and `Berry` refinement part of the ontology. For example, the refined ontology related to `StoneFruit` is presented in Fig. 3.

The case base has been limited to 124 tart recipes that are linked to 138 classes of the food hierarchy of $\mathcal{O}_{\text{std}}$.

Table 4. Average of the satisfaction score for CBR_{std} and CBR_{typ} for tart queries containing typical classes (S_T) or normal classes (S_N). The last column is the difference between the satisfaction score averages of CBR_{std} and CBR_{typ}, showing that CBR_{typ} improves the results for all the queries except two.

Set	id	Q	CBR_{std}	CBR_{typ}	$\Delta CBR_{std/typ}$
S_T	1	TartDish ∧ Apricot	0.88	1.35	0.47
	2	TartDish ∧ Mirabelle	1.03	1.29	0.26
	3	TartDish ∧ Peach	1.20	1.53	0.33
	4	TartDish ∧ Cherry	0.63	0.93	0.30
	5	TartDish ∧ Pomme	−0.29	0.59	0.88
	6	TartDish ∧ Pear	−0.40	1.67	2.07
	7	TartDish ∧ Blackberry	1.16	1.23	0.07
	8	TartDish ∧ Blueberry	0.81	1.14	0.33
	9	TartDish ∧ Raspberry	1.44	1.51	0.07
	10	TartDish ∧ RedCurrant	0.30	0.63	0.33
S_N	11	TartDish ∧ Date	0.18	0.18	0.00
	12	TartDish ∧ Litchi	−1.00	−0.69	0.31
	13	TartDish ∧ Mango	−0.33	−0.27	0.06
	14	TartDish ∧ Grape	−0.57	−0.43	0.14
	15	TartDish ∧ Lemon	−0.33	−0.26	0.07
	16	TartDish ∧ Melon	−1.60	−0.96	0.64
	17	TartDish ∧ Strawberry	0.82	0.66	−0.16

Queries. The experiment was made on queries about tarts. Each query is composed of `TartDish`, the dish type, and one ingredient belonging to a `PomeFruit`, `Berry` or `StoneFruit` subclass, which are the parts of $\mathcal{O}_{std}$ that have been refined. In order to evaluate (H1) and (H2), only the ingredients related to typical and normal classes are used in the queries.

6 Results and Discussion

The 10 first results returned by CBR_{std} and CBR_{typ} have been computed for the 17 possible queries involving a typical or a normal ingredient. Each system returns $17 \times 10 = 170$ answers. According to a random choice of cases when too many cases are retrieved at a given step of generalization, there are 108 common results between CBR_{std} and CBR_{typ}, so 232 different results in total. These 232 results were evaluated by 27 users (disjoint form the set of 14 users who participated to the data preparation, see Sect. 5.2), providing 1122 ratings.

Table 5. $\bar{s}$ relating to the type of adaptation: replacing a typical ingredient with the typical ingredient of the query (column "$T \leadsto \ldots$"), replacing a normal ingredient with the typical ingredient of the query (column "$N \leadsto \ldots$"), replacing a typical or a normal ingredient with the typical ingredient of the query (column "$T or N \leadsto \ldots$"), and replacing an atypical ingredient with the typical ingredient of the query (column "$A \leadsto \ldots$").

id	Q	$T \leadsto \ldots$	$N \leadsto \ldots$	$T or N \leadsto \ldots$	$A \leadsto \ldots$
1	TartDish ∧ Apricot	1.38	1.33	1.33	−1.33
2	TartDish ∧ Peach	1.42	1.47	1.43	−1.00
3	TartDish ∧ Mirabelle	1.38	1.33	1.36	−1.33
4	TartDish ∧ Cherry	1.43	0.53	1.13	−1.11
5	TartDish ∧ Pomme	1.49	−0.24	0.63	−1.01
6	TartDish ∧ Pear	1.59	−1.33	1.06	−1.13
7	TartDish ∧ Blueberry	1.60	0.98	1.08	0.00
8	TartDish ∧ Blackberry	1.68	1.06	1.21	1.11
9	TartDish ∧ Raspberry	1.37	1.65	1.51	1.20
10	TartDish ∧ RedCurrant	0.74	0.33	0.56	0.08

6.1 Validating (H): The Global User Satisfaction

Let $\bar{s}$ denote the average of user satisfaction scores. Table 4 shows, for each query, $\bar{s}$ for $\mathrm{CBR_{std}}$ results and $\bar{s}$ for $\mathrm{CBR_{typ}}$ results, as well as the difference between $\bar{s}$ of the two systems (column $\Delta\mathrm{CBR_{std/typ}}$). $\mathrm{CBR_{typ}}$ improves the results for 15 of the 17 queries. With $\bar{s} = 0.59$ for $\mathrm{CBR_{typ}}$ against $\bar{s} = 0.23$ for $\mathrm{CBR_{std}}$, $\mathrm{CBR_{typ}}$ better satisfies the user on average (difference of 0.36).

To demonstrate that user satisfaction is significantly higher for results of $\mathrm{CBR_{typ}}$ than for results of $\mathrm{CBR_{std}}$ (H), the Wilcoxon signed-rank test [18] is used to compare the medians of the user satisfaction scores. With a p-value of 0.0006731 ($p < 0.05$, stating that the results are significant according to the hypothesis), the difference in the results of $\mathrm{CBR_{typ}}$ compared to $\mathrm{CBR_{std}}$ is significant: (H) is supported.

6.2 Validating (H1) and (H2): User Satisfaction Related to the Adaptation Type

Table 5 gives $\bar{s}$ for adaptations with typical ingredients (see the table caption for explanations about the table description). Comparing the columns "$T \leadsto \ldots$" and "$N \leadsto \ldots$" shows that the results in which a typical ingredient replaces another typical ingredient are better compared to the results in which a typical ingredient replaces a normal ingredient. For 8 of the 10 queries, the results are better for "$T \leadsto \ldots$" and $\bar{s}$ of "$T \leadsto \ldots$" which is 1.41 is clearly better than $\bar{s}$ of "$N \leadsto \ldots$" which is only 0.71. In the same way, comparing the columns

"T or $N \rightsquigarrow \ldots$" and "$A \rightsquigarrow \ldots$" shows that the results in which a typical ingredient replaces a typical or a normal ingredient are better compared to the results in which a typical ingredient replaces an atypical ingredient. For all the queries the results are better for "T or $N \rightsquigarrow \ldots$" and $\bar{s}$ of "T or $N \rightsquigarrow \ldots$" which is 1.13 is clearly better than $\bar{s}$ of "$A \rightsquigarrow \ldots$" which is only -0.45. The p-value for the comparison of the results of "$T \rightsquigarrow \ldots$" and "$N \rightsquigarrow \ldots$" (resp. "$T$ or $N \rightsquigarrow \ldots$" and "$A \rightsquigarrow \ldots$") is 0.018. (resp. 0.002531). These p-values indicate that the results are significant and validate (H1).

Unfortunately, (H2) is not supported by the experiment. First, $\bar{s}$ for replacing an atypical ingredient with a normal ingredient is similar to replace a normal or a typical ingredient with a normal ingredient. Second, we have only 4 queries providing results that can be used to test (H2). Indeed, for 3 of the 7 queries about a normal ingredient, the generalization takes place in another part of the hierarchy because some classes are subclasses of several classes: `Date`, `Lemon`, and `Grape` have been respectively generalized in `DriedFruit`, `CitrusFruit`, and `AtypicalBerry`. Such generalizations return adaptations involving ingredients that are not in the refined parts of the hierarchy, and are not classified as atypical, normal nor typical.

7 Conclusion

This paper proposes an approach to refine a domain ontology using typicality, for a more accurate case retrieval. This refinement improves the case retrieval of a CBR system, improving at the same time the global results of the system. Subclasses of a class are separated in three sets of classes: atypical, normal and typical classes. These three sets are used to introduce intermediate classes in the ontology. The refinement of the ontology allows to generalize the query in a finer-grained manner. For example, searching for cases related to a given typical class will give priority to cases related to all the typical classes, and it is the same for normal classes. The resulting adaptations favor first the introduction of a typical instance (e.g. the peach the user wants to use, which is a part of the query) in cases using typical instances (e.g. the apricot in R_1), and second, the introduction of normal instances (e.g. litchis) in cases using normal (e.g. the mango in R_3) or typical instances (e.g. the apricot in R_1 or the cherries in R_2).

A version of a CBR system, CBR_{typ}, whose ontology was refined thanks to typicality was compared to a CBR system, CBR_{std}, using the ontology before the refinement. The experiment demonstrates the hypothesis that results of CBR_{typ} are more satisfying than results of CBR_{std} with a significant difference between user satisfaction of CBR_{typ} and user satisfaction of CBR_{std} and that, for adaptations returned when searching for a typical ingredient, it is better to prefer the replacement of this ingredient with a typical one, than with a normal one, and finally, with an atypical ingredient.

One of the possible extensions is to modulate the number of classes of typicality depending on the part of the ontology to refine. For example, in some parts of the ontology, the atypical and non atypical classes may be sufficient.

A limitation of the approach is the consideration of the typicality when one class can be generalized in two different classes. To manage this, future work will address the integration of the typicality degree in the similarity measure as a means of favoring some generalizations rather than others.

References

1. Richter, M.: The knowledge contained in similarity measures. Invited talk at the International Conference on Case-Based Reasoning (1995)
2. Cordier, A., Dufour-Lussier, V., Lieber, J., Nauer, E., Badra, F., Cojan, J., Gaillard, E., Infante-Blanco, L., Molli, P., Napoli, A., Skaf-Molli, H.: Taaable: a case-based system for personalized cooking. In: Montani, S., Jain, L.C. (eds.) Successful Case-based Reasoning Applications-2. SCI, vol. 494, pp. 121–162. Springer, Heidelberg (2014)
3. Smith, E.E., Medin, D.L.: Categories and Concepts. Cognitive Science Series. Harvard University Press, Cambridge (1981)
4. Beckner, M.: The Biological Way of Thought. Columbia University Press, New York (1959)
5. Rosch, E., Mervis, C.B.: Family resemblances: studies in the internal structure of categories. Cogn. Psychol. **7**(4), 573–605 (1975)
6. Barsalou, L.W.: Ideals, central tendency, and frequency of instantiation as determinants of graded structure in categories. J. Exp. Psychol. Learn. Mem. Cogn. **11**(4), 629 (1985)
7. Yeung, C.A., Leung, H.: Ontology with likeliness and typicality of objects in concepts. In: Embley, D.W., Olivé, A., Ram, S. (eds.) ER 2006. LNCS, vol. 4215, pp. 98–111. Springer, Heidelberg (2006)
8. Medin, D., Schaffer, M.: Context theory of classification learning. Psychol. Rev. **85**(3), 207–238 (1978)
9. Bareiss, E.R., Porter, B.E., Wier, C.C.: Protos: an exemplar-based learning apprentice. In: Gaines, B.R., Boose, J.H. (eds.) Machine Learning and Uncertain Reasoning, pp. 1–13. Academic Press Ltd., London (1990)
10. Dubois, D., Prade, H., Rossazza, J.P.: Vagueness, typicality, and uncertainty in class hierarchies. Int. J. Intell. Syst. **6**(2), 167–183 (1991)
11. Friedman, M., Ming, M., Kandel, A.: On the theory of typicality. Int. J. Uncertain. Fuzziness Knowl.-Based Syst. **03**(02), 127–142 (1995)
12. Weber-Lee, R., Barcia, R.M., Martins, A., Pacheco, R.C.: Using typicality theory to select the best match. In: Smith, I., Faltings, B. (eds.) Advances in Case-Based Reasoning. LNCS, vol. 1168, pp. 445–459. Springer, Heidelberg (1996)
13. Barsalou, L.W., Sewell, D.R.: Contrasting the representation of scripts and categories. J. Mem. Lang. **24**(6), 646–665 (1985)
14. Rosch, E.H.: On the internal structure of perceptual and semantic categories. In: Moore, T.E. (ed.) Cognitive Development and the Acquisition of Language, pp. 111–144. Academic, New York (1973)
15. Sanderson, M.: Test collection based evaluation of information retrieval systems. Found. Trends Inf. Retr. **4**(4), 247–375 (2010)
16. Quijano-Sánchez, L., Recio Garcia, J.A., Díaz-Agudo, B.: Using personality to create alliances in group recommender systems. In: Ram, A., Wiratunga, N. (eds.) ICCBR 2011. LNCS, vol. 6880, pp. 226–240. Springer, Heidelberg (2011)

17. Likert, R.: A technique for the measurement of attitudes. Arch. Psychol. **22**(140), 1–55 (1932)
18. Wilcoxon, F.: Individual comparisons by ranking methods. Biometrics Bull. **1**(6), 80–83 (1945)

CBR Meets Big Data: A Case Study of Large-Scale Adaptation Rule Generation

Vahid Jalali$^{(\boxtimes)}$ and David Leake

School of Informatics and Computing, Indiana University,
Bloomington, IN 47408, USA
{vjalalib,leake}@indiana.edu

Abstract. Adaptation knowledge generation is a difficult problem for CBR. In previous work we developed *ensembles of adaptation for regression* (EAR), a family of methods for generating and applying ensembles of adaptation rules for case-based regression. EAR has been shown to provide good performance, but at the cost of high computational complexity. When efficiency problems result from case base growth, a common CBR approach is to focus on case base maintenance, to compress the case base. This paper presents a case study of an alternative approach, harnessing big data methods, specifically MapReduce and locality sensitive hashing (LSH), to make the EAR approach feasible for large case bases without compression. Experimental results show that the new method, BEAR, substantially increases accuracy compared to a baseline big data k-NN method using LSH. BEAR's accuracy is comparable to that of traditional k-NN without using LSH, while its processing time remains reasonable for a case base of millions of cases. We suggest that increased use of big data methods in CBR has the potential for a departure from compression-based case-base maintenance methods, with their concomitant solution quality penalty, to enable the benefits of full case bases at much larger scales.

Keywords: Case-based reasoning · Ensemble of adaptations for regression · Locality sensitive hashing

1 Introduction

The growth of digital data is widely heralded. A 2014 article estimates that "[A]lmost 90 % of the world's data was generated during the past two years, with 2.5 quintillion bytes of data added each day" [1]. Individual organizations collect data sets on an unprecedented scale. For example, in 2013, a single health care network in the U.S. state of California was estimated to have over 26 petabytes of patient data from electronic health records alone [2]. Big data methods and resources have changed the practicality of using such large-scale data, with inexpensive cloud computing services enabling processing data sets of unprecedented scale. However, these are not a panacea: making good use of large-scale data remains a challenge (e.g., [3]).

© Springer International Publishing Switzerland 2015
E. Hüllermeier and M. Minor (Eds.): ICCBR 2015, LNAI 9343, pp. 181–196, 2015.
DOI: 10.1007/978-3-319-24586-7_13

Case-based reasoning's ability to reason from individual examples and its inertia-free learning make it appear a natural approach to apply to big-data problems such as predicting from very large example sets. Likewise, if CBR systems had the capability to handle very large data sets, that capability could facilitate the application of CBR to large data sources already identified as interesting to CBR, such as cases harvested from the "experience Web" [4], cases resulting from large-scale real-time capture of case data from instrumented systems [5], or cases arising from case capture in trace-based reasoning [6].

However, realizing the potential of CBR to have impact on big data problems will depend on CBR systems being able to exploit the information in case bases with size far beyond the scale now commonly considered in the CBR literature. The case-based reasoning community has long been aware of the challenges of scaling up CBR to large case bases. The primary response has been case-base maintenance methods aimed at reducing the size of the case base while preserving competence (e.g., [7,8]). Such methods have proven effective at making good use of case knowledge within storage limits. However, because compression methods delete some of the CBR system's knowledge, they commonly sacrifice some solution quality.

A key factor in success of CBR when applied to big data is efficient retrieval of cases. As CBR does not generalize beyond cases, it is extremely important to the success of a CBR system to be able to find required cases rapidly. In this paper we illustrate the practicality of applying big-data tools to increase the speed and scalability of CBR, using MapReduce and Locality Sensitive Hashing for finding nearest neighbors of the input query.

In previous work, we introduced and evaluated a method for addressing the classic CBR problem of acquiring case adaptation knowledge with *ensembles of adaptations for regression* (EAR) [9]. This work demonstrated the accuracy benefits of EAR [9–12], but also identified important efficiency concerns for large case bases. This paper presents a case study applying big data methods to addressing EAR's scale-up, leveraging techniques and frameworks well known to the big data community to enable large-scale CBR. It presents a new algorithm, BEAR,[1] applying the EAR approach in a MapReduce framework. The paper demonstrates that the use of big data methods substantially extends the size of case base for which the EAR approach is practical, to case bases of millions of cases even on a small Amazon Elastic MapReduce (EMR) cluster.[2]

The paper begins with a discussion of the relationship of big data and CBR, contrasting the "retain and scale up" approach of big data to the compression-based focus of case-base maintenance. It next introduces the EAR family of methods and the two big data methods to be applied, locality sensitive hashing [13] and MapReduce. With this foundation it introduces BEAR, a realization of EAR for big data platforms, and presents an experimental evaluation assessing BEAR's accuracy for a case base of two million cases. To assess the benefit of BEAR's ensemble approach it compares it to a baseline of a big data version

---

[1] Big data ensembles of adaptations for regression.
[2] http://aws.amazon.com/elasticmapreduce/.

of k-NN, using Locality Sensitive Hashing for implementing nearest neighbor search. It also shows that BEAR's approach helps alleviate the accuracy penalty that can result from using LSH instead of a traditional (exhaustive) approach for finding nearest neighbors, thus compensating for a potential drawback of using LSH. To assess the need for big data methods for BEAR's task and BEAR's scaleup potential, it also compares BEAR's scaleup performance to that of k-NN using a traditional (exhaustive) approach for finding nearest neighbors.

The evaluation supports the accuracy benefits of BEAR and that the speedup benefits of big data methods are sufficient to counterbalance the computational complexity of BEAR's rule generation and ensemble solution methods. Thus the use of big data methods may have benefits for CBR beyond simple speedups, by making practical the use of richer methods which can increase accuracy. Two million cases is large by the standards of current CBR practice, but true "big data" CBR will involve much larger data sets. The paper closes with a discussion of BEAR's potential for scaleup to such data sets.

2   Scaling CBR to Big Data

Big data has had a transformative effect on data management, enabling many enterprises to exploit data resources at previously unheard-of data scales. Large data sets such as electronic medical records collections may naturally be seen as containing cases; routine data capture in many domains could provide rich case bases. If cases can be retrieved sufficiently efficiently, CBR is an appealing method for large-scale reasoning because its lazy learning avoids the overheads associated with traditional rule mining approaches enables inertia-free adjustments to additional data, without the need for retraining.

However, CBR systems have seldom ventured into the scale of big data. For example, calculating metrics such as number of visitors or page views for a social media or e-commerce web site with hundreds of million users is a common practice at industry, but in current CBR research, experiments with tens of thousands of cases, or even much fewer, are common. Few CBR projects have considered scales up to millions of cases [14, 15], and to our knowledge, none have explored larger scales except a few exceptions such as a recent effort to apply big data methods focused on exact match only, rather than similarity-based retrieval [15].

When CBR research has addressed increased data sizes, the primary focus has been compression of existing data rather than scale-up. Considerable CBR research has focused on the efficiency issues arising from case-base growth. As the case base grows, the swamping utility problem can adversely affect case retrieval times, degrading system performance [16, 17]. Within the CBR community and the machine learning community studying instance-based learning, extensive effort has been devoted to addressing the swamping utility problem for case retrieval with case-base maintenance methods for controlling case-base growth, with the goal of generating case bases that are compact but retain coverage of as many problems as possible. Methods for developing compact competent case bases include selective deletion (e.g., [7, 18]), selective case retention (e.g., [19–21]), and competence-aware construction of case bases [8, 22–25]. Such methods generally trade off size against accuracy; they aim to retain as much competence as possible for a given amount of compression. This tradeoff has been seen as the price of making CBR feasible for domains in which the set of possible cases is large, but storage and processing resources are limited.

This paper argues that applying big data methods can change this calculus; that even for case bases on the order of millions of cases, big data methods can make the best case-base compression strategy *no compression at all*. If big data methods can enable CBR scale-up, dramatically increasing the feasibility of handling very large case bases, compression methods will be required only for extreme scale case bases—and, even for very large cases, might not be required at all in practice. Already, it has been observed that for common practical CBR tasks, even with conventional methods, case base size may not be an issue [26]; big data methods could bring CBR to bear on a new class of problems, at much larger scale.

3   Foundations of the Proposed Method

The case study in this paper focuses on applying CBR to numerical prediction tasks under big data settings, demonstrating the feasibility of big data approaches to provide good performance at scales on the order of millions of cases, with minimal quality loss. The method proposed in this paper builds on three currents of research. The first, from CBR, is the EAR family of methods [9] for case-based regression using ensembles of adaptations. The second, from big data is Locality Sensitive Hashing (LSH), a method for nearest neighbor search in big data platforms. The third is MapReduce, a popular framework for parallel processing of data.

3.1   The EAR Family of Methods

The acquisition of case adaptation knowledge is a classic problem for CBR. A popular approach to this problem, for numerical prediction (regression) tasks, is to generate adaptation rules automatically from the case base. The EAR family of methods solves numerical prediction problems using automatically-generated ensembles of adaptations to adapt prior solutions.

The EAR approach applies to any adaptation generation method, but it has been tested for a popular case-based rule generation method, the *Case Difference Heuristic*, which generates rules based on comparing pairs of cases. Given two cases A and B, with problem parts Prob(A) and Prob(B), and solution parts Sol(A) and Sol(B), the case difference heuristic approach assumes that problems with similar difference in their problem descriptions will have similar differences in their solutions. For example, for predicting apartment rental prices from a case base of rental properties and prices, if one apartment's monthly rent is $300 more than the rent of an otherwise highly similar apartment, and their difference is that the more expensive apartment has an additional bedroom, the

Algorithm 1. EAR's basic algorithm

Input:
Q: input query
n: number of base cases to adapt to solve query
r: number of rules to be applied per base case
CB: case base
Output: Estimated solution value for Q

> $CasesToAdapt \leftarrow$ NeighborhoodSelection(Q,n,CB)
> $NewRules: \leftarrow$ RuleGenerationStrategy($Q,CasesToAdapt,CB$)
> **for** c in $CasesToAdapt$ **do**
>   RankedRules $\leftarrow$ RankRules($NewRules,c,Q$)
> $ValEstimate(c) \leftarrow$ CombineAdaptations($RankedRules$, c, r)
> **end for**
> return CombineVals($\cup_{c \in CasesToAdapt} ValEstimate(c)$)

comparison might suggest a general rule: *When the previous apartment case has
one bedroom fewer, predict that the new apartment's rent will be $300 more than
the rent of the previous apartment* (We note that many possible rules could be
generated; the choice of rules is outside the scope of this paper).

More precisely, for cases A and B, the case difference heuristic approach
generates an adaptation rule applicable to a retrieved case C and problem P,
for which the difference in problems of A and B is similar to the difference
between the problem of C and P, i.e., for which diff(Prob(C),P) is similar to
diff(Prob(A),Prob(B)). The new rule adjusts Sol(C) to generate a new solution
N, such that diffSol(C),P is similar to diff(Sol(A),Sol(B)). For a more detailed
description, see Hanney and Keane [27].

The results of the case difference heuristic depend on the cases from which
rules are generated; the final results depend on the cases to which they are
applied. The EAR methods estimate the solution of a case by retrieving a set of
similar cases, adjusting their values by applying an ensemble of adaptation rules
and combining the adjusted values to form the final prediction. Algorithm 1
explains the overall approach of EAR. In Algorithm 1, *NeighborhoodSelection,
RuleGenerationStrategy,* and *RankedRules* respectively denote methods for
finding nearest neighbors, generating adaptation rules and adaptation retrieval
in EAR4. More details are provided in [9].

EAR has different variations based on the subsets of cases it uses as source
cases for solving input problems and the cases it selects as the basis for building
adaptation rules. Different variants use different combinations of local and global
cases. For example, EAR4, selects cases for both building solution and adaptation
rules from the local neighborhood of the input problem. In this paper we focus
on big data versions of EAR4 a family of EAR methods that generates both
solutions and adaptations from the local neighborhood of the input query.

EAR has been shown to provide significant gains in accuracy over baseline
methods [9]. However, because it depends on multiple case retrievals to generate

adaptation rules for multiple case neighborhoods, its application for large case bases, using conventional CBR techniques, can be expensive. We have developed compression-based methods to help alleviate this [11], but like all compression-based methods, these trade off accuracy for compression. This motivated us to explore the application of big data techniques to the EAR approach.

3.2   Locality Sensitive Hashing

Locality Sensitive Hashing (LSH) [13] was developed to decrease the time complexity of finding nearest neighbors for an input query in d-dimensional Euclidean space. LSH achieves this goal by approximating the nearest neighbor search process; it uses families of hashing functions for which the probability of collision is higher for cases which are similar (in terms of their input features). Since the introduction of LSH, various schemes have been proposed to improve various aspects of the core method [28–30].

LSH groups similar items into different buckets by maximizing the probability of collision for similar items. In contrast to nearest neighbor search, LSH does not require comparing a case with other cases to find its nearest neighbors. Instead, if an appropriate hashing function is used it is expected that a case and its nearest neighbors end up in the same bucket. LSH is an approximation method and it does not guarantee grouping a case and its nearest neighbors into the same bucket. However, though LSH sacrifices accuracy for efficiency, it has been demonstrated that LSH can be sufficiently accurate in practice [30].

Previous experiments have studied the performance of k-NN using locality-sensitive hashing to retrieve nearest neighbors (e.g. [29]), showing that LSH achieves higher efficiency compared to linear k-NN with the expected loss in accuracy. In this paper, we explore both the efficiency benefits of LSH for EAR's ensemble method, and the ability of EAR's ensemble method to provide good performance despite the approximations made by LSH.

3.3   MapReduce

MapReduce is a framework that enables parallel processing of data. The "map" step reads and filters data. Next, data is distributed among different nodes/reducers based on a particular field (key) where data is summarized and desired metrics are calculated for the subset of data in each reducer. Different implementations of the MapReduce framework are available. A popular open source implementations of MapReduce framework which is commonly used in industry is *Apache Hadoop*. Recent work by Beaver and Dumoulin [15] has applied MapReduce for CBR, but only for retrieval of exact match cases, rather than for similarity-based retrieval.

4   BEAR: A General Approach to Applying EAR Family Methods to Big Data

4.1   Overview

BEAR (Big-data Ensembles of Adaptations for Regression) is a realization of the EAR family of case-based regression methods in a big data platform, aimed at decreasing the cost of finding nearest neighbors, a process for which the computational expense may become serious issue for very large case bases. The EAR family of methods must identify nearest neighbors at three steps in their processing: to select source cases to adapt, to select candidate cases to build adaptation rules, and to retrieve adaptation rules. Among these three steps, retrieving adaptations is potentially the most challenging, because, for a case base of size n, the upper bound on the number of possible adaptation rules to generate is $O(n^2)$. However, EAR4 mitigates this by limiting the cases to participate in rule generation process to the cases in the local neighborhood of the input query. Therefore, in EAR4 it is likely that source case retrieval will be a more serious resource issue than the adaptation retrieval.

To overcome the challenges raised by the size of the case/rule base, we have investigated minimizing the size of the case base [11] and adaptation rule set [12] to improve performance (in terms of the required computing resources). Although these methods can be useful for reducing the case/rule base size, the process of case/rule base size reduction can still be time consuming and costly; they have not been applied to case bases with more than a few thousands cases. In contrast to these methods, BEAR aims to mitigate challenges brought by the size of the case base by leveraging existing frameworks and algorithms for processing big data to yield accurate estimates rapidly, using locality sensitive hashing on top of a MapReduce framework.

4.2   BEAR's Architecture

BEAR consists of two main modules: LSH for retrieving similar cases and EAR for rule generation and value estimation. The architecture of the system is designed to work in a MapReduce framework. In the map step cases and queries are read and hashed to different buckets using LSH. Cases and queries with identical hashed keys are sent to the same reducer node. In the reduce step two main activities are done: First, the nearest neighbors of each query (from the cases in the same reducer) are determined; Next, depending on the selected EAR method (i.e. EAR4), the adaptation rules are also generated.

For EAR4, which only uses local cases, adaptation rules are only generated and retrieved within the same bucket (for some other variations of EAR, e.g. EAR5, which uses global case information, adaptations would be generated within all buckets and the generated adaptations from different buckets unioned together to form the rule base from which adaptations are retrieved). The final estimates are generated by applying an ensemble of adaptation rules for adjusting

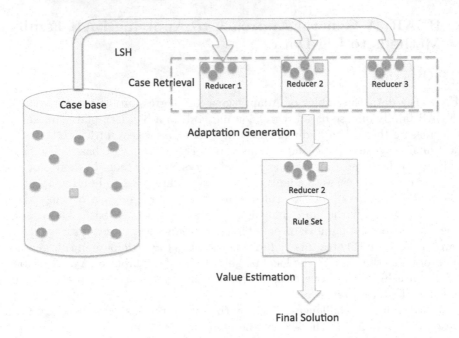

Fig. 1. Illustration of global BEAR process flow.

the base cases' values and combining those adjusted values. Figure 1 summarizes BEAR's process for estimating case solutions.

In Fig. 1, circles represent cases in the case base and the square represents the input query. Cases are hashed and transferred to different reducers based on their hashed keys. Next, adaptation rules are generated based on the cases hashed to the same reducer as the input query. Finally EAR4 is used to estimate the solution of the input query. Depending on the implementation of BEAR, there could be another step in its process flow (not depicted), using a similarity measure such as Euclidean distance to filter out cases in the same bucket as the input query based on a distance threshold or a predefined number of nearest neighbors.

The use of MapReduce offers the advantage of being able to process multiple queries simultaneously, enabling, for example, millions of queries to be processed in parallel. Also, even when single queries are processed sequentially, the use of MapReduce enables processing the cases in the case base in parallel on multiple nodes, rather than having to sequentially process all cases to select those whose LSH hashing keys match that of the query, which would not be scalable.

5   Evaluation

Our evaluation tests the execution time benefits associated with big data methods for scaled up case-based regression with ensembles of adaptations for regression, and studies whether BEAR's use of an ensemble of adaptations improves accuracy compared to applying a non-ensemble approach, when both methods use LSH for retrieval.

Because LSH is not guaranteed always to retrieve the optimal neighbors, we expect that using LSH rather than exhaustive search for nearest neighbors will somewhat degrade overall performance. Consequently, another question is whether the ensemble of adaptations approach, applied in the context of LSH, helps to mitigate this drawback.

This involves two types of tests. The first is a test of the accuracy of "traditional" k-NN, for which neighbors are selected exhastively, compared to that when LSH is used to select (an approximate set of) neighbors, and when LSH is used in conjunction with BEAR. The second, is ablation study to determine how much of the performance of BEAR can be ascribed to its ensemble method, as opposed to the fact that it uses adaptations, while k-NN does not. For the purposes of this case study, we test for a particular LSH implementation, described below, which we refer to as LSH1.

Specifically, our experiments address the following questions:

1. Q1: How does the accuracy of BEAR compare to that of a baseline of k-NN using LSH1 for finding nearest neighbors?
2. Q2: How does the ensemble approach of BEAR increase accuracy compared to applying single adaptations for adjusting base case values?
3. Q3: How does the accuracy of BEAR using LSH1 compare to that of exhaustive k-NN?
4. Q4: How does execution efficiency of BEAR compare to that of traditional (non-LSH) k-NN?

5.1 Experimental Design

We evaluated BEAR on four sample domains from the UCI repository [31]: Automobile (Auto), Auto MPG (MPG), Housing, and electric power consumption (Power). The goal for these domains is respectively to predict the auto price, fuel efficiency (miles per gallon), property value, and household global minute-averaged active power (in kilowatts). For all data sets records with unknown values are removed and feature values are normalized by subtracting feature's mean from the value and dividing the result by standard deviation of the feature's values. (Cases with missing features could be handled by standard feature imputation methods, but this is beyond the scope of our experiment.) In addition, for domains with non-numeric features, only numeric features are used. The accuracy is measured in term of Mean Absolute Error (MAE) in all experiments and ten-fold cross validation is used for conducting the experiments. For all domains parameters are tuned using hill climbing. In all experiments BEAR's performance is compared with that of an implementation of k-NN based on LSH which we refer to simply as k-NN from this point forward. Sample domains are chosen so that they cover both smaller and huge case bases. Table 1 summarizes the characteristics of the sample domains.

Table 1. Characteristics of the test domains

Domain name	# features	# cases	Avg. cases/solution	Sol. sd
Auto	13	195	1.1	8.1
MPG	7	392	3.1	7.8
Housing	13	506	2.21	9.2
Power	7	2,049,280	1.09	1.06

All records and features were used for the Housing (506) domain. Auto, MPG, and Power contained some records with unknown feature values, which were removed (46 out of 205 for Auto, 6 out of 398 for MPG and 25979 out of 2075259 for Power). For all domains only numeric attributes are used in the experiments to enable the application of p-stable locality sensitive hashing. All features of MPG and Housing were numeric, but 10 non-numeric features were removed from Auto and 1 from Power. We note that the numeric features are not required by the general BEAR method.

We note that LSH is a family of methods. Our implementation of BEAR uses Apache DataFu, originally introduced in [32], to support locality sensitive hashing. The corresponding class from Apache DataFu used in BEAR is *L2PStableHash*, with a 2-stable distribution and default parameter settings. It is important to note that because the focus of our experiments is primarily the comparison of BEAR to LHS-based k-NN, and both methods are based on the same version of LSH, the specific variant chosen is not significant to our results.

It also uses EAR4's Weka plugin's code [33], combined with some common functionality from Weka [34] to generate adaptation, retrieve and apply them and build the final prediction. The experiments are run on an EMR amazon cluster with one m3.xlarge master node and ten c3.2xlarge core nodes.

5.2   Experimental Results

Q1: How does the accuracy of BEAR compare to that of a baseline of k-NN using LSH1 for finding nearest neighbors? To address Q1, we conducted experiments to compare BEAR with k-NN using LSH1. In all experiments BEAR's estimations are generated using EAR4 to generate adaptation rules, retrieve adaptations and build final estimations based on nearest neighbors retrieved by LSH1. The experiments report estimation error in terms of Mean Absolute Error. Table 2 summarizes the results for four sample domains. In all domains BEAR outperforms k-NN by substantial margins. A one side paired t-test with 95 % confidence interval was used to assess the statistical significance of results achieved by BEAR in the smaller case bases (we excluded the Power domain from the statistical significance analysis because of the very large size of the case base). The null hypothesis is that the MAE of BEAR is greater than that of k-NN. The results of the t-test showed that $p<.01$, so the improvement of BEAR over k-NN is significant.

Table 2. Accuracy comparison for k-NN and BEAR

Domain name	MAE for k-NN	MAE for BEAR	% improvement over k-NN
Auto	2.04	1.18	42.14 %
MPG	2.62	2.06	21.40 %
Housing	3.73	2.84	23.98 %
Power	0.15	0.10	36.01 %

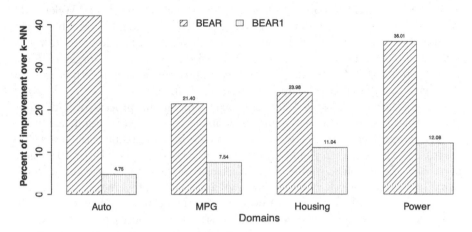

Fig. 2. BEAR vs. BEAR1

Q2: How does the ensemble approach of BEAR increase accuracy compared to applying single adaptations for adjusting base case values? To study the effect of applying an ensemble of adaptations on estimations' accuracy we implemented an ablated version of BEAR, *BEAR1* in which only one adaptation is applied to adjust case values. Figure 2 shows the percent of improvement in MAE over k-NN for BEAR and BEAR1. BEAR1 outperforms k-NN in all domains, but the improvement is less than that of BEAR over k-NN, which shows the benefit of ensemble approach of BEAR.

Q3: How does accuracy of k-NN and BEAR using LSH1 compare to that of traditional k-NN? Because LSH-based retrieval does not guarantee always selecting the true nearest neighbors to a case, some accuracy penalty may be expected. However, we hypothesize that BEAR's ensemble method helps alleviate the associated quality degradation. We tested this hypothesis by comparing the performance of traditional k-NN on the Auto, MPG, and Housing domains, with the previously-described testing scenario (because the relatively large size of the Power domain made traditional k-NN excessively expensive, we did not compare performance in the Power domain). MAE of traditional k-NN for auto and mpg domains was 1.31 and 2.14 respectively, compared to 1.18 and 2.06 for BEAR. However, in the housing domain, traditional k-NN slightly outperformed BEAR, with an MAE of 2.68, versus 2.84 for BEAR (approximately a 5% drop). Thus BEAR retains accuracy comparable to traditional k-NN.

Q4: How does execution efficiency of BEAR compare to that of traditional k-NN? Figure 3 shows the run time in seconds of traditional (non-LSH) k-NN and BEAR on different subsets of the Power domain, ranging from 20,000 to 820,000 randomly selected cases. The recorded run times are the total time for conducting ten fold cross validation. All experiments were run on a single machine with 16 GB memory and 2.8 GHz Intel Core i7 processor. Weka's [34] IBK package is used as the implementation of k-NN. For smaller case base sizes (e.g. 20,000 cases) k-NN is quite fast; a ten-fold cross validation test takes on the order of 1 second. For a case base approximately 31 times larger (615,000 cases) the test takes 4.5 h—approximately 16,000 times longer. When the size is increased to 820,000 cases, time increases to 21 h.

On the other hand, using LSH and parrallelizing the process over different nodes enables EAR4 to process same sizes of the case based in significantly less time. An interesting observation is that for a case base of 20,000 cases it actually takes less time for k-NN to yield the results than BEAR (it takes k-NN 10 seconds while takes BEAR 265 seconds). This is because of the communication overhead of MapReduce framework which makes applying big data techniques less efficient when applied to small case bases. However, k-NN run time increases very rapidly compared to BEAR as case base size increases to 205,000 cases, while the run time of BEAR increases at a much lower rate. Even when the entire power case base is used (over 2 million cases), BEAR takes less than 20 min to complete the experimental run on an EMR cluster with the configuration described in Sect. 5.1. We note that this corresponds to less than .1 second per problem solved, even on a small cluster. This supports the need to move to big data methods for practical large-scale CBR.

5.3   Overall Perspective: Scale-Up, Time, Space, and Accuracy

The experiments in this case study illustrate the ability of the BEAR approach, which combines ensemble adaptations with locality-sensitive hashing, both to remain efficient for large scale data and to provide substantial accuracy increases compared to non-ensemble adaptation of cases retrieved by LSH, and for k-NN using LSH. More generally, it illustrates the potential of CBR's reasoning capability (in the form of case adaptation) to provide strong benefits not present in big data/retrival-only methods. The largest test case base used in our experiments has two million cases, and was run on a small cluster (with ten core nodes). However, BEAR could easily be applied to substantially larger case bases with tens or even hundreds of millions of cases, and expected running times comparable to that reported in this paper, by increasing the computational resources to the level common in industrial settings (e.g. a cluster with hundred nodes or more).

In many treatments of case-based maintenance in CBR, having/maintaining a large number of cases is assumed to correspond to degraded retrieval and

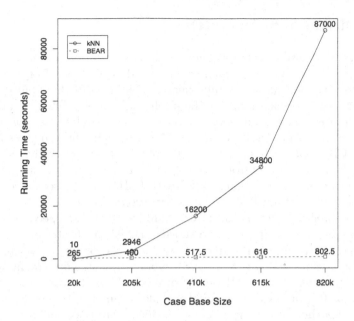

Fig. 3. Running time of k-NN and BEAR for different sizes of Power domain

processing time, potentially requiring sacrificing information by case deletion. However, leveraging big data platforms and techniques it is possible to avoid information loss, and consequently yield more accurate solutions, by retaining full case bases impractical for conventional methods and using them efficiently. The ability of big data to integrate both with flexible similarity-based retrieval and case adaptation is promising for the general ability of much scaled up CBR. This, in turn, could open the door to very large-scale CBR, with near-instant retrieval from case bases with millions of cases, plus the potential accuracy benefits of avoiding the need for case-base compression in many domains.

The previous experiments focus on the ability of big data methods to enable using full case bases. However, given the speed of those methods, for time-critical tasks it could even be feasible to sacrifice additional space for the sake of time. As a concrete example, for numerical prediction using BEAR in a domain with millions of cases (e.g. the Power domain), it would be possible to pre-process the data to generate the LSH keys for each case and store all cases with their corresponding hash keys in a NoSQL database. Because, in LSH, each record can be hashed with a set of hash families, this results in having case bases of size orders of magnitude greater than the original case base. However, with this NoSQL design, applying a method such as EAR4 on top of big data methods could enable processing thousands of queries in a matter of a few seconds even without MapReduce. Even for millions of queries, using MapReduce for query processing only and using the NoSQL database for case retrieval, average response time per query could still be in range of a few milliseconds.

6   Conclusion and Future Directions

In this paper we illustrated the practicality of a big-data version of ensembles of adaptation for regression, implemented in BEAR, which uses MapReduce and Locality Sensitive Hashing for finding nearest neighbors of the input query. We consider the results encouraging for the application of big data methods to the fuller CBR process, to exploit not only larger case bases but also collections of adaptation rules, without compression. Such methods might also present opportunities for CBR approaches to big data problems more generally, as an alternative to rule mining. In addition, the BEAR approach improves performance compared to the big data baseline k-NN with LSH1, and preserves comparable performance to that of much more costly traditional k-NN.

As future directions, we intend to compare accuracy and speed performance achieved by case base compression to those of BEAR, to better understand the tradeoffs between traditional and big data methods for CBR. Given BEAR's efficiency, we also intend to extend our methods to test more computationally expensive variations of the EAR family of methods as the case-base estimator module in BEAR. For example, generating rules from neighborhoods other than the local neighborhood of the input query—which requires consideration of many more cases—and adding contextual considerations in adaptation retrieval, have produced good small-scale results [10], but with high costs that raised concerns for their large-scale applicability by conventional CBR methods. The BEAR framework suggests a path for making practical such case-intensive methods.

Previous CBR research has applied big data methods to CBR when case retrieval relies on exact match (string-based) retrieval [15]; BEAR enables similarity-based matching. However, an important problem is how to apply these techniques to structured cases.

References

1. Kim, G.H., Trimi, S., Chung, J.H.: Big-data applications in the government sector. Commun. ACM **57**(3), 78–85 (2014)
2. Hoover, W.: Transforming health care through big data. Technical report, Institute for Health Technology Transformation (2013)
3. Greengard, S.: Weathering a new era of big data. Commun. ACM **57**(9), 12–14 (2014)
4. Plaza, E.: Semantics and experience in the future web. In: Althoff, K.-D., Bergmann, R., Minor, M., Hanft, A. (eds.) ECCBR 2008. LNCS (LNAI), vol. 5239, pp. 44–58. Springer, Heidelberg (2008)
5. Ontañón, S., Lee, Y.-C., Snodgrass, S., Bonfiglio, D., Winston, F.K., McDonald, C., Gonzalez, A.J.: Case-based prediction of teen driver behavior and skill. In: Lamontagne, L., Plaza, E. (eds.) ICCBR 2014. LNCS, vol. 8765, pp. 375–389. Springer, Heidelberg (2014)
6. Cordier, A., Lefevre, M., Champin, P.A., Georgeon, O., Mille, A.: Trace-based reasoning - modeling interaction traces for reasoning on experiences. In: Proceedings of the 2014 Florida AI Research Symposium, pp. 363–368. AAAI Press (2014)

7. Smyth, B., Keane, M.: Remembering to forget: a competence-preserving case deletion policy for case-based reasoning systems. In: Proceedings of the Thirteenth International Joint Conference on Artificial Intelligence, pp. 377–382. Morgan Kaufmann, San Mateo (1995)
8. Smyth, B., McKenna, E.: Building compact competent case-bases. In: Althoff, K.-D., Bergmann, R., Branting, L.K. (eds.) ICCBR 1999. LNCS (LNAI), vol. 1650, p. 329. Springer, Heidelberg (1999)
9. Jalali, V., Leake, D.: Extending case adaptation with automatically-generated ensembles of adaptation rules. In: Delany, S.J., Ontañón, S. (eds.) ICCBR 2013. LNCS, vol. 7969, pp. 188–202. Springer, Heidelberg (2013)
10. Jalali, V., Leake, D.: A context-aware approach to selecting adaptations for case-based reasoning. In: Brézillon, P., Blackburn, P., Dapoigny, R. (eds.) CONTEXT 2013. LNCS, vol. 8175, pp. 101–114. Springer, Heidelberg (2013)
11. Jalali, V., Leake, D.: Adaptation-guided case base maintenance. In: Proceedings of the Twenty-Eighth Conference on Artificial Intelligence, pp. 1875–1881. AAAI Press (2014)
12. Jalali, V., Leake, D.: On retention of adaptation rules. In: Lamontagne, L., Plaza, E. (eds.) ICCBR 2014. LNCS, vol. 8765, pp. 200–214. Springer, Heidelberg (2014)
13. Indyk, P., Motwani, R.: Approximate nearest neighbors: towards removing the curse of dimensionality. In: Proceedings of the Thirtieth Annual ACM Symposium on Theory of Computing. STOC 1998, pp. 604–613. ACM, New York (1998)
14. Daengdej, J., Lukose, D., Tsui, E., Beinat, P., Prophet, L.: Dynamically creating indices for two million cases: a real world problem. In: Smith, I., Faltings, B. (eds.) Advances in Case-Based Reasoning, pp. 105–119. Springer, Berlin (1996)
15. Beaver, I., Dumoulin, J.: Applying mapreduce to learning user preferences in near real-time. In: Delany, S.J., Ontañón, S. (eds.) ICCBR 2013. LNCS, vol. 7969, pp. 15–28. Springer, Heidelberg (2013)
16. Francis, A., Ram, A.: Computational models of the utility problem and their application to a utility analysis of case-based reasoning. In: Proceedings of the Workshop on Knowledge Compilation and Speed-Up Learning (1993)
17. Smyth, B., Cunningham, P.: The utility problem analysed: a case-based reasoning perspective. In: Proceedings of the Third European Workshop on Case-Based Reasoning, pp. 392–399. Springer, Berlin (1996)
18. Craw, S., Massie, S., Wiratunga, N.: Informed case base maintenance: a complexity profiling approach. In: Proceedings of the Twenty-Second National Conference on Artificial Intelligence, pp. 1618–1621. AAAI Press (2007)
19. Muñoz-Ávila, H.: A case retention policy based on detrimental retrieval. In: Althoff, K.-D., Bergmann, R., Branting, L.K. (eds.) ICCBR 1999. LNCS (LNAI), vol. 1650, pp. 276–287. Springer, Heidelberg (1999)
20. Ontañón, S., Plaza, E.: Collaborative case retention strategies for CBR agents. In: Ashley, K.D., Bridge, D.G. (eds.) ICCBR 2003. LNCS, vol. 2689, pp. 392–406. Springer, Heidelberg (2003)
21. Salamó, M., López-Sánchez, M.: Adaptive case-based reasoning using retention and forgetting strategies. Know.-Based Syst. 24(2), 230–247 (2011)
22. Zhu, J., Yang, Q.: Remembering to add: competence-preserving case-addition policies for case base maintenance. In: Proceedings of the Fifteenth International Joint Conference on Artificial Intelligence, pp. 234–241. Morgan Kaufmann (1999)
23. Angiulli, F.: Fast condensed nearest neighbor rule. In: Proceedings of the Twenty-second International Conference on Machine Learning, pp. 25–32. ACM, New York (2005)

24. Wilson, D., Martinez, T.: Reduction techniques for instance-based learning algorithms. Mach. Learn. **38**(3), 257–286 (2000)
25. Brighton, H., Mellish, C.: Identifying competence-critical instances for instance-based learners. In: Instance Selection and Construction for Data Mining, The Springer International Series in Engineering and Computer Science, vol. 608, pp. 77–94. Springer, Berlin (2001)
26. Houeland, T.G., Aamodt, A.: The utility problem for lazy learners - towards a non-eager approach. In: Bichindaritz, I., Montani, S. (eds.) ICCBR 2010. LNCS, vol. 6176, pp. 141–155. Springer, Heidelberg (2010)
27. Hanney, K., Keane, M.T.: The adaptation knowledge bottleneck: how to ease it by learning from cases. In: Leake, D.B., Plaza, E. (eds.) ICCBR 1997. LNCS, vol. 1266. Springer, Heidelberg (1997)
28. Gionis, A., Indyk, P., Motwani, R., et al.: Similarity search in high dimensions via hashing. VLDB **99**, 518–529 (1999)
29. Kulis, B., Grauman, K.: Kernelized locality-sensitive hashing for scalable image search. In: IEEE International Conference on Computer Vision ICCV (2009)
30. Datar, M., Immorlica, N., Indyk, P., Mirrokni, V.S.: Locality-sensitive hashing scheme based on p-stable distributions. In: Proceedings of the Twentieth Annual Symposium on Computational Geometry, SCG 2004, pp. 253–262. ACM, New York (2004)
31. Frank, A., Asuncion, A.: UCI machine learning repository (2010) http://archive.ics.uci.edu/ml
32. Hayes, M., Shah, S.: Hourglass: a library for incremental processing on hadoop. In: 2013 IEEE International Conference on Big Data, pp. 742–752 (2013)
33. Jalali, V., Leake, D.: Manual for EAR4 and CAAR weka plugins, case-based regression and ensembles of adaptations, version 1. Technical report TR 717, Computer Science Department. Indiana University, Bloomington (2015)
34. Witten, I., Frank, E., Hall, M.: Data mining: practical machine learning tools and techniques with Java implementations, 3rd edn. Morgan Kaufmann, San Francisco (2011)

Addressing the Cold-Start Problem in Facial Expression Recognition

Jose L. Jorro-Aragoneses[(⊠)], Belén Díaz-Agudo, and Juan A. Recio-García

Department of Software Engineering and Artificial Intelligence,
Universidad Complutense de Madrid, Madrid, Spain
{jljorro,belend}@ucm.es, jareciog@fdi.ucm.es

Abstract. In our previous research [5] we proposed a CBR approach
to infer the emotional state of the user through the analysis of a picture
taken from the front facing camera of her mobile device. We demonstrated that different people express emotions with different gestures
and got the best accuracy using a personal case base with self pictures
of the same user. However, in the cold start situation, where pictures
of the querying user are not available, the CBR system uses a generic
case base (GCB) made of pictures of anonymous people. Although the
performance using the GCB was acceptable on average there were several users with a very low accuracy. In this paper we compare our GCB
to other reference picture catalogues and evaluate our CBR approach
with state-of-the-art Facial Expression Recognition (FER) algorithms.
Results point out that our approach is only suitable for GCB including
semantically similar users. We use an ontology to group together users
with similar demographic and physiological information: sex, age and
ethnic group. We evaluate our CBR approach with small and specialized
case bases where pictures are semantically similar to the target population and demonstrate that it efficiently increases the accuracy in the
cold start situation and minimizes the noise in the case base.

1 Introduction

Emotional tagging of facial expressions is becoming a relevant topic for e-
commerce systems based on mobile devices. These devices include a wide range
of sensors able to capture the emotional state of the user, which is a very valuable feedback to infer her willing to consume a concrete product being proposed
by the e-commerce system. This information about the user's emotional state
has many potential applications and an unmeasurable value from the commercial
point of view. Although modern devices include sensors that can measure several
physiological variables of the user such as heart rate, temperature or even blood
pressure, the front camera is usually the best alternative to infer the emotional
response.

Usage of dynamically enriched information from the user context leads the
system to find better solutions that are adapted to the specific situations. In
our research we have focused on the difficult problem of dynamically acquiring

E. Hüllermeier and M. Minor (Eds.): ICCBR 2015, LNAI 9343, pp. 197–211, 2015.
DOI: 10.1007/978-3-319-24586-7_14

the *emotional context* of the user during a recommendation process. In [5] we described PhotoMood, a CBR system that uses gestures to identify emotions in faces, and presented preliminary experiments with MadridLive, a mobile and context aware recommender system for leisure activities in Madrid. In the experiments, the momentary emotion of a user is dynamically detected from pictures of the facial expression taken unobtrusively with the front facing camera of the mobile device.

The emotional state of the user inferred by PhotoMood was used as the feedback -like or dislike- for our recommender system for leisure activities. This CBR system uses as queries the pictures of the user taken while she is receiving recommendations from the system. By comparing these pictures to a case base of pictures tagged as positive or negative responses, we could infer the user's opinion about each activity being proposed. The published results stated that the use of a personal case base with pictures of the user achieved a higher performance than a generic case base with pictures from several users. Although a personal case base is the best alternative, new users have no pictures in the system and, therefore, the *cold-start* problem appears. The term *cold-start* is used in recommender systems to denote those users with little information in the system. This problem is very common and requires specific techniques to circumvent it. Due to the lower performance of the generic case base compared to the personal case base, we required an alternative approach to tag those users in cold-start. This paper tries to address this limitation by proposing the use of specific case bases with a small number of cases that have been semantically annotated and that groups pictures of users that are semantically similar. we can choose the most suitable set of cases to get an efficient and precise replacement of the personal case base for those users in cold-start. This intermediate approach rises the performance of our facial expression recognition (FER) system and achieves better results than a generic case base. We have performed an experimental evaluation that not only validates our proposal but also compares our CBR strategy to other state-of-the-art algorithms for facial emotion recognition.

Section 2 describes some related work on the FER field research. Next, in Sect. 3 we explain our previous work, algorithm and results on face emotional recognition based on the CBR paradigm. Section 4 introduces the cold-start problem and a first solution using generic case bases. We introduce some of our tools for visualization and organization of cases bases and we compare our approach with other FER approaches in these generic case bases. Section 5 describes a solution based on the semantic annotation of generic case bases to obtain specialized cases that share common gestures with the current query. We use age, sex and ethnic group as the features to annotate case bases. Experiments show that results on these specialized case bases are accurate and efficient. Section 6 concludes our work.

2   Related Work

Facial expression recognition (FER) is a research field very active in the area of computer vision. We can find a general overview of FER algorithms in [18].

FER algorithms are applied to different domains, such as recognition of students engagement [21], user experience feedback [13] or different recommender systems [19,20].

FER algorithms can be classified in several ways, although a common classification is based on the method to extract facial features [18]. According to this feature, FER algorithms can divided in two groups: appearance feature extraction and geometric feature extraction.

Appearance feature extraction consists on extracting changes in the appearance of the face, for example, skin texture. We find many works whose authors use this technique. Zhen and Zilu [22] use a FER algorithm based on sparse representation and Local Phase Quantization (LQP). They based their algorithm on the LPQ method of textures of images [14]. Other example is Taheri,Patel and Chellapa [17], where authors use a dictionary-based component separation algorithm (DCS). This method separates the neutral component and from expression component of an image. In Khanum, Mufti, Javed and Shafiq [7], authors detect the facial action elements (FAEs) and uses fuzzy logic combined to CBR to detect the emotion.

The second group of FER algorithms are the geometric feature extraction ones. It consists on extracting the location of elements of the face. An example of this type of algorithms is Kotsia and Pitas [8]. It uses the location of some points to detect deformations in the face and then detect the emotion from this deformation.

Our algorithm, PhotoMood [5,11], is classified in the geometric feature extraction. PhotoMood is a CBR system to detect emotions with facial recognition. In [11], we presented a similarity function based on the external contour of the mouth. In this paper, we improve the similarity function by adding additional gestures as it is explained in the following section.

3   PhotoMood: A CBR Approach to Face Emotion Recognition

This section explains the CBR algorithm implemented in PhotoMood. Photo-Mood [5,11] analyses the facial expression to detect the user emotion using two stages:

Image Pre-processing: PhotoMood obtains 46 coordinates of the user's facial gestures classified in 8 vectors:
- $\overline{v_1}$: External outline of mouth.
- $\overline{v_2}$: Internal outline of mouth.
- $\overline{v_3}$: Outline of right cheek.
- $\overline{v_4}$: Outline of left cheek.
- $\overline{v_5}$: Outline of right eye.
- $\overline{v_6}$: Outline of left eye.
- $\overline{v_7}$: Outline of right eyebrow.
- $\overline{v_8}$: Outline of left eyebrow.

CBR Process: Cases in the case base include pictures of users that have been annotated with the emotion tag t. From a query of gestures(Q) including these 8 vectors, the CBR system retrieves the most similar pictures and reuse their solution to estimate the user's emotion:

$$Q = <\overline{v_1^q}, \ldots, \overline{v_8^q}> \tag{1}$$

$$C = <\mathcal{D}_c, \mathcal{S}_c> \tag{2}$$

where

$$\mathcal{D}_c = <\overline{v_1^c}, \ldots, \overline{v_8^c}> \tag{3}$$

$$\mathcal{S}_c = \{t_1, \ldots, t_m\} \tag{4}$$

The system admits any number of emotions in the solution side of the cases. The system retrieves the most similar cases using the gesture vectors and reuse their emotion tags. In the recommender domain we use three tags *Like*, *Dislike* and *Surprise*, so m=3.

Retrieval. We used a K-Nearest Neighbour algorithm where the similarity value is computed by weighting the local similarity of each vector $\overline{v_i}$. Therefore each pair of vectors $<\overline{v_i^q}, \overline{v_i^q}>$ is compared and the resulting value is weighted with a value w_i that represents the relevance of the corresponding gesture in the global similarity computation:

$$Sim(Q, \mathcal{D}_c) = \sum_{i=1}^{8} w_i * Sim_i(\overline{v_i^q}, \overline{v_i^c}) \tag{5}$$

where

$$\sum_{i=1}^{8} w_i = 1 \tag{6}$$

We obtain the angle between each pair of points and the horizontal axis. Following the contour of a gesture, the arctangent between a pair of points is computed to obtain their angle. Each angle of the query Q is compared to the corresponding angle of the case $\mathcal{D}_c$ producing $|\overline{v_i}|$ (angle-level) similarity values:

$$Sim_i(\overline{v_i^q}, \overline{v_i^c}) = \frac{1}{|\overline{v_i}|} \sum_{j=1}^{z} 1 - \sqrt{(\arctan(\widehat{p_j^q p_l^q}) - \arctan(\widehat{p_j^c p_l^c}))^2} \tag{7}$$

where $l = (j+1) \mod z$

Reuse. To obtain the solution of the query, the system uses a weighted voting schema according to the similarity of the retrieves cases. The scoring function is:

$$score(t_i) = \sum sim(Q, \mathcal{D}_c) \; \forall \, c \, | \, \mathcal{S}_c = t_i \tag{8}$$

The solution assigned to the query is:

$$t_i = arg\,max\{score(t_i), \; i = 1, \ldots, m\} \tag{9}$$

Revise. The revise stage is external to the PhotoMood CBR module. The user
will be the responsible to modify the solution if it is wrong.

Retain. The system stores the cases that they were revised by the user.

3.1   Previous Results

In our previous experiments [5], we tested our CBR approach with a a personal
case base (PCB) of self pictures for each querying user, and with a generic case
base (GCB) that contains 300 images of anonymous people that we obtained
with search processes in Google Images[1]. We concluded that each gesture has
a different importance for each user and we used a Genetic Algorithm (GA) to
calculate the optimal set of weights for each user. Finally, we concluded that
the CBR system behaves better using the personal case base (PCB) and the
personal set of weights (pw) for each user. However, having a personal case base
for each user is not always possible. For example, when we have new users in
cold start. Besides, the computation of a personal set of weights for an user is a
computational expensive process. So, GCBs with very different people have been
characterized and evaluated. Next section describes how different GCBs behave
with different users in the cold start situation.

4   Generic Case Bases as a Solution to the Cold Start Problem

Cold start is one of the most challenging problems in recommender systems [10].
It is a potential problem in any knowledge based system based on information
about users or items. It appears when the system has not gathered enough information. For example, in the domain of product recommendation for new users
the system has no information about users' preferences in order to make recommendations. Although a relevant research has been conducted in this field the
cold-start problem is far from being solved and many different partial solutions
have been proposed (see [16] for a recent review).

In our previous research [5] we demonstrate that different people show emotions with different gestures and compute the specific set of weights that maximize the accuracy of our CBR classifier for each specific user. To do that, we
used a personal case base with self pictures of each user. We have experimented
with different configurations of case bases and sets of weights. As conclusion,
we pointed out that the system obtains the best results using a personal case
base for each user. However, in the cold start situation, where pictures of the
querying user are not available, the system uses a generic case base made of
pictures of anonymous people. Reasonable results were found when tagging two
basic emotions –like and dislike – using 46 feature points organized in 8 gesture
vectors. Experiments got an average precision in GCB of 89.34 % and an average

---

[1] We used Google Image Search with the queries "Happy face", "Unhappy faces" and
"Surprise faces".

precision for the personal case base of 92.16 %. Our generic Case Base in these preliminary experiments consists of 300 images of anonymous people that we obtained with search processes in Google Images.

Although the performance of the GCB in our previous work was acceptable on average, there were several users with a very low accuracy. After a comparison of our GCB with other reference picture catalogues used to evaluate state-of-the-art FER algorithms, we realised that our approach is only suitable for users similar to those ones in the case base. For example, the JAFFE [12] dataset contains pictures of Japanese women. Due to specific facial features of this population our GCB was unable to correctly classify our previous pictures (from European people) when used as queries in a cross-validation experiment. Moreover, after several experimental evaluations we noticed that the Genetic Algorithm used to optimize the similarity function decreased its performance when applied to large generic datasets including a wide range of users.

This experimentation with several generic case bases led us to conclude that the cold-start problem could not be addressed with only one generic case base. Nevertheless, the use of specialized case bases including pictures semantically similar to the target population could increase the accuracy of the system. We use the term *semantically similar* to denote those features that allows us to group users with similar facial expressions.

Initially, we tried to apply automatic machine learning approaches -such as clustering techniques- to obtain these groups of users with similar expressions. After several experimental evaluations results were unsatisfactory and we concluded that these groups had to be made according to semantic features of the users. As we will present in following sections we have used the information from the profiles of the users to map them to a specialized case base through a semantic mapping function based on an ontology. This ontology captures the *semantic similarity* between users according to features like age, ethnic group an gender.

Following subsections present this experimental evaluation. First we introduce the results when applying our approach to several reference datasets in Sect. 4.1. Next, we also compare different state-of-the-art FER algorithms to our CBR system 4.2. Finally, Sect. 5 presents our semantic approach to exploit specialized case bases.

4.1 Applying PhotoMood CBR to Reference Datasets

The first solution to the cold-start situation is the use of a GCB including a large number of pictures of different type of users, in order to cover as much as possible the solution space of facial expressions. We used 2 different reference datasets as case bases to evaluate our CBR approach:

1. *CK+*[6]*:* It contains 593 images from 123 individuals. These images have been divided in scenes where each scene contains images from neutral state to any emotion. The last image of each scene is the peak of the emotion (the image most representative of the emotion tagged in this scene). We only used the last image of each user that has been tagged with an emotion. Totally we

CK+			
	Dislike	Like	Surprise
Dislike	85.55%	10.98%	3.47%
Like	30.43%	66.67%	2.9 %
Surprise	6.02%	0.0%	93.98 %

JAFFE			
	Dislike	Like	Surprise
Dislike	93.42%	5.26%	1.32%
Like	45.16%	54.84%	0.0%
Surprise	86.67%	3.33%	10.0%

Fig. 1. Confusion matrix of the CK+ (top) and JAFFE (bottom) datasets using PhotoMood CBR.

have 327 images (45 angry, 18 contempt, 59 disgust, 25 fear, 69 happiness, 28 sadness and 82 surprise) to use like queries.

2. *JAFFE*[12]: It contains 213 face images from 10 subjects. These images are of Japanese women. These pictures are classified in: Neutral (30), Anger (30), Disgust (29), Fear (32), Happiness (31), Sadness (31) and Surprise (30).

As we explained these datasets are classified in 7 emotions, but PhotoMood was implemented to detect 3 emotions (like, dislike and surprise). Therefore we grouped similar emotions (for example sad and fear) to reduce the number of classes to the ones used by our system. We selected and grouped the emotions equivalent to those ones in PhotoMood. For the *Like* tag we included images tagged with the happy emotion. For the *Dislike* tag we grouped images whose tag is a dislike emotion (angry, contempt, disgust, fear and sadness). Finally, we included the *Surprise* tag alone as there is not a clear classification (Like or Dislike).

After several cross-validation evaluations we obtained poor results. We tested both datasets using leave-one-out and grouping the classes of pictures into the 3 emotions used in PhotoMood (*Like, Dislike* and *Surprise*). Results are shown in Fig. 1 as a confusion matrix. We can observe that there is no clear classification for any emotion.

In any CBR system there are two possible explanations for this lack of performance: (1) the additional knowledge of the system -this is the similarity function-has deficiencies. And (2) the deficiencies are in the case base. Beginning with the first possible explanation we developed introspection tools to analyse the performance of our similarity function. These tools are:

– Visualization of cases grouping. Ideally a good similarity function should group same class cases together. By using a graph distance visualization tool we should be able to identify clusters of cases belonging to the same emotion. This tool is shown in Fig. 2 applied to the CK+ dataset. In this figure each emotion is represented by a colour. We can observe that there are two clear clusters where the similarity function is working properly but the center of the image is too messy, denoting a bad classification.

– Distance matrix. Although the case-base visualization tool is very useful to informally evaluate the accuracy of the similarity function it depends on the graph visualization algorithm that tries to layout the cases depending on their distance. An efficient way to circumvent this drawback is the distance matrix. This NxN matrix (being N the size of the dataset) shows the similarity of every pair of cases by using a grey scale. A dark point denotes a high similarity and a whiter point reflects low similarity. By sorting the axis of the matrix according to the class (emotion) of the cases we should, ideally, find dark rectangles in the diagonal of the matrix (corresponding to the similarity of cases from the same emotion) and white areas in the surroundings. However, when visualizing the datasets with this technique we did not find this pattern. As we can observe in Fig. 3 there is only a good classification in the third class (*like* emotion) where there is a clear dark rectangle in the diagonal of the matrix and white areas in its vertical and horizontal sides.

– Similarity function introspection. The disappointing results of the similarity function led us to consider that it was not properly designed. As we previously explained it is a weighted average of several local similarities that compares 8 gestures of the user (eyebrow, outline of mouth, etc.). Perhaps this way to compute the similarity between emotions was not suitable for this domain so we developed an introspection tool to evaluate the accuracy of the similarity function for problematic pairs of cases. The resulting tool is shown in Fig. 4. In this case, pictures being compared belong to different emotions (anger and disgust). After analysing the accuracy of the similarity function with several pairs of pictures we concluded that the similarity function was performing reasonably well.

If the similarity function was performing well the only remaining explanation for the low accuracy of the CBR system is the other source of knowledge: the case-base. As we can clearly observe in Fig. 4 pictures are quite similar but classified with different emotions. No similarity function could distinguish among these emotions because they deeply depend on the facial expression of the user[2]. This analysis supports our premise that the personal case base is the best alternative to rise the performance of the CBR system. But when the user is in cold-start and there are no personal pictures we cannot use generic case bases because the classification deeply depends on the type of user. This is the main reason to develop an approach based on specialized case bases selected by using semantic knowledge that is presented in Sect. 5. But before developing such method we had to completely confirm that the CBR approach was valid by comparing it to state-of-the-art FER algorithms as described in the following section.

4.2 Comparison of PhotoMood with Other FER Approaches

Section 2 introduced several algorithms that exploit the textures in pictures to infer the emotional state for the user, i.e. [17,22]. We have compared our CBR

---

[2] We leave aside possible miss-classifications of pictures in the dataset.

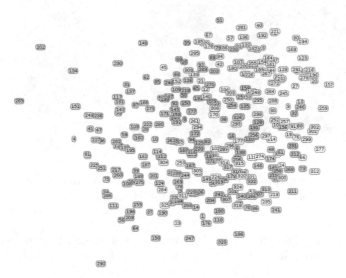

Fig. 2. Visualization of the CK+ dataset. Each colour representing a different emotion (Color figure online).

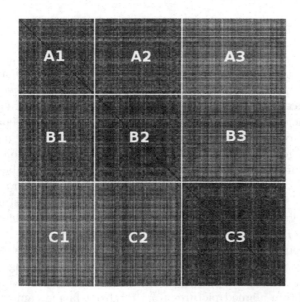

Fig. 3. Distance matrix of the CK+ dataset. A1, B2 and C3 areas representing a pictures of the same emotion.

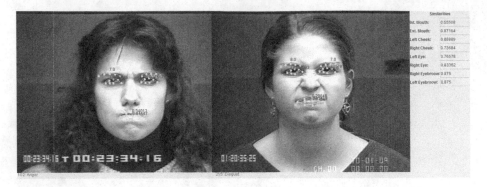

Fig. 4. Introspection tool to analyse the similarity function. Although similar pictures denote different emotions: anger (left) and disgust (right).

approach to these algorithms in order to validate our geometric similarity function. Figure 5 shows the results of this comparison for the CK+ dataset[3].

CK+			
	Like	Dislike	Surprise
SRC + LPQ [22]	100.00%	71.156%	100.00%
DCS-S1 [17]	98.55%	72.91%	98.70 %
PhotoMood	66.67%	85.55%	93.98 %

Fig. 5. Comparison with other FER approaches.

According to these results, our approach is similar in performance to state-of-the-art algorithms. It achieves a higher accuracy for the *dislike* class, but a lower performance for other classes. These inconclusive results are rooted in the low quality of the case base regarding the classification problem being addressed.

We can also compare these algorithms from the functional point of view. Texture-based algorithms require a large storage as they must store every single pixel of the picture. In our MadridLive scenario it is not a suitable solution for devices such as mobile phones with a limited storage capacity. Our approach only requires to store the 46 points defining the geometry of the gestures. Another relevant advantage of our CBR approach is related to the alignment of the faces in the pictures. Texture-based methods usually apply a *sparse representation* technique that requires a perfect alignment of every picture. This technique uses an image overlay training process to create a model representing the emotions of the user. However, this approach is not applicable in our scenario as it is impossible to obtain aligned pictures from the front facing camera of a mobile device. Being our method based on geometrical comparison it does allow to compare pictures with different face alignments.

---

[3] The accuracy is taken from [9], whereas the DCS-S1 accuracy is obtained from [17].

Once the CBR approach was validated next Section presents an improvement to solve the cold-start problem.

5   Use of Semantic Case Bases

After our analysis of the performance of the studied GCBs, we have found several drawbacks when trying to address the cold-start problem. First, to achieve an acceptable accuracy, the CBR system requires a large number of pictures that share gestures with the user query. Besides, to guarantee enough coverage for different types of users we would need to provide with a very large case base. The CBR process becomes inefficient and decreases accuracy due to the irrelevant noisy cases. However, we noticed that it is possible to increase the accuracy of the CBR system if the case base includes pictures from the same physiological group than the query user. It means that by using pictures of users with similar features like age, gender and ethnic group, we could rise the performance of our FER system. In this section we describe a proposal to solve the cold-start problem using small and specialized case bases according to the demographic and physiological features of the users.

Our first approach was the use of automatic clustering algorithms to find small related groups in GCB. This way we could find groups of pictures from users with similar features (Fig. 4). However after several experimental evaluations we concluded that this approach led us to generic groups of pictures meaningless from the point of view of expression recognition.

Our proposal is to manually create these clusters as specialized case bases that capture semantic features of the user being classified. We evaluate this proposal to check if the use of semantic case bases improves the accuracy and the efficiency of the FER system in a cold start situation. It is a two stage process:

- Find the proper specialized case base for a given user.
- Retrieve and reuse the most similar cases.

This approach requires additional knowledge about the user in order to select the most suitable specialized case base. MadridLive recommender [5] obtains this information from the user profile. During registration, users must state their age, gender and country. By adding additional semantic knowledge the system can be able to find the proper case base.

If users are described by age, gender and country, and our specialized case bases are described by the age range, gender and ethnic group of the users in the pictures there is a small vocabulary gap to be solved. Previous works [15] have pointed to ontologies as an efficient way to represent the semantic knowledge required to map between a query and case base description vocabulary. In our case, ontologies can formalize the additional knowledge required to obtain the proper case base of pictures for a query described in terms of the age, gender and country. Figure 6 shows a simplified view of our ontology. It illustrates how a user from Japan is mapped to a case base for Asiatic people[4].

---

[4] We have used the"Geographical Races" taxonomy proposed by Garn [4].

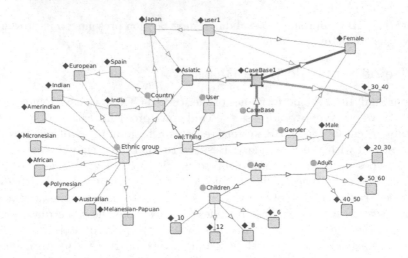

Fig. 6. Semantic annotation of specialized case bases

5.1   Experimental Results

Our approach is based on a semantic characterization of different generic case bases. We call Semantic Case Base (SCB) to these annotated GCBs. The SCB that is semantically most similar to the query is used to apply the CBR processes described in Sect. 3. The hypothesis we want to probe is that the use of a case base chosen this way repeats higher precision values. Our approach is considered as a solution to the cold start problem because the precision obtained by this specialized case base is close to the figures obtained by the user personal case base.

In the experiments to validate this hypothesis we have created 12 specialized case bases. Each case base contains images of a set of semantically similar users according to the following features:

– *Age,* classified in 2 categories, children (CH) and adults (AD).
– *Gender,* classified in 2 categories, men (M) and women (W).
– *Geographical area,* where we distinguish 3 main areas Africa (AF), Asia (AS) y Europa (EU).

By combining these categories we generated 12 case bases that were populated using Google Images:

– CHMAF: African male children. 10 Like and 9 Dislike.
– CHMAS: Asian male children. 19 Like and 20 Dislike.
– CHMEU: European male children. 20 Like and 17 Dislike.
– CHWAF: African female children. 21 Like and 16 Dislike.
– CHWAS: Asian female children. 19 Like and 19 Dislike.
– CHWEU: European female children. 20 Like and 20 Dislike.
– ADMAF: African male adults. 19 Like and 19 Dislike.

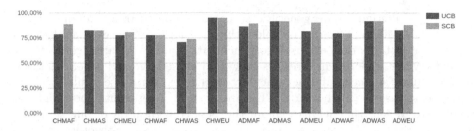

Fig. 7. Compare the use of specific case base between the union of case bases.

- ADMAS: Asian male adults. 14 Like and 12 Dislike.
- ADMEU: European male adults. 20 Like and 20 Dislike.
- ADWAF: African female adults. 10 Like and 15 Dislike.
- ADWAS: Asian female adults. 13 Like and 13 Dislike.
- ADWEU: European female adults. 20 Like and 20 Dislike.

In order to evaluate the validity of this approach we compared the performance of the system when using the most suitable semantic case base for each user to the performance when using a generic case base composed of all the semantic ones (referred as *union case base* (UCB)). Concretely, we have performed two tests:

- *Test 1:* Evaluation of every SCB using leave-one-out with cases from UCB.
- *Test 2:* Individual leave-one-out evaluation of every SCB only taking cases from itself.

Figure 7 shows the hit ratio when PhotoMood uses GCB and when it uses each semantic case base SCB. As we can observe the use of SCB always achieves at least the same performance than a generic case base, although there are many SCB improving the performance of GCB.

This experiment validates our hypothesis, that was statically confirmed by Wilcoxon test ($p-value < 0.05$), and allows us to conclude that the use of semantic case bases provides higher precision values. Besides, this approach solves the cold start problem as the precision obtained in this specialized case base is close to the values obtained in the user personal case base.

6   Conclusions

Mobility and context-awareness are two active research directions that open new potential to CBR systems [1–3]. In this paper we have addressed the cold-start problem in Facial Expression Recognition (FER). We propose a CBR system that is able to recognize emotions from pictures taken from the front facing camera of a mobile device.

In our previous work [5], we have concluded that personal case bases of user's pictures increase the performance of our CBR system. However, users in cold-start do not have enough pictures and an alternative case base must be adopted.

A plausible alternative are generic case bases composed of a large number of pictures from different users. To explore this approach we have evaluated several datasets of pictures that are used to benchmark state-of-the-art algorithms for FER. However, experimental results show that these large and generic datasets do not achieve an acceptable accuracy. In order to explain these results with the generic case bases we analysed our CBR process using a set of introspection tools, developed ad-hoc, that let us explore the best configuration of our similarity function. After a deep analysis and a concise comparison with other state-of-the-art FER algorithms, we can conclude that the proposed similarity function is correct and the only remaining explanation is the low quality of the generic case bases.

At this point, generic case bases are not an acceptable solution for those users in cold-start. Although cold-start users do not provide pictures to increase the performance of the system, they provide semantic information about their age, gender and country that can be exploited by our CBR system. Our proposal consists on the use of several specific case bases representing groups of users with similar features (those ones from the profile), and therefore, similar expressions. These specific case bases are labelled with semantic information that let the system choose the most suitable case base for each user according to the profile. However, there is a vocabulary gap between the descriptions of the profile and the labelling of the specific case bases that we solve by mean of an ontology. This ontology provides semantic information about these case bases. By using the ontology the CBR system chooses a semantic case base for each user that includes pictures from users with a similar facial expression.

Finally, this paper reports an experimental evaluation of the approach by comparing the semantic case bases to the generic ones. Results show that these case bases are a suitable solution for users in cold-start.

References

1. Adomavicius, G., Mobasher, B., Ricci, F., Tuzhilin, A.: Context-aware recommender systems. AI Mag. **32**(3), 67–80 (2011)
2. Benou, P., Bitos, V.: Context-aware query processing in ad-hoc environments of peers. JECO **6**(1), 88 (2008)
3. Braunhofer, M., Kaminskas, M., Ricci, F.: Location-aware music recommendation. IJMIR **2**(1), 31–44 (2013)
4. Garn, S.M.: Human Races, 3rd edn. Thomas, Springfield (1971)
5. Jorro-Aragoneses, J.L., Díaz-Agudo, B., Recio-García, J.A.: Optimization of a CBR system for emotional tagging of facial expressions. In: UKCBR (2014)
6. Kanade, T., Cohn, J.F., Tian, Y.: Comprehensive database for facial expression analysis. In: Proceedings of the Fourth IEEE International Conference on Automatic Face and Gesture Recognition, pp. 484–490 (2000)
7. Khanum, A., Mufti, M., Javed, M.Y., Shafiq, M.Z.: Fuzzy case-based reasoning for facial expression recognition. Fuzzy Sets Syst. **160**(2), 231–250 (2009)
8. Kotsia, I., Pitas, I.: Facial expression recognition in image sequences using geometric deformation features and support vector machines. IEEE Trans. Image Process. **16**(1), 172–187 (2007)

9. Lee, S.H., Plataniotis, K., Ro, Y.M.: Intra-class variation reduction using training expression images for sparse representation based facial expression recognition. In: IEEE Transactions on Affective Computing, p. 1 (2014)
10. Lika, B., Kolomvatsos, K., Hadjiefthymiades, S.: Facing the cold start problem in recommender systems. Expert Syst. Appl. **41**(4, Part 2), 2065–2073 (2014). http://www.sciencedirect.com/science/article/pii/S0957417413007240
11. Lopez-de-Arenosa, P., Díaz-Agudo, B., Recio-García, J.A.: CBR tagging of emotions from facial expressions. In: Lamontagne, L., Plaza, E. (eds.) ICCBR 2014. LNCS, vol. 8765, pp. 245–259. Springer, Heidelberg (2014)
12. Lyons, M., Akamatsu, S.: Coding facial expressions with gabor wavelets. In: Coding Facial Expressions with Gabor Wavelets. pp. 200–205 (1998)
13. Novak, D., Nagle, A., Riener, R.: Linking recognition accuracy and user experience in an affective feedback loop. IEEE Trans. Affect. Comput. **5**(2), 168–172 (2014)
14. Ojansivu, V., Heikkilä, J.: Blur insensitive texture classification using local phase quantization. In: Elmoataz, A., Lezoray, O., Nouboud, F., Mammass, D. (eds.) ICISP 2008. LNCS, vol. 5099, pp. 236–243. Springer, Heidelberg (2008)
15. Recio-García, J.A., Díaz-Agudo, B., González-Calero, P.A., Sánchez-Ruiz-Granados, A.: Ontology based CBR with jCOLIBRI. In: Proceedings of the 26th SGAI International Conference on Innovative Techniques and Applications of Artificial Intelligence, pp. 149–162. Springer, Cambridge (2006)
16. Son, L.H.: Dealing with the new user cold-start problem in recommender systems: A comparative review. Information Systems (0) (2014). http://www.sciencedirect.com/science/article/pii/S0306437914001525
17. Taheri, S., Patel, V., Chellappa, R.: Component-based recognition of facesand facial expressions. IEEE Trans. Affect. Comput. **4**(4), 360–371 (2013)
18. Tian, Y., Kanade, T., Cohn, J.F.: Facial expression recognition. Handbook of Face Recognition, pp. 487–519. Springer, New York (2011)
19. Tkalcic, M., de Gemmis, M., Semeraro, G.: Personality and emotions in decision making and recommender systems. In: First International Workshop on Decision Making and Recommender Systems, pp. 14–18. CEUR (2014)
20. Tkalcic, M., Kosir, A., Tasic, J.: Affective recommender systems: the role of emotions in recommender systems. In: Proceeding of the RecSys 2011 Workshop on Human Decision Making in Recommender Systems, pp. 9–13. Citeseer (2011)
21. Whitehill, J., Serpell, Z., Lin, Y.C., Foster, A., Movellan, J.: The faces of engagement: Automatic recognition of student engagementfrom facial expressions. IEEE Trans. Affect. Comput. **5**(1), 86–98 (2014)
22. Zhen, W., Zilu, Y.: Facial expression recognition based on local phase quantization and sparse representation. In: 2012 Eighth International Conference on Natural Computation (ICNC), pp. 222–225 (2012)

Flexible Feature Deletion: Compacting Case Bases by Selectively Compressing Case Contents

David Leake[(✉)] and Brian Schack

School of Informatics and Computing, Indiana University,
Bloomington, IN 47408, USA
{leake,schackb}@indiana.edu

Abstract. Extensive research in case-base maintenance has studied methods for achieving compact, competent case bases. This work has examined how to achieve good solution performance while limiting the number of cases retained, using approaches such as competence-based case deletion. Two fundamental assumptions of such approaches have been (1) that cases are approximately the same size and (2) that the only way to affect case base size is by deleting or retaining entire cases. However, in some domains different cases may contain different amounts of information, causing widely varying case sizes, and case solutions may themselves be compressible, with the ability to selectively delete portions of indices or solutions while still retaining varying levels of usefulness. In accordance with this more flexible view, this paper proposes a new maintenance approach, *flexible feature deletion*, which removes parts of cases, enabling compression of the case base by selective—and possibly non-uniform—size reduction of individual cases. It proposes and evaluates an initial set of feature deletion strategies. Experimental results support that when cases have varying size and compressible contents, flexible feature deletion strategies may enable better system performance than case-oriented strategies for the same level of compression.

Keywords: Case-base maintenance · Feature reduction · Case deletion · Case-base compression

1 Introduction

The performance of case-based reasoning systems depends on the coverage of their case bases and the quality of their cases. As the number of cases in the case base grows, increased retrieval costs [1,2] or storage constraints may require controlling case base size. Extensive case-based reasoning research has aimed to address this problem through case-base maintenance [3]. A key focus of this work has been on strategies for selecting cases to retain in the case base to maximize the competence achieved for a given number of cases. Approaches include strategies to guide deletion of cases from an existing case base [4], for determining when to retain a new case during problem-solving (e.g., [5]), and for ordering addition of cases from a candidate case set (e.g., [6,7]). All of these

© Springer International Publishing Switzerland 2015
E. Hüllermeier and M. Minor (Eds.): ICCBR 2015, LNAI 9343, pp. 212–227, 2015.
DOI: 10.1007/978-3-319-24586-7_15

strategies treat cases as single units, adding or deleting entire cases. We call such strategies "per-case" maintenance strategies.

Per-case strategies reflect two common implicit assumptions: (1) that all of the CBR system's cases will be of sufficiently uniform size that the size effects of deletion or addition do not depend on the chosen case, and (2) that the size of the internal contents of cases cannot be reduced. In domains for which each case must contain uniform knowledge, so that removal of any case information would severely impair the ability to use the cases, per-case strategies are the only appropriate choice. However, in some CBR domains, case contents are more flexible.

This paper questions the assumption of uniform case size in case-base maintenance, The assumption of uniform size means that, if cases are of different size, it is not possible, for example, to favor retention of smaller cases when those cases have comparable coverage. It also questions the assumption of maintenance only on a per-case basis, proposing that compression strategies can consider not only case deletion/addition but the deletion of components of particular cases. Rather than pre-determining a static set of features to be used throughout the life of the CBR system, the set of features to include in the case base could be adjusted based on requirements for storage, processing speed, and accuracy. There need be no requirement that all cases in the case base include the same set of features, just as there need not be uniform collections of components in the solution parts of cases, and the solutions need not be represented at the same level of granularity. This paper proposes a new, more flexible maintenance approach in which selective compression can be done at the level of the contents of individual cases, by removing selected features from either indexing or solution information. Thus this can be used to maintain both indexing features and features of a solution.

The motivation for adjusting case contents arises from domains in which cases are large and can be represented in multiple ways. For example, CBR has attracted interest for reasoning from imagery such as medical images (e.g., [8]). From any image, different features may be extracted, at different resolutions, and the amount of information required to represent different images might vary dramatically. In diagnostic domains, numerous features may carry information relevant to the diagnosis, with different pieces relevant to different degrees for different problems. When CBR is applied to design support, stored designs could selectively include different subsets of a full design or could include the design at different levels of detail. In a case-based planner generating highly complex plans, it is possible to retain the entire plan, or only key pieces, or to preserve full details for parts of the plans and high-level abstractions for others. Likewise, when CBR is applied to tasks such as aiding knowledge capture by supporting concept map construction [9], stored concept map cases could be retained at different levels of completeness. Exploiting this flexibility requires maintenance processes that can perform maintenance at a finer-grained level than simple retention or deletion of cases.

Feature compression is especially appropriate for complex domains in which cases are large, may contain extensive indexing or solution information, and in which partial information—for either indices or solutions—may still be useful. Feature compression prompts the question of when to delete an entire case versus when to achieve comparable space savings by abstracting, deleting, or otherwise compressing some of the features contained in one or more cases in the case base. There is no free lunch: Either method may entail accuracy losses, case deletion by removing what may be the most relevant solution for a problem; feature maintenance by affecting retrieval accuracy or quality of the solutions. The interesting question is how these methods compare.

This paper begins by discussing the range of applicability of flexible feature deletion and its relationship to standard case-base maintenance. It then defines a set of simple feature deletion strategies and evaluates their performance compared to per-case strategies for three domains, two with cases containing varying amounts of information and one with uniform size cases. A comparison of competence as a function of compression supports the value of flexible feature deletion for domains with variable-size cases.

2   When Feature Deletion is Appropriate

Feature compression is appropriate for a particular class of domains: Those in which a particular case can be represented with varying levels of information and still be useful. Even if indexing accuracy is reduced, the retrieved cases will provide value if they are still adaptable to usable solutions with an acceptable level of adaptation effort. Even if some poor retrievals result, they may be acceptable given savings in space; just as per-case maintenance usually involves a tradeoff of case base compactness against competence, feature deletion does as well.

The range of problems to which the case can be applied, and the reliability of its application, may vary with the specific information stored. Consequently, different feature deletion domains will exhibit different tradeoffs between per-case maintenance strategies and feature-maintenance strategies, as well as different tradeoffs between compression and quality. For some domains, such as a regression (numerical predication) task, feature deletion may only be possible for indexing information. In domains for which indexing information is based on many features, it may be possible to reduce case size by removing information about the values of some indices. Note that feature deletion of indices contrasts with the extensive work on selecting indexing vocabularies in the CBR literature, in that feature maintenance is aimed not at selecting an indexing vocabulary or maximizing retrieval accuracy, but instead at selectively compressing indexing information by deleting particular features, potentially from individual cases, with the recognition that some accuracy loss may result.

The deletion done by feature deletion is not necessarily limited to particular indexing dimensions (e.g., deleting the "age" attribute from all patient cases), but alternatively may delete specific attribute values (e.g., deleting the "age" attribute-value pair for specific patients, or only for a particular range of age values, such as those patients who fall into a default set for which age is not considered significant).

Feature deletion for indices could be especially relevant to situations in which extremely rich indexing information is available, such as a case-based agent to respond in a real-time strategy game, or a prediction system for driver behavior, for which the situation in which a plan was applied could be described with extremely rich detail—with fine-grained details which might be helpful to finding the perfect case, but not essential to finding a good case. Likewise, in a movie recommender domain, with movies characterized by their list of actors and the goal of recommending similar movies, a subset of the actors might be sufficient for good retrievals.

2.1 When to Apply Feature Deletion to Indices

Tasks are potential targets for feature deletion of indices if their cases have large indexing structures which can be reduced while retaining an acceptable level of indexing/similarity performance. Specifically, domains are appropriate if:

- *Indexing or similarity assessment depends on information about detail-rich situations from which many features could be generated.* If any low-level features of the current situation, or of a sequence of situations, might be available and potentially be relevant to deciding a response. In such domains, due to the potential for large amounts of indexing information, feature deletion could have significant effects on case base size.
- *Indexing or similarity assessment features are sufficiently closely related that acceptable accuracy is possible after removal of some features.* If features are closely related—even if they are not redundant—feature removal may have limited effects on system accuracy, helping to boost the amount of compression possible per unit of retrieval accuracy loss.

The CBR community has devoted substantial effort to methods for refining the indices used for cases, as well as on developing methods for assigning weights to features for similarity assessment. However, work in index/similarity refinement differs from feature deletion in a key way: The focus of index/similarity refinement is generally increasing retrieval accuracy, rather than compression of case data. Consequently, research on such methods does not address space/accuracy tradeoffs. Feature deletion is a primary focus of research on dimensionality reduction for CBR. However, such deletion is done uniformly across all cases; this work does not attempt selective deletion of a feature from some cases but not others.

2.2 When to Apply Feature Deletion to Solutions

Feature deletion is useful for domains in which the solution to a single problem can capture varying levels of information and still be useful. In such domains, parts of a large or complex solution may be removed or abstracted while still retaining the usefulness of a case, even if the level of usefulness varies with the specific information retained.

For example, as previously mentioned, in case-based planning, certain parts of a plan could be elided or abstracted to reduce storage. When a new planning problem is precisely covered by the retained material, there is no solution quality or efficiency loss. When it is not, the maintenance may result in increased adaptation cost to reconstruct the plan, or some competence could be lost—in weak-theory domains, plan failures could result if adaptation did not generate a perfect solution. However, partial deletion of case contents might still cause less competence loss than deletion of an entire case by per-case methods.

Case-based support for concept mapping [9] provides another example. Concept maps [10] are informal two-dimensional visual representations of concepts and their relationships, representing a particular user's conceptualization of a domain. The goal of support systems is to aid humans using electronic tools to build concept maps, by monitoring the concept map under construction, retrieving a past concept map relevant to the partial concept map they have constructed, and using it to suggest extensions to the concept map. Concept map cases contain rich structures of interconnected concepts, from which deletion of some parts may reduce the range of problems for which suggestions can be provided, but for which the remaining parts are still useful.

We note that for supporting concept map extension, any part of a concept map case may be viewed as the index or the solution, depending on which features are available as the input problem and the context of the retrieval [11]. Thus in the concept mapping domain, the same feature deletion process can be seen as simultaneously maintaining indices and solutions.

3   Bundling Features for Deletion

We can consider cases as composed of a set of primitive features which cannot be further decomposed. In what follows, for simplicity we will consider these to be attribute–value pairs. However, other representations are possible. Both indexing and solution information are defined by sets of features. For example, basic features could be combined to form complex structured cases, from which flexible feature deletion could remove multiple features corresponding to substructures.

Maintenance approaches for case-base compression can be seen as "bundling" different types of information together, to treat as a unit. Traditional per-case maintenance for case-base compression bundles together all features associated with a particular case and deletes the entirety of features associated with a particular case. In contrast, feature-bundled maintenance does an orthogonal bundling, deleting a single feature in all cases for which it appears. Flexible feature deletion can also apply an "unbundled" approach, simply deleting specific features from selected individual cases. To distinguish unbundled individual features from feature-based bundles, we call the individual features of a specific case "case-features." Fig. 1 illustrates the case-bundled, feature-bundled, and unbundled approaches.

Figure 2 summarizes eleven simple candidate strategies for selecting the next case or feature to delete, spanning case-bundled, feature-bundled, unbundled, and hybrid strategies, which we describe in more detail below. Random deletion strategies are included as a baseline. The simplicity of these strategies enables

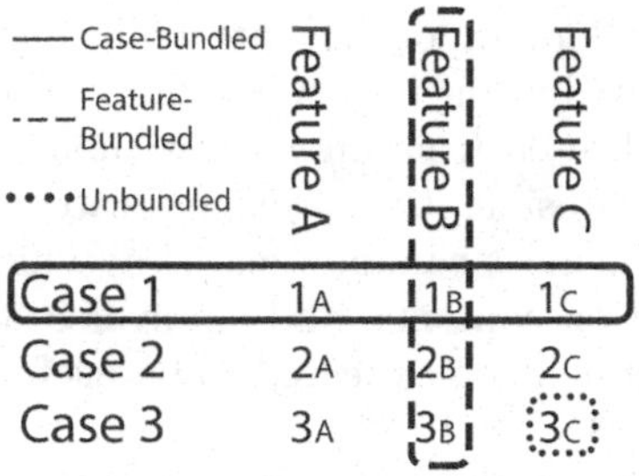

```
──── Case-Bundled      Feature A   ┌Feature B┐   Feature C
___ Feature-
     Bundled
•••• Unbundled
                     ┌─────────────────────────────┐
        Case 1       │  1A      I1B│      1C       │
                     └─────────────────────────────┘
        Case 2          2A      ¦2B│      2C
        Case 3          3A      ¦3B│     ⋮3C⋮
```

Fig. 1. Feature selection with case-bundled, feature-bundled, and unbundled strategies.

comparing case-bundled and feature-bundled strategies on an equal footing. Section 6 discusses future paths for more sophisticated flexible feature deletion strategies.

Strategy	Type of Bundling	Hybrid or Non-Hybrid
Random Case-Features	Unbundled	Non-Hybrid
Random Cases	Case-Bundled	Non-Hybrid
Large Cases	Case-Bundled	Non-Hybrid
Least Coverage	Case-Bundled	Non-Hybrid
Most Reachability	Case-Bundled	Non-Hybrid
Random Features	Feature-Bundled	Non-Hybrid
Rarest Features	Feature-Bundled	Non-Hybrid
Most Common Features	Feature-Bundled	Non-Hybrid
Rarest Cases / Least Coverage	Case-Bundled	Hybrid
Rarest Features / Least Coverage	Unbundled	Hybrid
Rarest Features / Large Cases	Unbundled	Hybrid

Fig. 2. Strategies for selecting the next case, feature, or case-feature to delete

1. **Case-Bundled Strategies**
   Case-bundled strategies follow the traditional CBR compression approach of removing entire cases, i.e., the bundle of features determined by the case. A key question for case deletion is the order in which to delete cases. A classic approach is to consider cases' *coverage* as the set of target problems that a case can solve, and *reachability* as the set of cases that can solve a given target problem [7]. Cases with higher coverage are considered more valuable to preserve; cases with low reachability are considered harder to replace. We consider simple strategies favoring each criterion. Another simple criterion is to include removing largest cases first (aiming to maximize size reduction).

2. **Feature-Bundled Strategies**
   Feature-bundled strategies ignore the boundaries of cases, replacing deletion of cases with deletion of common features across cases. For example, in a

movie recommendation domain, one feature might be the presence of a particular (little-known) individual; if that was unimportant to recommendations, that feature could be deleted from all cases without impairing recommendation performance. We consider the baseline strategy of random deletion, a strategy of removing the most common features (which might be expected to have the least information content), and an inverse strategy of removing the rarest features (which might be expected to be useful in fewer instances).

3. **Unbundled Strategies**
   Unbundled strategies ignore the boundaries of both cases and features. Deletion need not be done uniformly on a per-case or per-feature basis; individual features may be deleted from some cases and retained in others. For example, in the movie domain, the feature corresponding to a particular actor could be deleted only from selected cases (e.g., those in which the actor had a walk-on role). We consider only one basic unbundled strategy, removing random features of random cases.

4. **Hybrid Strategies**
   We also consider three hybrid strategies, each combining two strategies with equal weight (weightings could also be tuned). The strategies are Large Cases / Least Coverage, Rare Features / Least Coverage, and Rare Features / Large Cases. Combining two case-bundled strategies, as in Large Cases / Least Coverage, yields a case-bundled strategy, and combining two feature-bundled strategies yields a feature-bundled strategy. However, combining two differently-bundled strategies (e.g., Rare Features / Least Coverage) yields an unbundled strategy in which the scorings of the constituent parts are used to determine case-features to delete.

We note that different strategies have substantially different computational cost. Case size and feature rarity can be calculated rapidly because they do not require problem solving. However, coverage depends on the ability of a case to solve the problems associated with other cases, so requires more costly testing involving other cases in the case base.

4   Evaluation

To help understand the relationship of per-case and flexible feature deletion strategies, we tested the compression/competence tradeoff for the strategies in Fig. 2, across three domains. Our evaluation addresses two questions:

1. For given compression, how does the retrieval accuracy of the strategies compare?
2. How does the retrieval time change as the number of case-feature pairs decreases, and does this depend on the retrieval strategy?

We hypothesize that at higher compressions, accuracy will tend to decrease for all strategies, but that non-case-bundled maintenance strategies will outperform case-bundled strategies. We also hypothesize that, as the total number of features decreases, retrieval time will decrease as well, with decreases roughly independent of strategy used.

4.1   Test Data

Tests used three data sets, from movie, legal, and travel domains. Movie data was drawn from the Internet Movie Database (IMDb)[1], in which each case is a film or television show, and each feature is an actor in that film or show. The sample contained 100,000 case-feature pairs in 74,720 cases with 38,374 features.

Legal data was extracted from the LegiScan database on the 113th session of the United States Congress[2]. Each case is a bill, and each feature is a sponsor or co-sponsor of a bill. The sample contained 50,000 case-feature pairs in 7,785 cases with 552 features.

Travel data was the CBR Wiki travel package case base[3]. Each case is a travel package and features are the types, prices, regions, etc. This case base contains 14,700 case-feature pairs in 1,470 cases, with 2,902 distinct feature-value pairs.

All features for the IMDb and law domains are Boolean; features correspond to the presence of a particular actor in a film or sponsor of a bill. The features for the travel domain are key-value pairs, which were treated as Boolean features based on whether a particular pair was present.

4.2   Indexing and Similarity Criteria

In the experiments, when features were deleted from case content, the corresponding indices were deleted as well, keeping indices and case content synchronized.

Case similarity was calculated by Jaccard similarity of case-features. For calculating competence, problems were considered to be solved successfully if the system was able to retrieve a case for which the Jaccard similarity of features exceeded 50 %. Additional tests were run for a scenario assuming minimal shared coverage, in which cases were considered to cover only with the closest adjacent case in the original case base, so successful retrieval was defined as the system retrieving the same case retrieved during the initial leave-one-out testing. Results were similar in both conditions. For reasons of space, only the results for traditional similarity are reported here.

4.3   Hybrid Strategies

The hybrid strategies in the experiments rank cases by summing normalized scores corresponding to each of their constituent strategies. The score assigned to a case for Large Cases is the size of that case divided by the size of the largest case in the case base. The coverage score assigned to a case for Least Coverage is the coverage of the case divided by the maximal case coverage. The score for Rare Features is based on the commonality of the feature, defined as the number of cases that contain that feature divided by the number of cases containing the maximally common feature in the case base; rarity of a feature f is $1 - commonality(f)$.

---

[1] http://www.imdb.com/interfaces
[2] https://legiscan.com/
[3] http://cbrwiki.fdi.ucm.es/mediawiki/index.php/Case_Bases

4.4   Evaluation Procedure

The evaluation first establishes baseline performance by leave-one-out testing for the entire case base. Next, performance is tested for compression to nine different case base sizes, ranging from 90 % to 10 % of the case base. For each test, the entire original case base is used as test problems, and a test problem is considered solved if there exists in the compressed case base a case (other than the test case) within the 50 % similarity threshold.

When compressing the case bases, if the desired number of case-feature pairs does not fall exactly on a boundary between cases, then the single case in which this division falls is unbundled to delete features within a case.

4.5   Experimental Results

Figures 3, 4 and 5 show accuracy after each round of maintenance. The graphs compare the eleven strategies across the IMDb, law, and travel domains. For readability, the graphs are divided into three parts with the same horizontal and vertical scales. The third graph compares the best four strategies from the other two graphs. Each type of bundling has a different type of connecting line. Solid lines indicate case-bundled strategies, dashed lines indicate feature-bundled strategies, and dotted lines indicate unbundled strategies.

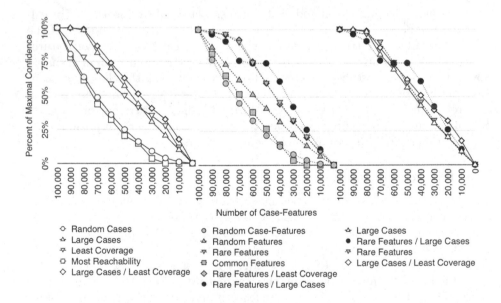

Fig. 3. Competence retention for varying compression levels for the IMDb case base

Figure 3 shows results for the IMDb data, for which the best four strategies were Large Cases, Rare Features / Large Cases, Rare Features, and Large Cases /

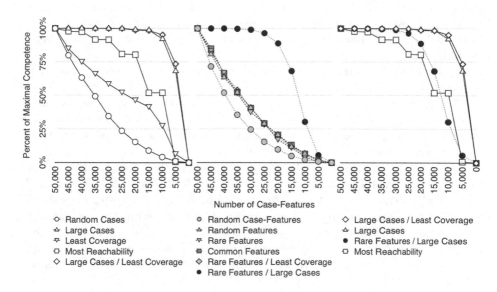

Fig. 4. Competence retention for varying compression levels for the law case base

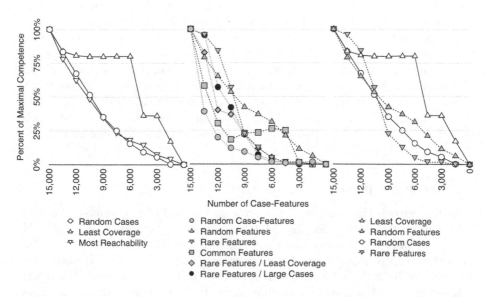

Fig. 5. Competence retention for varying compression levels for the travel case base

Least Coverage. Three of the best strategies consider the size of the cases, which supports having maintenance consider not only the benefit of retaining a case (its solution coverage) but also its storage cost. Two of the best strategies are hybrid strategies, and two are non-case-bundled. The worst strategy was Most Reachability.

Given the established importance of coverage, that Least Coverage is outperformed by Large Cases on the IMDb and law data sets might seem surprising, but this is explained by the substantial case size variation in these domains. For example, the IMDb case base includes multi-episode soap operas such as *The Bill*, which span hundreds of actors but also include numerous relatively unknown actors who never appear widely.

Figure 4 shows results on the law data, for which the best four strategies were Large Cases / Least Coverage, Large Cases, Rare Features / Large Cases, and Most Reachability. Note that these overlap with three of the best strategies on the IMDb case base but in a different order. The worst strategy was Random Case-Features. The law data set has a much smaller number of features than the IMDb data set, and we hypothesize that its features are more likely to have comparable importance, making random deletion more likely to remove significant content.

Figure 5 shows results for the travel data. Because all cases are initially the same size, the strategies Large Cases and Large Cases / Least Coverage do not apply, so are omitted from the graphs. However, the hybrid strategy Rare Features / Large Cases is still applicable, because as the Rare Features strategy deletes features, only cases with those features will be compacted, resulting in different case sizes. The best strategies were Least Coverage, Random Features, Random Cases, and Rare Features. That deleting cases with least coverage is best is consistent with the key role coverage has has been ascribed in case-base maintenance research. That Random Features is second is surprising, but could be explained if many features in this domain have comparatively low information content. Although Rare Features is one of the top four strategies, its performance is quite poor, which could correspond to rare features tending to be important for distinguishing relevant cases. As with the other two data sets, two of the best strategies were non-case-bundled. However, in contrast, none of the best strategies were hybrid.

4.6   Retrieval Speed

Figure 6 compares the retrieval times after each round of maintenance for each of the four best strategies for the IMDb case base, for retrieval from a MySQL database. It also includes Random Case-Features as a baseline. The Average line shows the mean retrieval time of the five strategies in the graph. All tests were run on a MacBook Pro with a 2.5 GHz Intel Core i5 processor and 8 GB of RAM.

Random Case-Features, the baseline, gave the best retrieval times, and Rare Features, the only feature-bundled strategy, gave the worst. Both of the case-bundled strategies, Large Cases and Large Cases / Least Coverage yield similar retrieval times, but the two unbundled strategies, Rare Features / Large Cases

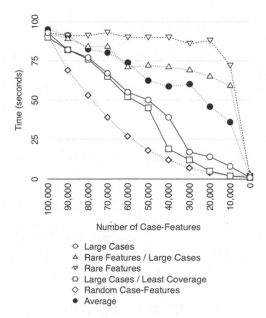

Number of Case-Features

△ Large Cases
△ Rare Features / Large Cases
▽ Rare Features
⊟ Large Cases / Least Coverage
◇ Random Case-Features
● Average

Fig. 6. Comparison of the retrieval times after each round of maintenance between the four best strategies on the cinema data set

and Random Case-Features, yield very different retrieval times. Most of the strategies have a fairly linear decline, but Rare Features declines slowly until the 10,000 case-feature mark where it drops abruptly. Because the retrieval function uses Jaccard similarity, retrieval time depends on the number of case features in the intersection between cases. However, the rarest features would seldom fall into any intersections, which explains why removing them has the least effect on retrieval time.

5 Related Work

Case-based reasoner maintenance [3] is an active area of CBR research. Much of this work develops methods to compress the case base, such as competence-based case deletion [4], deletion methods taking class boundaries into account by considering local complexity [12], optimizing the tradeoff between size and accuracy [13], deletion aimed at preserving diversity [14], and strategies for case retention and forgetting (e.g., [5, 15–17]. Such methods differ from flexible feature deletion that they retain or delete entire cases without adjusting case contents.

Research on maintenance of case contents has generally focused on quality improvement rather than case base compression (e.g., [18, 19]). However, research on case abstraction research, in aiming to compact the case base by removing concrete cases subsumed by abstractions [20], can be seen as in the spirit of replacing cases with more compact versions.

Flexible feature deletion applies to indexing features as well as cases. Maintenance of indexing features has been extensively studied in CBR, applying methods such as feature deletion, addition, and reweighting, but again with the goal of improving retrieval accuracy rather than decreasing the amount of storage required for the indices themselves (e.g., [21–24]). Feature set reduction has been combined with case selection, to improve accuracy while compressing the case base [25].

6   Future Research Questions for Feature Deletion

The feature deletion approach raises a rich range of questions for fully exploiting its potential. A key question is how to develop knowledge-based feature deletion rules, especially for flexible feature deletion for complex structured cases. Other questions include how feature deletion strategies should interact with the indexing and adaptation knowledge containers, how feature deletion can preserve case integrity, and how feature deletion should be reflected in case provenance and explanation.

– *Coupling Feature Deletion with Index Maintenance:* As case contents are deleted, the relevance of case indices may change. Consequently, feature deletion may need to be accompanied by index maintenance to assure that as cases are compressed the system still retrieves the most similar cases. Feature weight information might be used to suggest features which could be deleted with limited harm.
– *Benefiting from the Relationship of Feature Deletion to Case Adaptation:* Feature deletion can be seen as a form of "before the fact" adaptation of cases, in which the adaptation is driven not by a new problem to solve, but by a combination of (1) compression goals, and (2) performance goals. Richer feature deletion methods could draw on a CBR system's adaptation procedures to perform operations beyond simple deletion of case components, such as abstractions or substitutions of alternatives requiring less space. Enabling such methods requires reasoning about the competence effects of replacing a case with various candidate adapted versions, as well as performance effects (whether replacing a case with a given compressed version will decrease problem-solving speed), and the balance to strike between them.
– *Maintaining Case Integrity Despite Feature Deletion:* Another question is the relationship of feature deletion to the cohesiveness of a case. From the early days of CBR, an argument for CBR has been that cases can implicitly capture interactions among case parts. Deleting portions of a case risks some of that cohesion, making it a concern to address in feature deletion strategies. That case adaptation faces the same risks but is effective supports optimism for some levels of compression, and research on hierarchical CBR has supported the usefulness of sometimes considering subparts of complete cases individually. However, how much compression can be done without excessive harm to case integrity, and how to manage the process to avoid such harm, are interesting questions.

– *Reflecting Feature Deletion in Provenance and Explanation:* Because feature deletion results in stored cases which differ from the cases originally captured, it (like case adaptation) may weaken the ability to justify proposed solutions by past experience. Likewise, changes from the original cases may make it difficult to apply provenance-based methods for predicting solution characteristics such as solution accuracy and trust (e.g., [26]). Addressing these complications might require maintaining records of case maintenance process as part of the provenance trace used for explanation, as well as reasoning about (and presenting to users) information about the parts of the case which have been affected by feature maintenance.

7   Conclusion

This paper has proposed a new case-base maintenance approach, flexible feature deletion, which questions the assumptions that cases are of uniform size and that maintenance must treat cases as unitary objects. Flexible feature deletion enables selective deletion of case contents rather than restricting deletion to the case level. It has illustrated tasks for which flexible feature deletion may be desirable, such as domains in which reasoning can be done with different amounts of information, in which flexible feature deletion enables selectively compressing different parts of different cases. Its experimental results show that case-based maintenance may need to change when case contents are non-uniform; in such settings feature-based strategies may give better accuracy than per-case strategies, and that total case-base size and retrieval times may not always be aligned, giving a space/time tradeoff which it may be possible to exploit.

The paper focuses primarily on knowledge-light maintenance strategies. Interesting future directions are to refine the strategies tested here with additional knowledge, for example, leveraging case adaptation knowledge, and to explore when other knowledge-light techniques for compression of cases and feature bundlings could yield useful maintenance strategies.

References

1. Francis, A., Ram, A.: Computational models of the utility problem and their application to a utility analysis of case-based reasoning. In: Proceedings of the Workshop on Knowledge Compilation and Speed-Up Learning (1993)
2. Smyth, B., Cunningham, P.: The utility problem analysed: a case-based reasoning perspective. In: Smith, I., Faltings, B. (eds.) EWCBR-1996. LNCS, vol. 1168, pp. 392–399. Springer, Heidelberg (1996)
3. Wilson, D., Leake, D.: Maintaining case-based reasoners: dimensions and directions. Comput. Intell. **17**(2), 196–213 (2001)
4. Smyth, B., Keane, M.: Remembering to forget: a competence-preserving case deletion policy for case-based reasoning systems. In: Proceedings of the Thirteenth International Joint Conference on Artificial Intelligence, San Mateo, pp. 377–382, Morgan Kaufmann (1995)

5. Muñoz-Ávila, H.: A case retention policy based on detrimental retrieval. In: Althoff, K.-D., Bergmann, R., Branting, L.K. (eds.) ICCBR 1999. LNCS (LNAI), vol. 1650, pp. 276–287. Springer, Heidelberg (1999)

6. Zhu, J., Yang, Q.: Remembering to add: competence-preserving case-addition policies for case base maintenance. In: Proceedings of the Fifteenth International Joint Conference on Artificial Intelligence, pp. 234–241, Morgan Kaufmann (1999)

7. Smyth, B., McKenna, E.: Building compact competent case-bases. In: Althoff, K.-D., Bergmann, R., Branting, L.K. (eds.) ICCBR 1999. LNCS (LNAI), vol. 1650, pp. 329–342. Springer, Heidelberg (1999)

8. Wilson, D., O'Sullivan, D.: Medical imagery in case-based reasoning. In: Perner, P. (ed.) Case-Based Reasoning on Images and Signals. Studies in Computational Intelligence, vol. 73, pp. 389–418. Springer, Heidelberg (2008)

9. Leake, D., Maguitman, A., Reichherzer, T.: Experience-based support for human-centered knowledge modeling. Knowl. Based Syst. **68**, 77–87 (2014)

10. Novak, J., Gowin, D.: Learning How to Learn. Cambridge University Press, New York (1984)

11. Leake, D., Maguitman, A., Reichherzer, T., Cañas, A., Carvalho, M., Arguedas, M., Brenes, S., Eskridge, T.: Aiding knowledge capture by searching for extensions of knowledge models. In: Proceedings of the Second International Conference on Knowledge Capture (K-CAP), New York, pp. 44–53, ACM Press (2003)

12. Craw, S., Massie, S., Wiratunga, N.: Informed case base maintenance: a complexity profiling approach. In: Proceedings of the Twenty-Second National Conference on Artificial Intelligence, pp. 1618–1621, AAAI Press (2007)

13. Lupiani, E., Craw, S., Massie, S., Juarez, J.M., Palma, J.T.: A multi-objective evolutionary algorithm fitness function for case-base maintenance. In: Delany, S.J., Ontañón, S. (eds.) ICCBR 2013. LNCS, vol. 7969, pp. 218–232. Springer, Heidelberg (2013)

14. Lieber, J.: A criterion of comparison between two case bases. In: Haton, J.-P., Keane, M., Manago, M. (eds.) EWCBR-1994. LNCS, vol. 984, pp. 87–100. Springer, Heidelberg (1995)

15. Ontañón, S., Plaza, E.: Collaborative case retention strategies for CBR agents. In: Ashley, K.D., Bridge, D.G. (eds.) ICCBR-2003. LNCS, vol. 2689, pp. 392–406. Springer, Heidelberg (2003)

16. Romdhane, H., Lamontagne, L.: Forgetting reinforced cases. In: Althoff, K.-D., Bergmann, R., Minor, M., Hanft, A. (eds.) ECCBR 2008. LNCS (LNAI), vol. 5239, pp. 474–486. Springer, Heidelberg (2008)

17. Salamó, M., López-Sánchez, M.: Adaptive case-based reasoning using retention and forgetting strategies. Knowl. Based Syst. **24**(2), 230–247 (2011)

18. Racine, K., Yang, Q.: Maintaining unstructured case bases. In: Leake, D.B., Plaza, E. (eds.) ICCBR 1997. LNCS, vol. 1266, pp. 553–564. Springer, Heidelberg (1997)

19. Salamó, M., López-Sánchez, M.: Rough set based approaches to feature selection for case-based reasoning classifiers. Pattern Recogn. Lett. **32**(2), 280–292 (2011)

20. Bergmann, R., Wilke, W.: On the role of abstraction in case-based reasoning. In: Smith, I., Faltings, B. (eds.) EWCBR-1996. LNCS, vol. 1168, pp. 28–43. Springer, Heidelberg (1996)

21. Arshadi, N., Jurisica, I.: Feature selection for improving case-based classifiers on high-dimensional data sets. In: Proceedings of the Eighteenth International Florida Artificial Intelligence Research Society Conference (FLAIRS-2005), pp. 99–104, AAAI Press (2005)

22. Fox, S., Leake, D.: Learning to refine indexing by introspective reasoning. In: Veloso, M., Aamodt, A. (eds.) ICCBR-1995. LNCS, vol. 1010, pp. 431–440. Springer, Heidelberg (1995)

23. Muñoz-Avila, H.: Case-base maintenance by integrating case-index revision and case-retention policies in a derivational replay framework. Comput. Intell. **17**(2), 280–294 (2001)

24. Zhang, Z., Yang, Q.: Towards lifetime maintenance of case base indexes for continual case based reasoning. In: Giunchiglia, F. (ed.) AIMSA 1998. LNCS (LNAI), vol. 1480, pp. 489–500. Springer, Heidelberg (1998)

25. Li, Y., Shiu, S., Pal, S.: Combining feature reduction and case selection in building CBR classifiers. IEEE Trans. Knowl. Data Eng. **18**(3), 415–429 (2006)

26. Leake, D.B., Whitehead, M.: Case provenance: the value of remembering case sources. In: Weber, R.O., Richter, M.M. (eds.) ICCBR 2007. LNCS (LNAI), vol. 4626, pp. 194–208. Springer, Heidelberg (2007)

A Case-Based Approach for Easing Schema Semantic Mapping

Emmanuel Malherbe[1,2]($\boxtimes$), Thomas Iwaszko[2], and Marie-Aude Aufaure[2]

[1] Multiposting S.A.S.U, 3 Rue Moncey, 75009 Paris, France
emalherbe@multiposting.fr
[2] CentraleSupélec - M.A.S. Laboratory, Grande Voie des Vignes,
92295 Chatenay-Malabry, France
tiwaszko@gmail.com, marie-aude.aufaure@centralesupelec.fr

Abstract. Given two text schemas, how can one map each item from the first schema to an item of the second one, in order to have the best semantic correspondence? While this latter criterion has to be defined by a human, in this paper we present a new application of the Case-Based Reasoning (or CBR) methodology, that helps the user in finding a good match thanks to a score function taking into account previous results (i.e. solved cases). Such a technique makes the schema mapping easier and less error-prone. The proposed solution is in use in an industrial context, at the Multiposting firm.

Keywords: Case based reasoning · Schema mapping · Language processing · Data mining

1 Introduction

Nowadays, business related websites make information about jobs and companies available to anyone. Such websites that present complex data use specific nomenclatures or schemas to structure data and represent it visually. This abundance of schemas describing a same area of interest raises industrial issues. Indeed, it becomes increasingly difficult to aggregate data from different sources, or to understand how distinct schemas are related one to each other. The communication and cooperation between heterogeneous schemas has become a crucial issue for applications in the domain of e-recruitment, but also e-business or life sciences [20].

A possible approach to resolve this issue is to match schemas manually and pinpoint equivalences between concepts/items. This process is called name-based schema mapping and focuses only on textual contents and the associated semantic meaning. We want every item of an initial schema to be mapped to an item of a destination schema. Given an item of an initial schema, the problem is thus to find its semantic equivalent among the items of the destination schema. At the Multiposting firm, where employees face the incoherence between schemas in the e-recruitment domain, this type of problem has to be solved very often.

© Springer International Publishing Switzerland 2015
E. Hüllermeier and M. Minor (Eds.): ICCBR 2015, LNAI 9343, pp. 228–243, 2015.
DOI: 10.1007/978-3-319-24586-7_16

Schemas in the e-recruitment domain include descriptions of: contract types, job locations, job sectors, diploma requirements.

In this paper, we show how by relying on a corpus of schemas that were matched manually, one can match match new pairs of schemas more easily. In other words, we show that schema mapping greatly benefits from the use of Case-Based Reasoning (CBR). The contributions presented in this article are threefold:

– we formalize the problem of schema-matching and show that it boils down to multiple independent sub-problems
– we instantiate the CBR process to those sub-problems in order to solve them and we perform first benchmarks
– we propose changes to improve the CBR system performance

Prior to the Sects. 4, 6, 7 that describe respectively each one of these steps, in Sect. 2 we present the industrial context; in Sect. 3 we discuss motivations and the scientific background for this work.

2 Industrial Context

Recruitment websites (such as Monster, Linkedin or Indeed) require recruiters to fill out many fields based on a specific schema/nomenclature. Those fields include but are not limited to: job sector, contract type, years of experience, geographical area. When posting a job offer on n websites, the same kind of task has to be done n times.

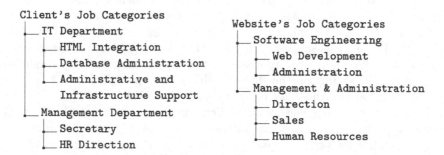

Fig. 1. Example of the schema matching problem: given the text schemas $A^{example}$ in the left-hand side and $B^{example}$ in the right-hand side, what is, for each item in $A^{example}$ the best semantic equivalent in schema $B^{example}$?

Multiposting is a firm based in Paris and specialized in on-line recruitment. Various clients worldwide and its advanced techniques allow the firm to post about 3 millions announces a year on the Internet. To face the increasing number of recruitment websites Multiposting aims at helping recruiters by spreading their job offer; posting it on several job boards at a time. In order to relief the recruiter from a redundant and time-consuming task, the firm proposes a unique interface to be filled once with job offer's information (description, contract type, experience required...) specified with respect to the client company nomenclatures. To spread the job campaign quickly, Multiposting needs to adapt the information to all existing job websites including new ones. This is done by mapping the recruiter's schema into the website corresponding one, as described in Fig. 1. By doing so, the recruiter only needs to fill its internal fields (job sector, contract type, ...), and does not need to fill the specific fields of each recruitment website.

Multiposting has to constitute and maintain over the years a huge database, containing numerous semantic equivalences between schemas/nomenclatures of different websites and companies' system. Before the tool called smartmapping was developed, the employees of Multiposting used to define equivalences between schemas manually with no computer assistance at all, item by item. This manual process was difficult, as dozens of schemas needed to be mapped one to another each week, due to the constant increase of Multiposting clients and supported websites. This justifies the effort to design a new semi-automatized tool for schema mapping. The resulting software tool has been implemented and is now used daily through an interface presented in Sect. 7.3.

3   Related Work

The problem of schema mapping has received a lot of attention over the past decade, with the growth of Internet and database integration (Madhavan *et al.* [9]), a common application being databases schemas or website XML schemas. There exists several approaches to schema mapping: to focus on instances, on the structure of the schema, on the data type constraints, on the description or the name of the columns. Mapping reuse is also a possibility, as discussed by Rahm *et al.* [3,10]. In this case, schemas are linked thanks to a golden standard schema. However the golden standard approach is time consuming, and can result in a loss of quality depending on the choice of the golden standard.

Name-based approaches have been proposed by several authors (Shvaiko *et al.* [11], Jean-Mary *et al* [7]). The similarity between names are direct, the similarity measures uses the n-grams of words, or the Levenshtein distance: it assumes that names are similar for both schemas, and will not work for two strictly different words with same meaning that we tackle by our algorithm. The use of WordNet for extended semantic similarity seems to give good results, as presented by Manakanatas *et al.* [18]. They propose a similarity measure by separating names into tokens, and leverage WordNet similarity between them. They also reuse already identified mapping, but only when names to be mapped are exactly the same as for previous mapping, which is very restrictive. Our approach goes further in the mapping reuse by learning the similarity measure from previous mappings.

Madhavan *et al.* [5] proposes to train a machine learning algorithm to learn the similarity between names; however, it relies on inferring negative cases (wrong mapping) on every schema mapping of our data-set. This inference would generate negative cases of uneven quality, and it would lead to tens of millions of cases for our corpus (see Table 1), whereas our method is much lighter with 200,000 positive cases.

Textual CBR is a sub-field of CBR and relies on texts to gather information to solve problems (Richter *et al.* [12]). It often leverage the bag of words (BOW) representation of documents, such as in Information Retrieval, but differs from this field as it focuses on solving a problem. A promising alternative to the BOW is proposed by Dufour-Lussier *et al.* [13] by extracting structured features from texts. It requires long documents and is thus inapplicable in the case of schemas, as the labels are only a few words long.

Textual CBR can have the advantage to gather information from a large amount of documents, but of uneven quality, requiring a particular focus on the adaptation of solutions. In the system of Lamontagne *et al.* [14], the cases are emails, for a given question they adapt a solution from a retrieved email case by selecting relevant portions and modify them before giving the final answer. This approach can't be extended to schema as names are too short. An alternative for building a solution is to display ranked suggestions to user and request validation such as in [19] and in the approach proposed in this paper.

Sani *et al.* [15] study the terms relatedness, that could be applied to name-based schema mapping. Their CBR uses a word generalization, based on higher semantic concepts. There is no existing word generalization for recruitment sector, when dealing with contract types, job sectors or qualifications. A way to get this word generalization would be by computing co-occurrences in a relevant corpus, and could also lead to a higher level co-occurrence computation such as done by Chakraborti *et al.* [16]. Again, a relevant corpus for recruitment with documents mixing contract types, qualification names or job industries does not exist, leading us to tackle the name-based schema mapping issue by leveraging a corpus of mappings.

4   Formalizing the Problem of Schema Mapping

Our goal in this section is to give a more formal description of the text schema mapping problem, using a representation of schemas relying on the vectorization of documents.

4.1   Prerequisites: Tools for Representing Textual Documents

To represent documents, we use the bag-of-words model described in the text retrieval literature by Salton *et al.* [1]. Before computing such a representation, we process textual documents thanks to common methods, detailed in [4]. We first remove accents and enforce lowercase, before separating the text into words, also denoted tokens. We eventually stem these words to keep only the semantic roots; we used the French stemmer of Porter [21] because the most part of the real-world data-set we use is in French.

After processing documents, we use the bag-of-words model with binary weights, which again is a common technique [1]. This allows us to represent documents as binary vectors $d = (d_1, \ldots d_n)$, $d \in \{0,1\}^n$, where n is the number of words in the vocabulary V. The set V contains every word that occurs in any of all considered documents.

Using binary weights means that we ignore the number of times a word occur, we focus only on its appearance/its absence. Thus, for $1 \leq i \leq n$, d_i is a binary value depending on whether the word of V indexed by i appears in the modeled text or not. From now on, the term document designates such vectors $d \in \{0,1\}^n$.

Documents Similarity: As explained in the previous subsection, documents are vectors representing documents. To compare documents one to another, we compute the cosine similarity often used in the data mining literature [1,4]. We denote f this similarity function, whose definition is:

$$f(d, d') = cos(\theta) = \frac{\langle d, d' \rangle}{\|d\| \, \|d'\|} \quad \in [0, 1] \tag{1}$$

Where $\langle d, d' \rangle$ is the scalar product between two vectors, $\langle d, d' \rangle = \sum_{i=1}^{n} d_i \times d'_i$ And $\|d\| = \sqrt{\langle d, d \rangle}$ the corresponding norm of vector d. The measure $f(d, d')$ approaches the percentage of words in common in the two documents.

4.2 Representation of Schemas and Items

We firstly introduce here the notion of item, which is a single entity of the schema (a node if one considers the schema as an tree). Items are for instance contract types, job sectors or experience levels. To define an item, we will leverage the *name of the node*, which is a textual document. We also consider the name of the *parent node*, to capture the hierarchy of the schema. This parent name gives the semantic context of the item. For clarity, we will only focus on such 2-level items, but higher levels of the hierarchy could be taken into account, by considering for instance the name of the *grandparent node*. An *item* is thus defined as a pair of documents $a = (a_1, a_2)$, constituted by the entity name a_1, and the name a_2 of its parent. In the case where there is no parent for the considered node, the document a_2 is equal to the null vector $\mathbf{0}$, which represents the empty text.

This definition of items allows to represent a schema as a set of items. For instance, the left-hand side schema in the Fig. 1 is represented as the following items set:

$$A^{example} = \{(\text{``}HTML\ Integration\text{''}, \text{``}IT\ Department\text{''}),$$
$$(\text{``}Database\ Administration\text{''}, \text{``}IT\ Department\text{''}),$$
$$(\text{``}Administrative\ and\ Infrastructure\ Support\text{''}, \text{``}IT\ Department\text{''}),$$
$$(\text{``}HR\ Direction\text{''}, \text{``}Management\ Department\text{''}),$$
$$(\text{``}Secretary\text{''}, \text{``}Management\ Department\text{''})$$

And similarly for the right-hand side schema of Fig. 1:

$$B^{example} = \{(\text{``}Web\ Development\text{''},\ \text{``}Software\ Engineering\text{''}),$$
$$(\text{``}Administration\text{''},\ \text{``}Software\ Engineering\text{''}),$$
$$(\text{``}Direction\text{''},\ \text{``}Management\ \&\ Administration\text{''}),$$
$$(\text{``}Sales\text{''},\ \text{``}Management\ \&\ Administration\text{''})\},$$
$$(\text{``}Human\ Resources\text{''},\ \text{``}Management\ \&\ Administration\text{''})$$

In this paper, for clarity, *schema* will equivalently denote the formal schema (on the top) and the corresponding items set. Thus, $B^{example}$ will be referred as a schema. Schemas will be written in capital letters, such as A, and we can write $a \in A$ where a is an item as defined above.

4.3 Formal Problem Statement

Given two schemas A, B the general problem tackled in this article aims at associating one by one each item $a \in A$ to its "equivalent" item in the schema B, according to the meaning of a. We solve this global problem by considering successively the items $a \in A$; thus our schema to schema mapping boils down to solving several sub-problems, each of them being denoted by a triplet $pb = (a, A, B)$. Sub-problem pb consists in finding the item in schema B that is the closest to the meaning of a in schema A. A will be referred as the *initial schema* and B as the *destination schema*. We assume the problem have systematically a solution: it is not true in theory, but it is an industrial constraint that every item in A finds an equivalent in B. In practice, the schemas to be mapped describe similar knowledge and the assumption is verified.

We can compare the problem and its solution (a, A, B, b) to the definition of mapping by Bouquet *et al.* [17]. Similarly to them, we will define in Eq. 4 of Sect. 5 a *score* that estimate the quality of the mapping, which correspond to what they refer to a degree of trust to the mapping. On the other hand, it is worth noticing that contrary to them, we explicit the schemas as part of the problem, and we consider our problem as *asymmetric*.

For instance, let us consider the problem of finding a semantic equivalent of the first item in the left in Fig. 1 among the schema in the right. If we write the first item as $a^{example} = (\text{``}HTML\ Integration\text{''},\ \text{``}IT\ Department\text{''})$, this problem is formalized as $pb = (a^{example}, A^{example}, B^{example})$. The solution of this target problem would then be $(\text{``}Web\ Development\text{''},\ \text{``}Software\ Engineering\text{''})$. In the following, we denote $b^{example}$ this item, where $b^{example} \in B^{example}$.

5 Instantiating the CBR Process to Schema Mapping

To tackle the problem, we use the CBR methodology [2], with a large case base populated manually (see figures of Table 1). It is important to notice that the proposed CBR system provides a computer-assisted schema matching and is not fully automatic.

Formally, we denote C the case base. A case $c \in C$ is simply a pair: $c = (pb, sol)$ where $pb = (a, A, B)$ describes the mapping of a single element and $sol \in B$ the previously retained solution.

The CBR methodology is applied "as is" with the four standard steps: Retrieve, Reuse, Revise, Retain. In our case, the last step *Retain* is easy: it simply consists in saving the newly constructed case in the base. That's why below, we give details only on the three first steps of the CBR methodology.

5.1   The Retrieve Step

First of all, we define a function for comparing two items one to another. Given two arbitrary items a, a', the similarity between them is defined such as:

$$g(a, a') = \frac{f(a_1, a_1') + f(a_2, a_2')}{2} \quad \in [0, 1] \tag{2}$$

All values computed by this function belong to the interval $[0, 1]$, with 1 corresponding to a perfect match. The first term $f(a_1, a_1')$ computes the similarity between the leaves of each item, whereas the second term compares the parents. For instance, with items introduced in Sect. 4.3, we have:

$$
\begin{aligned}
g(a^{example}, b^{example}) = & f(\text{``HTML Integration''}, \text{``Web Development''})/2 \\
& + f(\text{``IT Department''}, \text{``Software Engineering''})/2 \\
= & \, 0
\end{aligned}
$$

This example also shows us the need to go beyond a *direct match* between items, which leads here to a zero similarity whereas items' meanings are related.

Then, to determine what is the useful case for solving a new problem that occurs, we introduce an inter-problem similarity function. It is simply defined as:

$$sim(pb, pb') = g(a, a') \tag{3}$$

When we have $sim(pb, pb') = 1$, it implies that: $a = a'$. One notes that this similarity only takes into account the item a, forgetting schemas A, B when retrieving useful cases. We will propose similarity functions in the Sect. 7 that tackle this limit.

Thanks to this function, for every new problem pb' that occurs, we can determine which existing case $c = (pb, sol)$ stored in the case base is the most useful to help us solve pb', retrieve it and exploit the information provided by its solution, as explained in the next subsection.

5.2   The Reuse Step

In this step, we want to exploit the most similar case found during the *Retrieve* step. However, if we consider a source problem $pb' = (a', A', B')$, we do not necessarily have $B = B'$. Thus, it is not possible to take the corresponding source solution sol' as our current solution because $sol' \notin B$; we need an adaptation to find a solution in B, which constitutes our current possible solutions.

Furthermore, any CBR system aims at improving its "problem solving capacity" over time. In our context, proposing a ready-made solution for each mapping would likely fail to enhance the system's performance, because of the fact that distinguishing a good from a bad mapping is a matter of semantics and it thus seems not reasonable to fully automatize the schema matching.

As a consequence, the revise step is necessary in our system, but it is greatly eased by the reuse of retrieved cases: we can compute from them a score function estimating the relevancy of the possible solutions. To do so, we retrieve not one but the k most similar cases from the base, in order to use more information to rank possible mappings. Let us denote C' the subset of C that contains the k most similar cases, with respect to the similarity function g. In our experiments $k = 100$ gave satisfying results. For clarity, every case $c' \in C'$ will be written $c' = (pb', sol')$, to differ from the target problem pb.

Given the target problem $pb = (a, A, B)$, a ranking of the possible solutions $b \in B$ can be computed using the score function $score(pb, b) \in \mathbb{R}$ defined as:

$$score(pb, b) = \max_{c' = (pb', sol') \in C'} sim\left(pb, pb'\right) \times g\left(b, sol'\right) \qquad (4)$$

The score function returns a value in the interval $[0, 1]$. The idea is to find a source problem pb' similar to pb such that the corresponding solution sol' is similar to item b - the first similarity being evaluated by the function sim (problems similarity) and the second one by the function g (items similarity).

5.3 The Revise Step: Ranking Suggestions to the User

Previously, we stated that the user has to input new mapping information continuously to improve the system over time. That is why our *Revise* step consists in asking the right mapping to the user, and most important: make this task as fast and easy as possible thanks to the previously computed scores. Given the problem $pb = (a, A, B)$, items b of B are ranked with respect to $score(pb, b)$, and the user chooses manually which element of B corresponds to a. The ranking of all elements of B helps a lot, especially when the cardinality of B is big.

6 Review of the First Results

6.1 Experimental Data-Set

As explained in Sect. 2, schemas mapped at Multiposting are nomenclatures of e-recruitment websites. They are in different languages, mainly French, English, German, Spanish and Polish. They represent different kinds of information, and for a good part of them Multiposting employees have manually assigned them to categories such as "contract type" of "job function". Moreover, many nomenclatures are specialized, such as university degrees, an e-recruitment for managers or for disabled people. However this diversity of schemas should be handled by our algorithm, retrieving only useful cases.

Characteristics of our data-set are listed in the Table 1. The schemas are annotated with 6 different categories. Note that we also have an "unknown"

Table 1. Characteristic from our data-set, by schema category: number of terms, *average* number of items per schema, number of item pairs mapped, of schema pairs mapped and number of schemas.

Schema category	All	Contract	Study	Experience	Location	Sector	Function
Number of terms	99875	713	1736	288	36960	5120	12541
# items per schema	87	7	23	7	3152	55	227
Schema mappings	3,356	182	96	71	18	340	691
Item mappings	215,701	1,330	839	629	2,017	19,926	88,581
Schemas concerned	2,257	184	94	69	21	205	296

schema category in our data-set for uncategorized schema. The same table also gives statistics for the full case base, resulting of the aggregation of all categories.

6.2   Performance Criteria

To evaluate the performance of our CBR approach, and compare the variants, we used a cross validation on the data-set. The Leave-one-Out Cross Validation is commonly used for testing a CBR algorithm [6], and in more general for machine learning algorithms. To cross validate our system, at each fold, we take out a schema mapping from the case base, and we run the algorithm on the excluded mapping to see the ranks of the items selected.

From this cross validation, we deduce the relevancy of the suggestions. To do so we compute the precision at k which is a metric from information retrieval systems [8]:

$$precision@k = \mathbb{E}(\text{solution is in the k top ranked items}) \in [0, 1]$$

This value is computed by counting the number of test mappings for which the algorithm had the validated suggestion in the top k, divided by the number of test mappings. One notes that the precision at k increases with k; in our experiments, we computed $precision@k$ for $k = 1..10$.

6.3   Evaluation of the First Results

We computed the $precision@k$ from the cross-validation. It is displayed in Fig. 2, with k as abscissa. From this figure, we can conclude that the case-based reasoning applies to our problem; for instance, in 70 % of the runs, the validated mapping has been ranked in the 5 first suggestions. However, the performance seems still quite low: we note that the algorithm firstly suggested a correct mapping in only 36 % of the case. As a consequence, we cannot imagine a totally automatic CBR taking the mapping with highest $score(pb)$ as solution.

Figure 3 shows the precision of the first suggestion, with schemas mappings grouped with respect to their category. The geographical location and contract

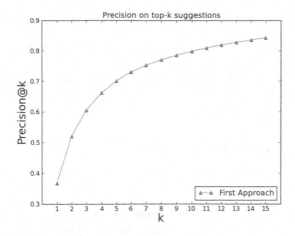

Fig. 2. Precision on top k suggestions for this first approach.

type are the categories with highest precision; these problems are more simple, it mainly maps synonyms - the location could even be mapped automatically by taking the first suggestion as semantic equivalent. On the contrary, job sectors and job functions are more difficult to map: the problem is to find semantic equivalents, that can be synonyms, but also hyponyms or hypernyms, for instance. As it could be expected, the sectors, which are less precise than functions, are better suggested by our CBR.

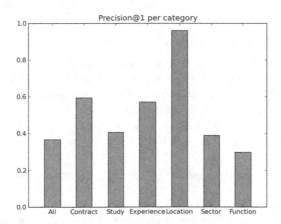

Fig. 3. Precision on the first suggestion, by schema categories.

7   System Improvements and Extensive Benchmarking

In this section, we explore some improvement and their effect on the system's performances. In the previous section, we had to make arbitrary choices concerning the definition of the inter-problem similarity, and the score function used in the ranking. In the following, we propose more sophisticated definitions in order to make the CBR system better.

7.1   Alternative Inter-problem Similarity

First of all, let us introduce an operation to compare not only items but whole schemas one to another. As stated in Sect. 4, we did not exploit the schemas A or B to define the inter-problem similarity. We denote $cat(A) \in \{0,1\}^n$ the vector representating the bag-of-words of the text one would obtain by concatenating all documents of schema. $cat(A)$ conveys thus crucial information, which are the terms contained in A. For example, with the taxonomies introduced in Fig. 1, $cat(A^{example})$ is the representation in the vector space of the document:

"$HTML\ Integrator,\ IT\ Department, Administrative\ and\ Infrastructure\ Support,$

$Database\ Administration, HR\ Direction,\ Management\ Department, Secretary$"

and $cat(B^{example})$ is the representation in the vector space of:

"$Management\ \&\ Administration,\ Software\ Engineering, Web\ Development,$

$Direction, Administration, Sales, Human\ Resources$"

Thanks to this new definition, we can compute by $f(cat(A), cat(A'))$ how the two schemas A and A' are similar by evaluating the terms in common. We then refine the inter-problem similarity to take into account the schema B, which is the set of possible solution items:

$$sim(pb, pb') = g(a, a') \times f(cat(B), cat(B')) \tag{5}$$

This similarity will be referred as *"right vocabulary"* similarity, because $cat(B)$ represents the vocabulary of B. The similarity function sim defined in the first solution (Sect. 4) will now be referred as *"no vocabulary"* similarity. For instance, with example problems defined in Sect. 4:

$$sim(pb^{example}, pb'^{example}) = g(a^{example}, b^{example}) \times f(cat(A^{example}), cat(B^{example}))$$
$$= 0 \times 0.15$$

Where right schemas share common terms, *Management* and *Administration*.

Similarly, we can try other variants of this similarity as well:
- a variant that considers the schema A, referred by *"left vocabulary"* similarity:

$$sim(pb, pb') = g(a, a') \times f(cat(A), cat(A')) \tag{6}$$

- a variant that considers both schemas A, B, referred by *"both vocabularies"* similarity:

$$sim(pb, pb') = g(a, a') \times f(cat(A), cat(A')) \times f(cat(B), cat(B')) \tag{7}$$

Advantages of Alternative Similarity Functions: Those variants are proposed so that the system takes into account the context, when performing a text schema mapping.

The first solution proposed in this article uses a very basic inter-problem similarity based only on the inter-item similarity g of items a, a' that we want to map. This does not take into account the context of problem to be compared one to another. Indeed, the schemas B, B' in which we want to find a semantic equivalent to a, a' respectively may be totally different even if a and a' are very similar.

In practice, this phenomenon occurs when two problems with almost identical items a, a' are applied to schemas of items that model documents in different languages, e.g. English and in French. Similarly, mapping an item to a list of job sectors, or to a list of university courses are also two very different problems because of the context, that is the global difference between B and B'.

7.2 Alternative Score Function

Given an item $a \in A$ to map, in the end it is the human user that validates/ chooses which item $b \in B$ is equivalent. Even though the proposed CBR system helps, it is still possible that human operators make mistakes.

Since the score function proposed previously (see Eq. 4) is based on the single case that maximizes inter-problem similarity, a single bad mapping can influence further responses negatively. In other words, a single flawed case can change score values dramatically and the system may fail in helping the user in the very long run. On the other hand, we can hope the best solution b for problem pb to have several similar cases in the base; we thus proposed to consider not one, but 5 similar cases to define the score used for ranking the solutions. We define an extended score function, such as:

$$
\begin{aligned}
score'(pb, b) = & \max_{c'=(pb',sol')\in C'} sim\left(pb, pb'\right) \times g\left(b, sol'\right) \\
& + \; 2^{nd}\max_{c'=(pb',sol')\in C'} sim\left(pb, pb'\right) \times g\left(b, sol'\right) \quad (8) \\
& + \; \dots \\
& + \; 5^{th}\max_{c'=(pb',sol')\in C'} sim\left(pb, pb'\right) \times g\left(b, sol'\right)
\end{aligned}
$$

Advantages of the New Score Function: This new score function computes a value considering not only the most similar case, but five most similar cases.

Using several cases at a time, the system becomes a lot less human error-prone. The new score function return meaningful results, even if a bad mapping is unintentionally saved along the way.

7.3 Overall Evaluation

Figure 4 shows results when we try variants of similarity measures defined in Eqs. 3, 5, 6 and 7, with a simple revision step with the scoring function of Eq. 4.

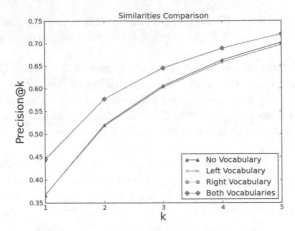

Fig. 4. Precision of the system using three different inter-problem similarity functions proposed in Sect. 7.1. The curves for the *"no vocabulary"* and *"left vocabulary"* similarities are almost indistinguishable, such as the curves for *"right vocabulary"* and *"both vocabularies"* similarities.

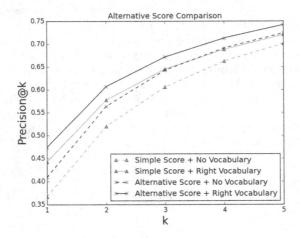

Fig. 5. Precision of the system using two different score functions defined in Sects. 5.3 and 7.2, with and without right vocabulary inter-problem similarity.

At first sight, we see an improvement when the right vocabulary is taken into account, with more than 10 % increase for the accuracy at the first suggestion. Second, we note that taking into account the left vocabulary is not useful: curve obtained when the left vocabulary is included in the similarity has exactly the same shape. Therefore the problem of schema mapping breaks down to sub-problems that are independent, as we supposed in the first place. The problem depends more on items that are possible equivalent; on the right vocabulary.

Figure 5 shows results for the extended revision step, with the simple score function defined in Eq. 4 and with the extended score defined in Eq. 8. We only

Table 2. Average rank of the solution in the computer-assisted system, computed by cross-validation or on user's real-world usage.

Evaluation	Cross validation		User's usage on new mappings
System	Initial system	Tuned system	Tuned system
Average rank	2.56	2.14	1.72

compared algorithms with simple *"no vocabulary"* similarity (Eq. 3), and *"right vocabulary"* similarity (Eq. 5). In both case, the extended revision step increase accuracy by about 5 % in general. Thus, to take into account several similar cases seems relevant. This could be because of the poor quality of some cases, that could lead to biased suggestions.

We decided to implement the system tuned with the alternative score function and the *"right vocabulary"* similarity. In the interface, items on the right side represent the schema B of possible mappings, and are ranked with respect to $score'(pb)$. In order to ease the visualization, the items are associated to a color (gradients of blue) proportional to $score'(pb, b) / \max_b score'(pb, b)$. The users are satisfied by this tool.

After three months of use they have significantly reduced the time spent on schema mapping: 62 % of items has been mapped by a clic on one of the top suggestion, 81 % on the top 3, whereas in the unranked case the user had to look at dozens of items (87 on average, see Table 1). We computed the average rank of the user's choice, sometimes denoted by mean rank [8]. The results are presented in Table 2, along with the average rank computed by cross validation for the initial system described in Sect. 5, and the tuned system proposed in this section (alternative score and *"right vocabulary"* similarity). The figures show how helpful the ranking is, and we evaluate the saving time per item mapping at 5 seconds. As there are about 30 schema mapped every week with 87 items per schema on average, the time saved by the employees using our system is approximately 3 and a half hours every week.

8   Conclusions

In this paper, we proposed a new method for easing text schema mapping based on the CBR methodology. For this, we defined a novel similarity measure between two schema mapping problems, taking into account the destination schema and the hierarchy of items to be mapped. The revise step involves a scoring function that aims at ranking the solutions for manual validation. The scoring function we proposed is an improvement for all problems facing a large case base with some poor quality/flawed cases. Compared to the state of the art of name-based schema mapping we studied more in depth on learning semantics rules from the corpus. Moreover, training a baseline model of text mining on document vectors involved in each problem would be much more computational expensive than our approach. All the real-world experiments we conducted show the utility of storing previous mapped cases and reusing them in a CBR system. The system is now implemented and used daily by Multiposting's employees, easing the task of schema mapping and reducing drastically the time spent on this task.

In our future work, we plan to take automation of the system a step further. As suggested by accuracy measures on the first solution, the CBR mapping decisions are not accurate enough to be considered valid every time; however, we could take the first result automatically in most of the cases, and limit manual validation only for unsure mappings. By applying a threshold on a confidence measure that assess how sure is the top ranked item by the CBR, we expect to automatize half of the mappings by this approach, reducing even more the human resources dedicated to this task.

References

1. Salton, G., Wang, A., Yang, C.S.: A vector space model for information retrieval. J. Am. Soc. Inf. Sci. **18**(11), 613–620 (1975)
2. Watson, I.: Case-based reasoning is a methodology not a technology. Knowl. Based Syst. **12**(5–6), 303–308 (1999)
3. Do, H.H., Rahm, E.: COMA - a system for flexible combination of schema matching approaches. In: Proceedings of the 28th VLDB Conference (2002)
4. Ruthven, I., Lalmas, M.: A survey on the use of relevance feedback for information access systems. Knowl. Eng. Rev. **18**(2), 95–145 (2003)
5. Madhavan, J., Bernstein, P.A., Doan, A.H.: Corpus-based schema matching. In: ICDE (2005)
6. Gu, M., Aamodt, A.: Evaluating CBR systems using different data sources: a case study. In: Roth-Berghofer, T.R., Göker, M.H., Güvenir, H.A. (eds.) ECCBR 2006. LNCS (LNAI), vol. 4106, pp. 121–135. Springer, Heidelberg (2006)
7. Jean-Mary, Y.R., Shironoshita, E.P., Kabula, M.R.: Automatic ontology matching with semantic verification. J. Web Seman. **7**(3), 235–251 (2009)
8. Buttcher, S., Clarke, C.L.A., Cormack, G.V.: Information Retrieval: Implementing and Evaluating Search Engines. MIT Press, Cambridge (2010)
9. Madhavan, J., Bernstein, P.A.: Generic schema matching, ten years later. In: Proceedings of the 37th VLDB Conference (2011)
10. Massmann, S., Raunich, S., Aumuller, D., Arnold, P., Rahm, E.: Evolution of the COMA match system. In: Ontology Matching, p. 49 (2011)
11. Shvaiko, P., Euzenat, J.: Ontology matching: state of the art and future challenges. IEEE Trans. Knowl. Data Eng. **25**(1), 158–176 (2013)
12. Richter, M.M., Weber, R.O.: Textual CBR. In: Richter, M.M., Weber, R.O. (eds.) Case-Based Reasoning, pp. 375–409. Springer, Heidelberg (2013)
13. Dufour-Lussier, V., Le Ber, F., Lieber, J., Nauer, E.: Automatic case acquisition from texts for process-oriented case-based reasoning. Inf. Syst. **40**, 153–167 (2014)
14. Lamontagne, L., Lee, H.-H.: Textual reuse for email response. In: Funk, P., González Calero, P.A. (eds.) ECCBR 2004. LNCS (LNAI), vol. 3155, pp. 242–256. Springer, Heidelberg (2004)
15. Sani, S., Wiratunga, N., Massie, S., Lothian, R.: Should Term-Relatedness be used in text representation? In: Delany, S.J., Ontañón, S. (eds.) ICCBR 2013. LNCS, vol. 7969, pp. 285–298. Springer, Heidelberg (2013)

16. Chakraborti, S., Wiratunga, N., Lothian, R., Watt, S.N.K.: Acquiring word similarities with higher order association mining. In: Weber, R.O., Richter, M.M. (eds.) ICCBR 2007. LNCS (LNAI), vol. 4626, pp. 61–76. Springer, Heidelberg (2007)
17. Bouquet, P., Euzenat, J., Franconi, E., Serafini, L., Stamou, G., Tessaris, S.: D2. 2.1 Specification of a common framework for characterizing alignment. In: University of Trento (2004)
18. Manakanatas, D., Plexousakis, D.: A tool for semi-automated semantic schema mapping: design and implementation. In: DISWEB (2006)
19. Daniels, J., Rissland, E.: What you saw is what you want: using cases to seed information retrieval. In: Leake, D.B., Plaza, E. (eds.) ICCBR-1997. LNCS, vol. 1266, pp. 325–336. Springer, Heidelberg (1997)
20. Bellashsene, Z., Bonifati, A., Rahm, E., et al.: Schema Matching and Mapping. Springer, Heidelberg (2011)
21. Porter, M.F.: Snowball: a language for stemming algorithms (2001)

Great Explanations: Opinionated Explanations for Recommendations

Khalil Muhammad(✉), Aonghus Lawlor, Rachael Rafter, and Barry Smyth

Insight Centre for Data Analytics, School of Computer Science and Informatics,
University College Dublin, Dublin, Ireland
khalil.muhammad@insight-centre.org

Abstract. Explaining recommendations helps users to make better decisions. We describe a novel approach to explanation for recommender systems, one that drives the recommendation ranking process, while at the same time providing the user with useful insights into the reason why items have been recommended and the trade-offs they may need to consider when making their choice. We describe this approach in the context of a case-based recommender system that harnesses opinions mined from user-generated reviews, and evaluate it on TripAdvisor hotel data.

Keywords: Recommender systems · Case-based reasoning · Explanations · Opinion mining · Sentiment analysis

1 Introduction

Recommender systems are a familiar part of the digital landscape helping millions of users make better choices about what to watch, wear, read, and buy. But generating suggestions is just the start. Explaining recommendations can make it easier for users to make decisions, increasing conversion rates and leading to more satisfied users [1–5]. Usually explanations provide a post-hoc rationalisation for the suggested items. But our work is motivated by a more intimate connection between recommendations and explanations, which poses the question: can the recommendation process itself be guided by structures generated to explain the suggestions to users?

We describe a case-based hotel recommender based on cases that are mined from the opinions in user-generated reviews; see also [6–8]. The central contribution of this work is a technique for generating personalised, feature-based explanations that can be used as part of an explanation interface in a recommender system but also during recommendation ranking. We provide examples based on real-world TripAdvisor data and discuss the results of an initial evaluation to explore the structure and utility of the resulting explanations.

2 Related Work

There is a history of using explanations to support reasoning in intelligent systems with approaches based on heuristics [9], CBR [10–12], and model-based

© Springer International Publishing Switzerland 2015
E. Hüllermeier and M. Minor (Eds.): ICCBR 2015, LNAI 9343, pp. 244–258, 2015.
DOI: 10.1007/978-3-319-24586-7_17

techniques [13] for example. More recently explanations have been used to support the recommendation process ([1–5]) by justifying recommendations to users. Good explanations promote trust and loyalty, increase satisfaction, and make it easier for users to find what they want.

Early work explored the utility of explanations in collaborative filtering systems with [1] reviewing different models and techniques for explanation based on MovieLens data. They considered a variety of explanation interfaces leveraging different combinations of data (ratings, meta-data, neighbours, confidence scores etc.) and presentation styles (histograms, confidence intervals, text etc.) concluding that most users recognised the value of explanations.

Bilgic and Mooney [14] used keywords to justify items rather than disclosing the behaviour of similar users. They argued that the goal of an explanation should not be to "sell" the user on the item but rather to help the user to make an informed judgment. They found users tended to overestimate item quality when presented with similar-user style explanations. Elsewhere, keyword approaches were further developed by [2] in a content-based, collaborative hybrid recommender capable of providing explanations such as: *"Item A is suggested because it contains features X and Y that are also included in items B, C, and D, which you have also liked."*; see also the work of [15] for related ideas based on user-generated tags instead of keywords. Note that this style of explanation justifies the item with reference to other items, in this case items that the user had previously liked.

Explanations can also relate one item to others. For example, Pu and Chen [3] build explanations that emphasise the tradeoffs between items. For example, a recommended item can be augmented by an explanation that highlights alternatives with different tradeoffs such as *"Here are laptops that are cheaper and lighter but with a slower processor"* for instance; see also related work by [16].

Here we focus on generating explanations that are feature-based and personalized (see also [17]), highlighting features that are likely to matter most to the user. But, like the work of [3,16], our explanations also relate items to other recommendation alternatives to help the user to better understand the trade-offs and compromises that exist within a product-space; see also [18]. However, our work also leverages the opinions in user-generated reviews as its primary source of item and recommendation knowledge. A unique feature of our approach is that explanations are not generated purely to justify recommendations but also to influence their ranking in the recommendation set.

3   Mining Experiential Cases

Our approach is summarised in Fig. 1 which we will describe with reference to TripAdvisor hotels and reviews. The *opinion mining* component extracts features and sentiments from reviews to produce hotel cases. This also generates user profiles from the reviews a user has submitted (or, for example, from the reviews they have previously viewed or marked as useful). The recommendation engine takes a user query (and profile) and retrieves a set of matching hotels and

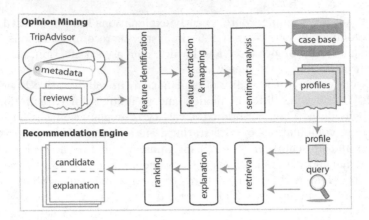

Fig. 1. An overview of the experiential product recommendation architecture.

then, generating explanations for each of these candidates, uses these explanations to rank the hotels for recommendation. It is this combination of opinion mining and explanation-based ranking that sets this work apart from others.

3.1   Opinion Mining

To identify and extract features from reviews we use the methods of [7,8]; we will refer to these (e.g. the *carpets* or the quality of *orange juice* at breakfast) as *review features*. While [7] use these as the basis for case descriptions, we find that they are less suitable for our needs, especially as the basis of explanations. For this reason we harness higher-level features available in the meta-data for hotels and map the review features back to these higher-level features. Since we will be focusing on TripAdvisor data, we map these review features back to a set of known *amenities* (e.g. *room quality, bar/restaurant* etc.); we refer to these features as *item features*. In this way we use this amenity meta-data as the primary features of our cases while still leveraging the opinions expressed in reviews to associate sentiment information with these amenities.

Mining Review Features. As with [8] we mine *bi-gram* features and *single-noun* features; see also [19,20]. For example, bi-grams which conform to one of two basic part-of-speech co-location patterns are considered — a noun followed by a noun, such as *shower screen* (NN), or an adjective followed by a noun, such as *twin room* (AN) — excluding bi-grams whose adjective is a sentiment word (e.g. *excellent, terrible* etc.) in the sentiment lexicon [19]. Separately, single-noun features are validated by eliminating nouns that are rarely associated with sentiment words in reviews as per [19], since such nouns are unlikely to refer to product features; these extracted features are the *review features*.

Mapping Review Features to Item Features. Taking all review texts, we apply k-means clustering, using sentence co-occurence, to associate review features with item features (amenities). While beyond the scope of this work suffice it to say that this provides a mapping between review features, such as *orange juice* and item features such as *breakfast*.

Evaluating Feature Sentiment. Again, as per [7], for a review feature f_i in a review sentence S_j, we determine whether there are any sentiment words in S_j. If not, f_i is marked *neutral*, otherwise we identify the sentiment word w_{min} with the minimum word-distance to f_i. Next we determine the part-of-speech (POS) tags for w_{min}, f_i and any words that occur between w_{min} and f_i. The POS sequence corresponds to an *opinion pattern*. We compute the frequency of all opinion patterns recorded after a pass of all reviews; a pattern is *valid* if it occurs more than average. For valid patterns we assign sentiment to f_i based on the sentiment of w_{min} and subject to whether S_j contains any negation terms within 4 words of w_{min}. If there are no negation terms then the sentiment assigned to f_i in S_j is that of the sentiment word in the sentiment lexicon; otherwise this sentiment is reversed. If an opinion pattern is not valid then we assign a *neutral* sentiment to each of its occurrences within the review set; see [21] for a fuller description.

Generating Experiential Cases. For each item/hotel H_j we have review features $\{f_1, ..., f_m\}$ mined from $reviews(H_j)$. Each f_i is mapped to a item feature F_i and we aggregate the review feature's mentions and sentiment scores to associate them with the corresponding F_i. So $F(H_j)$ is the set of item features $\{F_1, ..., F_n\}$ of hotel H_j. We can compute various properties of F_i: the fraction of times it is mentioned in reviews (its *importance*, see Eq. 1) and the degree to which it is mentioned in a positive or negative light (its *sentiment*, see Eq. 2, where $pos(F_i, H_j)$ and $neg(F_i, H_j)$ denote the number of times that feature F_i has positive or negative sentiment in reviews for H_j, respectively). Thus, each hotel can be represented as a *case*, $case(H_j)$, which aggregates item features, importance and sentiment data as in Eq. 3.

$$imp(F_i, H) = \frac{count(F_i, H)}{\sum_{\forall F_k \in F(H_j)} count(F_k, H_j)} \tag{1}$$

$$sent(F_i, H_j) = \frac{pos(F_i, H_j)}{pos(F_i, H_j) + neg(F_i, H_j)} \tag{2}$$

$$case(H_j) = \{[F_i, sent(F_i, H_j), imp(F_i, H_j)] : F_i \in F(H_j)\} \tag{3}$$

3.2   The Recommendation Engine

The recommendation engine returns a set of items (hotels) based on some query and user profile. Previous work has described related approaches to recommendation using opinions and sentiment [6,7] but here we describe a very different approach, one that bases recommendation on the ability to generate compelling explanations. The core of this is a novel approach to generating opinionated explanations and a way to score these explanations for recommendation ranking. We will discuss this in detail in the next section of this paper.

4 Generating Opinionated Explanations

Before describing our explanation approach it is important to understand the setting: we assume the target user U_T is presented with a set of hotel recommendations $\{H_1...H_k\}$ based on some user query which might include features such as star rating, price and location, and our task is to generate an explanation for each H_i. To simplify the explanation process let us say for now that we will build an explanation that will highlight two types of features: (1) reasons why they might choose the hotel; and (2) reasons why they might avoid the hotel.

4.1 A Basic Explanation Structure

Our basic explanation comes in two parts. The *pro* part is a set of (positive) hotel features that are reasons to choose the hotel. The *con* part is a set of (negative) features that can be considered as reasons to avoid the hotel. More formally, a feature F_i of hotel H_T is a *pro* if and only if it has a majority of positive sentiments ($sent(F_i, H_T) > 0.7$ in the case of our TripAdvisor data) and if its sentiment is *better than* at least one of the alternative hotels, H' (that is, $betterThan(F_i, H_T, H') > 0$); see Eqs. 4 and 5. Obviously this does not guarantee a pro will be a strong reason to choose H_T — it might only be better than a small fraction of the alternatives — but it is a possible reason to choose the hotel. Likewise a feature is a *con* if it has a negative sentiment ($sent(F_i, H_T) < 0.7$) and if it is *worse than* at least one alternative case; see Eqs. 6 and 7.

$$pro(F_i, H_T, H') \leftrightarrow sent(F_i, H_T) > 0.7 \wedge betterThan(F_i, H_T, H') > 0 \qquad (4)$$

$$betterThan(F_i, H_T, H') = \frac{\sum_{H_c \in H'} 1[sent(F_i, H_T) > sent(F_i, H_c)]}{|H'|} \qquad (5)$$

$$con(F_i, H_T, H') \leftrightarrow sent(F_i, H_T) <= 0.7 \wedge worseThan(F_i, H_T, H') > 0 \qquad (6)$$

$$worseThan(F_i, H_T, H') = \frac{\sum_{H_c \in H'} 1[sent(F_i, H_T) < sent(F_i, H_c)]}{|H'|} \qquad (7)$$

Then, we can construct a basic explanation as a set of pros and a set of cons as in Eqs. 8 and 9; for example, $Pros(H_T, H')$ is a set of tuples, each tuple comprising a pro feature and its *betterThan* score and likewise for $Cons(H_T, H')$

$$Pros(H_T, H') = \{(F, v) : pro(F, H_T, H') \wedge v = betterThan(F, H_T, H')\} \qquad (8)$$

$$Cons(H_T, H') = \{(F, v) : con(F, H_T, H') \wedge v = worseThan(F, H_T, H')\} \qquad (9)$$

4.2 Personalised Explanations

The approach described in Sect. 4.1 treats each hotel feature equally, but in reality different features will matter to different users. If we wish to create compelling explanations then we will need to focus on those features that matter to the target user. For this, we assume we have access to user profiles made up of the same type of features as cases, each with a relative importance value to reflect the importance (imp) of the feature to the user as in Eq. 10. A more detailed account of the user profiling is beyond the scope of this work but briefly we create profiles just as we create hotel cases, as mentioned previously, by mining opinions from the user's reviews and mapping these review features to item features. Then we can calculate $imp(F_i, U)$ in a similar manner to how we calculated $imp(F_i, H)$: as the number of occurrences of F_i in $Reviews(U)$ divided by the total number of feature occurrences in $Reviews(U)$.

$$Profile(U) = \{[F_i, imp(F_i, U)] : F_i \in Reviews(U)\} \tag{10}$$

Now we can modify the way we generate the pros (or cons) of an explanation so that in addition to capturing the feature and its $betterThan$ (or $worseThan$) scores we can also include an importance score for the target user U_T as in Eqs. 13 and 14.

$$pro(F, U_T, H_T, H') \leftrightarrow$$
$$sent(F, H_T) > 0.7 \wedge betterThan(F, H_T, H') > 0 \wedge imp(F, U_T) > 0 \tag{11}$$

$$con(F, U_T, H_T, H') \leftrightarrow$$
$$sent(F, H_T) < 0.7 \wedge worseThan(F, H_T, H') > 0 \wedge imp(F, U_T) > 0 \tag{12}$$

$$Pros(U_T, H_T, H') =$$
$$\{(F, v, m) : pro(F, U_T, H_T, H') \wedge v = betterThan(F, H_T, H') \wedge m = imp(F, U_T)\} \tag{13}$$

$$Cons(U_T, H_T, H') =$$
$$\{(F, v, m) : con(F, U_T, H_T, H') \wedge v = worseThan(F, H_T, H') \wedge m = imp(F, U_T)\} \tag{14}$$

In this way, for a target user U_T and hotel H_T, as well as a set of alternative hotels H', we can construct an explanation for H_T relative to H' that emphasises those pros and cons that matter to U_T. An example explanation structure is shown in Fig. 2, for a user *Peter Parker* and a *Clontarf Castle Hotel* in Dublin. Based on the user's profile we can see that he is interested in a number of listed features including *Bar/Lounge, Free Breakfast, Airport Transport, Restaurant, Leisure Centre, Shuttle Bus, Swimming Pool, and Room Service*, in order of decreasing importance score. In *Clontarf Castle* some of these features have been positively reviewed in the past (high sentiment scores) and so are

		Feature	Importance	Sentiment	BetterThan
		Bar/Lounge*	0.25	0.71	60%
	Pros	Free Breakfast	0.22	0.79	10%
Hotel:		Free Parking*	0.18	0.95	90%
Clontarf Castle		Restaurant*	0.15	0.86	70%
		Shuttle Bus	0.06	0.75	10%
User:		Feature	Importance	Sentiment	WorseThan
Peter Parker		Room Service	0.50	0.46	20%
	Cons	Airport Transport*	0.21	0.20	90%
		Leisure Centre*	0.11	0.31	75%
		Swimming Pool	0.10	0.45	33%

Fig. 2. An example of a raw explanation structure showing pros and cons that matter to the user along with associated importance, sentiment, and better/worse than scores.

listed as pros (e.g. *Bar/Lounge and Restaurant*) while others have been more negatively reviewed (e.g. *Airport Transport and Swimming Pool*) and are listed as cons. In each case we can see the proportion of alternative recommendations that this hotel is better or worse than with respect to a particular pro or con, respectively. For example, *Clontarf Castle* has been reviewed very favourably for its *Free Parking* (sentiment of 0.95) and it is better for this than 90% of the alternative recommendations. In contrast its *Leisure Centre* appears to be lacking (sentiment of only 0.31) and it is worse than 75% of the alternatives. Of course there are also some features that matter to the user but that do not appear in the hotel's reviews and so these are not in the explanation.

4.3   Compelling Explanations

The explanation structure so far can be made up of a large number of features. In fact, as we shall see later, in our TripAdvisor dataset basic explanations tend to include an average of 6–7 pros and 2 or 3 cons. That is a lot of features to present to the user especially since not all of them will be very compelling. Many of the pros might be better than only a small fraction of the other recommendations. One option is to filter features based on how strong a reason they may be to choose or reject the target hotel case. We define a *compelling* feature to be one that has a *betterThan* (pro) or *worseThan* (con) score of > 50% instead of just > 0. Thus, a compelling pro is one that is better than a majority of alternative recommendations and a compelling con is one that is worse than a majority of alternatives. A compelling pro may be a strong reason to choose the target hotel; a compelling cons is a strong reason to avoid it.

We define a *compelling explanation* as a non-empty explanation which contains only compelling pros and/or compelling cons. For instance, referring back to Fig. 2, we have marked compelling features with an asterisk after their name; so, the compelling explanation derived from this basic explanation would include *Bar/Lounge, Free Parking, Restaurant* as pros and *Airport Transport* and *Leisure Centre* as cons. These are all features that matter to the user and they distinguish the hotel as either better or worse than a majority of alternatives.

4.4   Using Explanations to Rank Recommendations

A unique element of this work is our proposal to use explanations to rank recommendations. To do this we need to score explanations to reflect how strongly they are likely to be when convincing the user to choose (or reject) a given hotel; hotels with the strongest explanations should appear at the top of the ranking. To do this we use a straightforward scoring function to measure the strength of an explanation as the weighted sum of its pros minus the weighted sum of its cons as shown in Eq. 15.

$$
\begin{aligned}
strength(U_T, H_T, H') = \\
\sum_{f \in Pros(U_T, H_T, H')} betterThan(f, H_T, H') \times imp(f, U_T) - \\
\sum_{f \in Cons(U_T, H_T, H')} worseThan(f, H_T, H') \times imp(f, U_T)
\end{aligned}
\tag{15}
$$

We can consider two versions of this scoring function, one that is applied to basic recommendations and one that is applied to compelling explanations. In each case the core calculation remains the same but only the features change. For example, applying the metric to the compelling features in Fig. 2 we calculate score of 0.15 based on a pro-score of 0.42 and a cons score of 0.27 (that is, $0.42 - 0.27 = 0.15$). Using this scoring function we can now rank-order hotels for recommendation in descending order of explanation strength.

4.5   Presenting Explanations to the User

So far we have said nothing about how these explanations might be presented to the user. For completeness, in Fig. 3 we illustrate one example for Clontarf Castle. The explanation is in the pop-up on the main hotel photo. We show the compelling version of the explanation with 3 pros and 2 cons.

The pros and cons are ordered based on their importance to the target user. The horizontal (sentiment) bar next to each shows the relative sentiment associated with the feature and beneath each is an indication of the *betterThan* or *worseThan* score, as appropriate. Evidently, Clontarf Castle is superior to a significant majority of alternatives in terms of its *Bar/Lounge*, *Free Parking*, and *Restaurant*, all of which are important to the target user, but it loses out to a majority of alternatives in terms of its *Airport Transportation* and *Leisure Centre*.

The user can request a more detailed explanation to reveal the full set of explanation features. By hovering over a sentiment bar the user can see a summary of the opinions extracted from reviews about that feature; this is shown for the *Bar/Lounge* feature in Fig. 3. And by clicking on the text that references alternatives the user will be brought to a list of the relevant alternatives;

Fig. 3. An example explanation showing pros and cons that matter to the target user along with sentiment indicators (horizontal bars) and information about how this item fares with respect to alternatives.

for example, if the user selected the "worse than 75 % of alternatives" for the Leisure Centre feature she would be brought to a list of these superior alternatives. In this way explanations also serve as a navigation structure to help users navigate between these very alternatives. This is just one approach to presenting explanations to the user and future work will consider interface issues further.

5   Evaluation

There are 4 important aspects to our approach to generating opinionated explanations for recommendation: (i) we separately emphasise the pros and cons of each item; (ii) we use information about features that matter to the user to personalise these explanations; (iii) we link the explanation to recommendations which offer better or worse feature options; (iv) we propose to use these explanation structures for the ranking of recommendations themselves. In combination we believe that these aspects make for a novel and potentially powerful approach to explanations for recommender systems and we provide some evaluation data to support this in what follows.

5.1   Data and Methodology

We use a TripAdvisor dataset as a source of users, reviews, and hotels. This dataset contains 1,000 users who have each written at least 10 hotel reviews for 2,370 hotels that they had booked. These reviews are used for user profiles. In addition we had more than 220,000 reviews by almost 150,000 reviewers available for the hotel cases.

For each target user U_T we select a hotel that they have booked, H_B, and collect a set of 10 related hotels from TripAdvisor. These additional hotels are those that TripAdvisor recommends as *related hotels*; we understand that TripAdvisor generates these using a combination of location, similar users, and meta-data.

Our intention is to simulate a typical session in which U_T has located a hotel of interest H_B, and a set of alternatives suggested by TripAdvisor. The booked hotel and the alternatives represent a set of recommendations for U_T. For each such session we generate an explanation for each of the 11 recommended hotels for U_T; in fact we will generate a basic explanation and a compelling explanation for each hotel. We analyse various properties of these explanations in addition to their utility for ranking the hotels for recommendation.

5.2   Pros vs. Cons, Better vs. Worse

First we investigate the number of pros and cons and their *betterThan/ worseThan* scores. Starting with basic explanations, Fig. 4(a) shows the average number of pros and cons generated per explanation (left y-axis) and also the average *betterThan/worseThan* scores (right y-axis). We can see that on average we are recommending about 5.8 pros versus only 2.2 cons reflecting the strong positive bias amongst reviews.

Interestingly we see a significant difference between the average *betterThan* score for pros (0.42) compared to the average *worseThan* score for cons (0.63). In other words, for a typical hotel, its pros will typically be better than about 42 % of the alternatives in the recommendation session. In contrast, when it comes to the cons, it is usually the case that the hotel in question does worse than most of the alternatives in the recommendation session.

Figure 4(b) shows corresponding results for compelling explanations. Incidentally about 97 % of the basic explanations are compelling. Now we can see that the average number of pros and cons is more balanced; there are 1.76 pros vs 1.55 cons. The average *betterThan* and *worseThan* scores for these explanations are 70 % and 75 %, respectively. These explanations are simpler to interpret

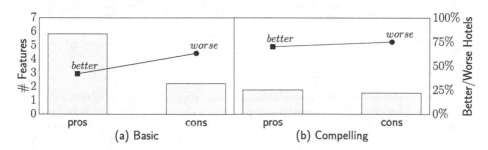

Fig. 4. The average number of pros and cons and the average *betterThan* and *worseThan* scores per explanation for basic explanations and compelling explanations.

(having fewer features) and more compelling in the sense that their features are better/worse that a strong majority of alternatives. Intuitively this combination of simplicity and compellingness should make them effective when it comes to helping users to decide, be it to accept or reject a given recommendation.

5.3   Using Explanations to Rank Recommendations

Earlier we described how to compute the *strength* of an explanation as a function of its pros and cons (see Eq. 15) and we proposed to use this score to rank hotels for recommendations. To evaluate how well this might work we need a ground-truth against which to judge our hotel recommendations. We propose the average rating that is available alongside each TripAdvisor hotel for this; a similar approach has been used by [6,7].

For each recommendation session we re-rank the recommended hotels (including the booked hotel) according to the strength of their basic and compelling recommendations and note the average position of the booked hotel. Then we compare the average rating of the booked hotel to the average ratings of the hotels above and below the booked hotel in the ranking. Ideally we would like to see all hotels above the booked hotel to have better average ratings and all hotels ranked below the booked hotel to have lower average ratings.

It noteworthy that we can expect this to be a tough test. After all the user chose the booked hotel for a reason and so we can expect it to be a highly rated one, all things being equal. Related to this, it is also worth noting that the average ratings for hotels in the recommendation sessions tend to be very high — TripAdvisor is unlikely to suggest poorly rated hotels — and so rating diversity can be low within sessions providing little opportunity for measurable ranking improvement. To deal with this we ordered our sessions based on variance of average user ratings (across the hotels in each session) and selected the top 20 % (200 sessions) that had the highest average user rating variance.

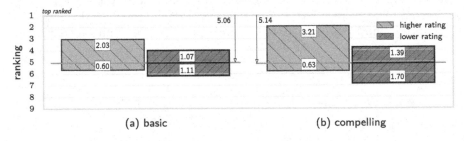

Fig. 5. A comparison of recommendation ranking results for basic and compelling explanations, showing only the first 9 rank positions. The solid horizontal lines indicate the average rank of the booked hotel. In each pair of bars, the green (left) show the average number of better-rated hotels above and below the booked hotel using our rankings. And in each pair the red bars (right) show the number of lower rated hotels above and below the booked hotel.

The results are shown in Fig. 5 as a bar chart that needs some explanation. First the horizontal lines that bound the charts at the top and the bottom represent the position of the top ranked hotel (position 1 in the ranking) and the position of the bottom ranked hotel (position 10 in the ranking). Between these boundaries there are two separate bar charts for the rankings based on the strength of: (a) basic explanations; (b) compelling explanations. The horizontal origin-line for each bar-chart is positioned between the top and bottom boundaries to reflect the average position of the booked hotel in each session. For basic explanations the booked hotel is ranked on average at position 5.06 and for compelling explanations the booked hotel is ranked a little lower at 5.14.

Next, each bar-chart contains 2 bars to reflect the number of recommendations that have a higher average rating than the booked hotel (the left-hand bar) and the number of recommendations with a lower average user rating than the booked hotel (the right-hand bar). The vertical position of these bars relative to the origin-line indicates whether these higher or lower rated hotels appear above or below the booked hotel in the ranking.

For example, for the compelling explanations (Fig. 5(b)) we see: the booked hotel ranked at position 5.14; an average of 3.21 recommendations above it with higher average ratings; only 0.63 hotels with higher ratings ranked below it (left-hand bar). This is good because it means that by ranking hotels by the strength of their explanations we are able to produce a ranking that tends to push a large majority (84 %) of higher rated hotels above the booked hotel. Next we look at the bar corresponding to lower rated hotels (right hand bar). Most of these lower rated hotels (1.70) are ranked below the booked hotel, but some (1.39) are ranked above. Again this is positive as it means that our explanation-based ranking tends to rank most of the poorer quality hotels below the booked one, although sometimes a lower rated hotel is ranked above the booked hotel. The results are broadly similar when we look at the basic explanations, although slightly fewer higher ranked hotels appear above the booked hotel.

6   Discussion and Limitations

To sum up, we have described a novel approach to generating explanations for opinionated recommender systems that can be used not only to help justify recommendations to users, but also to influence the recommendation ranking. In the space available we have left out many details and a number of items remain open for discussion, for example:

1. In our evaluation we base our profiles on reviews that have been authored by users but this introduces a significant cold-start problem in practice since most users are not active reviewers. Nevertheless there are many other ways to generate profiles such as mining opinions from reviews that users have rated, liked, or simply read. Moreover, even when user profiles remain lacking in features we could use those features we have to identify *similar users* and harness their profiles (and the features that matter to them) when generating explanations for the target user.

2. We have also said relatively little about how explanations might be presented to users, other than by showing one concrete example. Again this is a matter for future work where we will consider a variety of recommendation interfaces and styles, each emphasising different aspects of explanations. It will be interesting to see which styles users will find most helpful and compelling, and whether these do in fact support more satisfactory choices.

3. While the evaluation results on ranking are far from conclusive, they do suggest that using explanations for ranking can deliver a high quality ordering of recommendations. Indeed, when we compute the average rank correlation between the ground-truth (average TripAdvisor rating) ordering and the basic or compelling based orderings, we find correlation values of approximately 0.62 indicating a reasonable correlation between our explanation-based rankings and the ground-truth; this is yet another sign that the explanation-based approach is effective for recommendation ranking.

4. Finally, we have limited our research, thus far, to focusing on hotel reviews from TripAdvsor. However, there is nothing in the work that suggests this should be a limitation. In fact earlier work by [6–8] has applied similar opinion mining techniques to good effect to other types of user reviews such as those found on Amazon for consumer electronic products.

7   Conclusions

This work builds on recent research in the case-based reasoning community by bringing together ideas from CBR, opinion mining, and recommender systems. Its main contribution is a novel approach to explanation that can also be used to influence recommendation ranking. Rather than relying on similarity as a proxy for user relevance we base recommendation decisions on the ability to explain/justify recommendations to the user; this bears a resemblance to the work [22] which proposed the use of adaptation knowledge as a part of the case retrieval and ranking process, arguing that *adaptability* served as a more reliable metric for retrieval than traditional notions of similarity.

This is very much a work in progress. We have described our approach to generating explanations and provided some point examples on how such explanations might be used in practice. We analysed the structure of these explanations based on a TripAdvisor dataset of hotel reviews. We demonstrated that it is feasible to generate compelling explanations as part of the recommendation process, and that these explanations could be used for effective ranking.

Future work will focus on live-user trials of this approach. This will include experimenting with different presentation formats for our explanation structures to investigate whether users find them more or less useful, and whether there is evidence to suggest that such explanations do lead to better decisions in practice.

Acknowledgments. This work is supported by the Insight Centre for Data Analytics under grant number SFI/12/RC/2289.

References

1. Herlocker, J.L., Konstan, J.A., Borchers, A., Riedl, J.: Explaining collaborative filtering recommendations. In: Proceedings of The ACM Conference on Computer Supported Cooperative Work, pp. 241–250, ACM (2000)
2. Symeonidis, P., Nanopoulos, A., Manolopoulos, Y.: Providing justifications in recommender systems. IEEE Trans. Syst. Man Cybern. Part A Syst. Hum. **38**(6), 1262–1272 (2008)
3. Pu, P., Chen, L.: Trust-Inspiring explanation interfaces for recommender systems. Knowl. Based Syst. **20**(6), 542–556 (2007)
4. Coyle, M., Smyth, B.: Explaining search results. In: Proceedings of The 19th International Joint Conference on Artificial Intelligence, pp. 1553–1555, Morgan Kaufmann Publishers Inc (2005)
5. Friedrich, G., Zanker, M.: A taxonomy for generating explanations in recommender systems. AI Mag. **32**(3), 90–98 (2011)
6. Dong, R., Schaal, M., O'Mahony, M.P., McCarthy, K., Smyth, B.: Mining features and sentiment from review experiences. In: Delany, S.J., Ontañón, S. (eds.) ICCBR 2013. LNCS, vol. 7969, pp. 59–73. Springer, Heidelberg (2013)
7. Dong, R., O'Mahony, M.P., Smyth, B.: Further experiments in opinionated product recommendation. In: Lamontagne, L., Plaza, E. (eds.) ICCBR 2014. LNCS, vol. 8765, pp. 110–124. Springer, Heidelberg (2014)
8. Dong, R., Schaal, M., O'Mahony, M.P., Smyth, B.: Topic extraction from online reviews for classification and recommendation. In: Proceedings of The 23rd International Joint Conference on Artificial Intelligence (2013)
9. Buchanan, B.G., Shortliffe, E.H.: Rule Based Expert Systems: The Mycin Experiments of the Stanford Heuristic Programming Project. The Addison-Wesley Series in Artificial Intelligence, vol. 3. Addison-Wesley Longman Publishing Co., Inc., Boston (1984)
10. Sørmo, F., Cassens, J., Aamodt, A.: Explanation in case based reasoning perspectives and goals. Artif. Intell. Rev. **24**(2), 109–143 (2005)
11. McSherry, D.: Explaining the pros and cons of conclusions in CBR. In: Funk, P., González Calero, P.A. (eds.) ECCBR 2004. LNCS (LNAI), vol. 3155, pp. 317–330. Springer, Heidelberg (2004)
12. Doyle, D., Cunningham, P., Bridge, D.G., Rahman, Y.: Explanation oriented retrieval. In: Funk, P., González Calero, P.A. (eds.) ECCBR 2004. LNCS (LNAI), vol. 3155, pp. 157–168. Springer, Heidelberg (2004)
13. Druzdzel, M.J.: Qualitative verbal explanations in Bayesian belief networks. Artif. Intell. Simul. Behav. Q. Spec. Issue Bayesian Netw. **94**, 43–54 (1996)
14. Bilgic, M., Mooney, R.J.: Explaining recommendations: Satisfaction vs. Promotion. In: Proceedings of Beyond Personalization 2005: A Workshop on the Next Stage of Recommender Systems Research at The 2005 International Conference on Intelligent User Interfaces, pp. 13–18 (2005)
15. Vig, J., Sen, S., Riedl, J.: Tagsplanations: explaining recommendations using tags. In: Proceedings of The 13th International Conference on Intelligent User Interfaces, pp. 47–56, ACM Press (2008)
16. Reilly, J., McCarthy, K., McGinty, L., Smyth, B.: Explaining compound critiques. Artif. Intell. Rev. **24**(2), 199–220 (2005)
17. Tintarev, N., Masthoff, J.: The effectiveness of personalized movie explanations: an experiment using commercial meta-data. In: Nejdl, W., Kay, J., Pu, P., Herder, E. (eds.) AH 2008. LNCS, vol. 5149, pp. 204–213. Springer, Heidelberg (2008)

18. McSherry, D.: Similarity and compromise. In: Ashley, K.D., Bridge, D.G. (eds.) ICCBR 2003. LNCS, vol. 2689, pp. 291–305. Springer, Heidelberg (2003)
19. Hu, M., Liu, B.: Mining opinion features in customer reviews. In: Proceedings of The 19th National Conference on Artificial Intelligence, pp. 755–760, AAAI Press (2004)
20. Justeson, J.S., Katz, S.M.: Technical terminology: some linguistic properties and an algorithm for identification in text. Nat. Lang. Eng. **1**(1), 9–27 (1995)
21. Moghaddam, S., Ester, M.: Opinion digger: an unsupervised opinion miner from unstructured product reviews. In: Proceedings of The 19th ACM International Conference on Information and Knowledge Management, pp. 1825–1828, ACM Press (2010)
22. Smyth, B., Keane, M.: Adaptation-Guided retrieval: questioning the similarity assumption in reasoning. Artif. Intell. **102**(2), 249–293 (1998)

Learning and Applying Adaptation Operators in Process-Oriented Case-Based Reasoning

Gilbert Müller$^{(\boxtimes)}$ and Ralph Bergmann

Business Information Systems II, University of Trier, 54286 Trier, Germany
{muellerg,bergmann}@uni-trier.de
http://www.wi2.uni-trier.de

Abstract. This paper presents a novel approach to the operator-based adaptation of workflows, which is a specific type of transformational adaptation. We introduce the notion of workflow adaptation operators which are partial functions transforming a workflow into a successor workflow, specified by workflow fractions to be inserted and/or deleted. The adaptation process itself chains adaptation operators during a local search process aiming at fulfilling the query as best as possible. Further, the paper presents an algorithm that learns workflow adaptation operators from the case base automatically, thereby addressing the common problem of adaptation knowledge acquisition. An empirical evaluation in the domain of cooking workflows was conducted which demonstrates convincing adaptation capabilities without a significant reduction of the workflows' quality.

Keywords: Process-oriented case-based reasoning · Operator-based adaptation · Workflows

1 Introduction

Process-aware information systems (PAISs) [7] support the operational business of an organization based on models of their processes. PAISs include traditional workflow management systems as well as modern business process management systems. In the recent years, the use of workflows has significantly expanded from the original business area towards new application fields such as e-science, medical healthcare, information integration, and even cooking [10,24,25]. Process-oriented case-based reasoning (POCBR) [20] covers research on case-based reasoning (CBR) for addressing problems in PAISs. Recent research deals with approaches to support modeling, composition, adaptation, analysis, monitoring and optimization of business processes or workflows [2,12,13,18,21,22,26]. Workflow adaptation addresses the adaptation of a retrieved workflow from a repository (case base) to fulfill the specific needs of a new situation (query). In POCBR, adaptation methods that originate from case adaptation in CBR are proposed for this purpose. In our previous work we have investigated case-based adaptation [17], compositional adaptation [22], as well as the use of generalized cases [23] for adaptation.

© Springer International Publishing Switzerland 2015
E. Hüllermeier and M. Minor (Eds.): ICCBR 2015, LNAI 9343, pp. 259–274, 2015.
DOI: 10.1007/978-3-319-24586-7_18

In this paper, we present a novel operator-based approach [3] for adapting workflow cases represented as graphs. The *workflow adaptation operators* we propose in Sect. 3 are partial functions specifying ways of adapting a workflow towards a successor workflow. Like in STRIPS, the operators are specified by two workflow sub-graphs, one representing a workflow fraction to be deleted and one representing a workflow fraction to be added. The adaptation process (see Sect. 5) transforms a retrieved workflow into an adapted workflow by chaining various adaptation operators. This process can be considered a search process towards an optimal solution w.r.t. the query. Most importantly, we also propose an algorithm to learn such workflow adaptation operators automatically from the case base, thereby extending previous work on learning adaptation knowledge [4,9,27] towards POCBR (see Sect. 4). Thus, the knowledge acquisition bottleneck for adaptation knowledge is avoided. Finally, we experimentally evaluate our approach in the domain of cooking (see Sect. 6). We can show that with the learned workflow adaptation operators, a high percentage of the changes requested for a retrieved workflow can be fulfilled without significantly reducing the quality of the adapted workflows.

2 Foundations

We now briefly introduce relevant previous work in the field of POCBR.

2.1 Workflows

Broadly speaking, a *workflow* consists of a set of *activities* (also called *tasks*) combined with *control-flow structures* like sequences, parallel (AND) or alternative (XOR) branches, as well as repeated execution (LOOPs). In addition, tasks exchange certain *data items*, which can also be of physical matter, depending on the workflow domain. Tasks, data items, and relationships between the two of them form the *data flow*.

We illustrate our approach in the domain of cooking recipes (see example workflow in Fig. 1). A cooking recipe is represented as a workflow describing the instructions for cooking a particular dish [24]. Here, the tasks represent the cooking steps and the data items refer to the ingredients being processed by the cooking steps. An example cooking workflow for a pasta recipe is illustrated in Fig. 1. Based on our previous work [2,22,23] we now introduce the relevant formal workflow terminology.

Definition 1. *A workflow is a directed graph $W = (N, E)$ where N is a set of nodes and $E \subseteq N \times N$ is a set of edges. Nodes $N = N^D \cup N^T \cup N^C$ can be data nodes N^D, task nodes N^T, or control-flow nodes N^C. In addition, we call $N^S = N^T \cup N^C$ the set of sequence nodes. Edges $E = E^C \cup E^D$ can be control-flow edges $E^C \subseteq N^S \times N^S$, which define the order of the sequence nodes or data-flow edges $E^D \subseteq (N^D \times N^S) \cup (N^S \times N^D)$, which define how the data is shared between the tasks.*

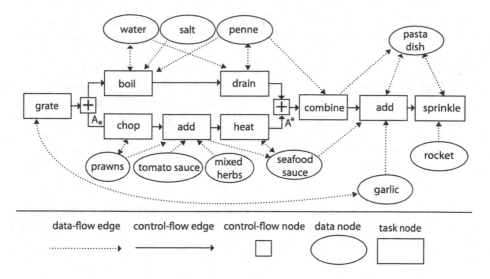

Fig. 1. Example of a block-oriented cooking workflow

The control-flow edges E^C of a workflow induce a strict partial order on the sequence nodes S. Thus, we define $s_1 < s_2$ for two sequence nodes $s_1, s_2 \in S$ as a transitive relation that expresses that s_1 is executed prior to s_2 in W. We further define $n \in]x_1, x_2[$ iff $x_1 < n < x_2$, describing that node n is located between x_1 and x_2 in W w.r.t. the control-flow edges.

We further denote that two data nodes $d_1, d_2 \in N^D$ are data-flow connected $d_1 \bowtie d_2$ if there exists a task that consumes data node d_1 and produces data node d_2. Moreover, $d_1 \bowtie^+ d_2$ denotes that $d_1, d_2 \in N^D$ are transitively data-flow connected:

$$d_1 \bowtie d_2, \text{iff } \exists t \in N^T : ((d_1, t) \in E^D \wedge (t, d_2) \in E^D) \tag{1}$$

$$d_1 \bowtie^+ d_2, \text{iff } d_1 \bowtie d_2 \vee \exists d \in N^D : (d_1 \bowtie d \wedge d \bowtie^+ d_2) \tag{2}$$

2.2 Block-Oriented Workflows

We now restrict the workflow representation to block-oriented workflows [22], i.e., workflows in which the control-flow structures form blocks of nested workflows with an opening and closing control-flow element. These blocks must not be interleaved.

Definition 2. *A* block-oriented workflow *is a workflow in which the control-flow nodes* $N^C = N^{C_*} \cup N^{C^*}$ *define the control-flow blocks. Each control-flow block has an opening node from* N^{C_*} *and a related closing node from* N^{C^*} *specifying either an AND, XOR, or LOOP block. These control-flow blocks may be nested but must not be interleaved and must not be empty.*

Figure 1 shows an example block-oriented workflow, containing a control-flow block with an opening AND control-flow node A_* and a related closing AND control-flow node A^*.

Further, we introduce a terminology of *consistent block-oriented workflows*. According to Davenport, "[...] a process is simply a structured, measured set of activities designed to produce a specific output [...]" [5]. In the following, these specific workflow outputs are denoted as $W^O \subseteq N^D$. In the cooking domain, the specific output is the particular dish produced, i.e., "pasta dish" in Fig. 1. Hence, for a *consistent workflow*, we require that each ingredient must be contained in the specific output, as otherwise the ingredient as well as the related tasks would be superfluous.

Definition 3. *A block-oriented workflow is* consistent, *iff each produced ingredient is contained in the specific output of the workflow. Thus, each ingredient must be transitively data-flow connected to the specific output* W^O, *i.e.,* $\forall d \in N^D \exists o \in W^O d \ltimes^+ o$.

2.3   Semantic Workflow Similarity

To support retrieval and adaptation of workflows, the individual workflow elements are annotated with ontological information, thus leading to a *semantic workflow* [2]. In particular, all task and data items occurring in a domain are organized in taxonomy, which enables the assessment of similarity among them. We deploy a taxonomy of cooking ingredients and cooking steps for this purpose. In our previous work, we developed a semantic similarity measure for workflows that enables the similarity assessment of a case workflow w.r.t. a query workflow [2].

The core of the similarity model is a local similarity measure for semantic descriptions $sim_\Sigma : \Sigma^2 \rightarrow [0, 1]$. In our example domain the taxonomical structure of the data and task ontology is employed to derive a similarity value that reflects the closeness in the ontology. It is combined with additional similarity measures that consider relevant attributes, such as the quantity of an ingredient used in a recipe (see [2] for more details and examples). The similarity $sim_N : N^2 \rightarrow [0, 1]$ of two nodes and two edges $sim_E : E^2 \rightarrow [0, 1]$ is then defined based on sim_Σ applied to their assigned semantic descriptions. The similarity $sim(QW, CW)$ between a query workflow QW and a case workflow CW is defined by means of an admissible mapping $m : N_q \cup E_q \rightarrow N_c \cup E_c$, which is a type-preserving, partial, injective mapping function of the nodes and edges of QW to those of CW. For each query node and edge x mapped by m, the similarity to the respective case node or edge $m(x)$ is computed by $sim_N(x, m(x))$ and $sim_E(x, m(x))$, respectively. The overall workflow similarity with respect to a mapping m, named $sim_m(QW, CW)$ is computed by an aggregation function (e.g. a weighted average) combining the previously computed similarity values. The overall workflow similarity is determined by the best possible mapping m

$$sim(QW, CW) = \max\{sim_m(QW, CW) \,|\, \text{admissible mapping } m\}.$$

This similarity measure assesses how well the query workflow is covered by the case workflow. In particular, the similarity is 1 if the query workflow is exactly included in the case workflow as a subgraph. Hence, this similarity measure is not symmetrical.

2.4   Partial Workflows and Streamlets

We aim at reusing workflow parts within the representation of adaptation operators. Therefore, we now introduce the definition of partial workflows according to Müller and Bergmann [22] and the new definition of so-called streamlets.

Definition 4. *For a subset of tasks $N_p^T \subseteq N^T$, a partial workflow W_p of a block-oriented workflow $W = (N, E)$ is a block-oriented workflow $W_p = (N_p, E_p \cup E_p^{C+})$ with a subset of nodes $N_p = N_p^T \cup N_p^C \cup N_p^D \subseteq N$. $N_p^D \subseteq N^D$ is defined as the set of data nodes that are linked to any task in N_p^T, i.e., $N_p^D = \{d \in N^D | \exists t \in N_p^T : ((d,t) \in E^D \vee (t,d) \in E^D)\}$. $N_p^C \subseteq N^C$ is the maximum set of control-flow nodes such that W_p is a correct block-oriented workflow. W_p contains a subset of edges $E_p = E \cap (N_p \times N_p)$ connecting two nodes of N_p supplemented by a set E_p^{C+} of additional control-flow edges that retain the execution order of the sequence nodes, i.e., $E_p^{C+} = \{(n_1, n_2) \in N_p^S \times N_p^S | n_1 < n_2 \wedge \not\exists n \in N_p^S : ((n_1, n) \in E_p^C \vee (n, n_2) \in E_p^C \vee n \in]n_1, n_2[)\}$.*

In general, control-flow nodes are part of a partial workflow if they construct a workflow w.r.t. the block-oriented workflow structure. The additional edges E_p^{C+} are required, to retain the execution order $s_1 < s_3$ of two sequence nodes if for $s_1, s_2, s_3 \in S$ holds $s_2 \in]s_1, s_3[$ but $s_2 \notin N_p$. Figure 2 illustrates a partial workflow W_p of the workflow W given in Fig. 1. One additional edge is required in this example, depicted by the double-line arrow since "grate" and "add" are not linked in W.

Based on Definition 4, we now introduce *streamlets* that represent a partial workflow constructed by all tasks linked to a certain data node $d \in N^D$. Hence, a streamlet describes the partial workflow comprising the tasks processing a certain data node d. Thus, it is the smallest fraction of a workflow regarding d. Streamlets will become the smallest fraction of a workflow to be modified by workflow adaptation operators.

Definition 5. *A streamlet $W_d = (N_d, E_d)$ for data $d \in N^D$ in workflow W is defined as a partial workflow for the subset of tasks connected to d, i.e. $\{t \in N^T | (t,d) \in E^D \vee (d,t) \in E^D\}$. The data node d is referred to as the head data node of W_d. Further, let the tasks in W_d that do not produce d be defined as anchor tasks A_d for d, i.e., $A_d = \{t \in N_d^T | \not\exists (t,d) \in E_d^D\}$.*

An example streamlet is illustrated in Fig. 2. In general, anchor tasks (see double-lined rectangle, task node "add") are those tasks that consumes the head data node (see double-lined circle, data node "garlic") but that do not produce it. Hence, these tasks mark the positions where the head data node is linked into

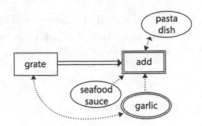

Fig. 2. Example of a streamlet W'

the overall workflow and used as an input to other tasks (e.g. after adding garlic it is used together with the seafood sauce as part of the pasta dish). Please note that a streamlet may contain more than one anchor, e.g., salt can be added to the boiled water and to the pasta sauce.

3   Workflow Adaptation Operators

In CBR, we usually distinguish between substitutional, transformational, and derivational adaptation approaches [4,27]. Substitutional and transformational adaptation approaches make use of adaptation knowledge represented as adaptation rules or adaptation operators. Adaptation rules describe how differences between the problem description in the query and the retrieved case (rule's precondition) can be compensated by certain changes of the solution in the retrieved case (rule's conclusion). Adaptation operators [3] however, do not explicitly represent differences between query and retrieved case, but they are partial functions specifying ways of adapting a case towards a successor case. The adaptation process in CBR transforms a retrieved case into an adapted case by chaining various adaptation operators. Consequently, workflow adaptation is performed by applying chains of adaptation operators $W \overset{o_1}{\to} W_1 \overset{o_2}{\to} \ldots \overset{o_n}{\to} W_n$ to the retrieved workflow W, thereby computing the adapted workflow W_n. This process can be considered a search process towards an optimal solution w.r.t. the query.

3.1   Definition of Workflow Adaptation Operators

For applying operator-based adaptation to POCBR, a notion of workflow adaptation operators is required defining them as partial functions that transform workflows. We loosely follow the representation idea of STRIPS operators and define an adaptation operator by specifying an insertion and a deletion streamlet. Operator preconditions are not explicitly specified, but result implicitly from those streamlets. In a nutshell, a workflow adaptation operator, if applicable, removes the deletion streamlet from the workflow and adds the insertion streamlet instead. In Fig. 3 an example of an operator is shown, specifying that in a cooking workflow prawns can be replaced by tuna, while at the same time the preparation step chop needs to be replaced by drain. Besides operators that exchange workflow streamlets, they could also just insert or just delete a streamlet of a workflow. We now give a formal definition of workflow adaptation operators, specifying their representation and operational semantics.

Definition 6. *Let $\mathcal{W}$ be the set of all consistent block-oriented workflows. A workflow adaptation operator is a partial function $o : \mathcal{W} \mapsto \mathcal{W}$ transforming a workflow $W \in \mathcal{W}$ into an adapted workflow $o(W) \in \mathcal{W}$. The adaptation operator o is specified by an insertion streamlet o_I and a deletion streamlet o_D, each of which can also be empty. Based on the presence of the two streamlets, we distinguish three types of adaptation operators.*

- *An* insert *operator consists only of an insertion streamlet with the anchor tasks A_d. The application of o to W inserts o_I into W except for the anchor tasks A_d at the positions of the best matching anchor tasks A_d. The operator is only applicable (precondition A) iff in W tasks matching A_d exist and if the resulting workflow is consistent.*
- *A* delete *operator consists only of a deletion streamlet o_D with the head data node d and the anchor tasks A_d. The application of o to W deletes the streamlet W_d from W except for the anchor tasks A_d. The operator is only applicable (precondition B) iff there exists a workflow streamlet W_d in W which is sufficiently similar to o_D and if the resulting workflow after deletion is consistent.*
- *An* exchange *operator consists of an insertion and a deletion streamlet. The application of o to W deletes o_D from W (except for the anchor tasks) and subsequently inserts o_I (except for the anchor tasks) at the position of the best matching anchor tasks. The operator is only applicable iff both previously defined preconditions A and B are fulfilled and if the resulting workflow after deletion and insertion is consistent.*

The conditions of identical head node and the minimum similarity between the streamlet and the deletion streamlet serve as a precondition to check whether the streamlet W_d is similar enough to the deletion streamlet. This ensures that operators are only applied if a similar streamlet is present in the workflow. For the insertion, a matching anchor is needed to ensure that the streamlet can be added to a suitable position of the workflow merging in the right data node.

An example of a workflow exchange adaptation operator o is given in Fig. 3. The head data nodes are marked by a double-circled data object, i.e., tuna or prawns, respectively. Further, the anchor tasks are marked by double-lined rectangles. These anchor tasks are used during adaptation, to identify the position of the streamlet within the entire workflow, i.e., the position at which the insertion streamlet is inserted. Hence, the example adaptation rule describes that prawns can be exchanged by tuna, if in W a streamlet W_d similar to o_D is present. This also enforces that tasks have to be changed as well, because the chop task also has to be exchanged by a drain task.

3.2   Details of the Operator Application

We now give some more details about how operators are applied, making the previous definition more precise. The space limitation prevents us from a detailed

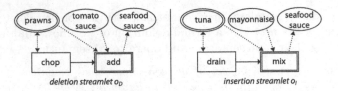

Fig. 3. Example adaptation operator

description of the algorithm. Hence, we illustrate our approach primarily by an example application of the exchange operator shown in Fig. 3 to the workflow given in Fig. 1.

To determine the applicability of a delete or exchange operator o for a workflow W, the definition requires that there exists workflow stream W_d in W for the head node d of o_D that is sufficiently similar to the deletion streamlet o_D. We implement this condition by a similarity threshold Δ_S, i.e., we require that $sim(W_d, o_D) \geq \Delta_S$. Further, we require that the output data nodes of the anchor tasks of o_D are the same as the output data nodes of the mapped tasks in W, ensuring that the data node d is removed from the same successive data nodes (e.g. see "seafood sauce" in Fig. 3). To remove the streamlet W_d from W (see Fig. 4 for an example) a partial workflow is constructed, containing all tasks of W except of the non-anchor tasks contained in W_d, i.e., $N^T \setminus (N_d^T \setminus A_d)$.

To add the insertion streamlet o_I to W, tasks in W must be identified that match the anchor tasks A_d of o_I. For this purpose, the partial workflow constructed from o_I for the anchor tasks A_d is considered. This partial workflow contains the anchor tasks as well as all connected data nodes. To match the anchor, the similarity between this partial workflow and W is determined. If this similarity exceeds the threshold Δ_S, the matching tasks in W are determined by the computed admissible mapping function. Further, we require that the output data nodes of the anchor tasks A_d are the same as the output data nodes of the mapped tasks in W. After a successful anchor mapping, the insertion streamlet is added at the position of the best matching anchor. This means

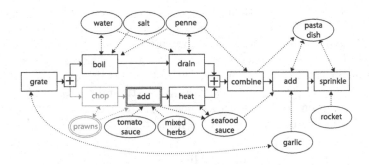

Fig. 4. Streamlet W_d removed from W

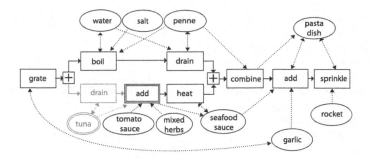

Fig. 5. Streamlet R_I added to W

that all edges, tasks (except of anchors), and data nodes (if not already present) of o_I are inserted into the workflow W. Then, the inserted streamlet o_I is connected with an additional control-flow edge that link the tasks of the streamlet into the workflow in front of the best matching anchor in W. In the illustrated scenario, the streamlet o_I is inserted in front of the first "add" task (see Fig. 5). In the special case that the insertion streamlet contains more than one anchor, the streamlet is split. To this end, for each anchor, the tasks are determined that belong to this anchor, i.e., previous tasks (w.r.t. to control-flow edges), which are transitively data flow connected until another anchor is reached. The procedure is then applied for each part of the split streamlet.

4   Automatic Learning of Workflow Adaptation Operators

The basic idea behind the automatic learning of workflow adaptation operators (see Algorithm 1) is to explore the knowledge already present in the case base [4,9,27]. To achieve this, pairs of similar workflows in the case base are compared, i.e., a query workflow W_q and a case workflow W_c that have a similarity value higher than a given threshold Δ_W, i.e. $sim(W_q, W_c) \geq \Delta_W$. The adaptation operators are then constructed from the differences in the data nodes between those two workflows. Hence, they describe which data nodes have to be exchanged, inserted or deleted in order to transform the set of data nodes N_q^D to N_c^D. The differences are determined by accessing the mapping produced during the similarity computation (see Sect. 2.3). More precisely, the data nodes of a query workflow are mapped to those of the case workflow. Mapped data nodes are then assumed to be replaceable data nodes. However, only mappings are regarded between data nodes that have a similarity value higher than a threshold Δ_D. This ensures that data nodes that are not similar to each other are not considered as replaceable. For each mapping, two streamlets are constructed based on the corresponding query data node and case data node. We thereby assume that these two streamlets can be exchanged by each other, i.e., the query streamlet represents the deletion streamlet and the case streamlet represents the insertion streamlet. For the remaining data nodes (the ones with no mapping with a similarity value larger than Δ_D), insert or delete operators are created.

Algorithm. LEARN_OPERATORS(CB);
Input: Case base CB
Output: Set of operators $operators$
$operators = \emptyset$;
forall the $W_q \in CB$ **do**
 forall the $\{W_c \in CB | sim(W_q, W_c) \geq \Delta_W \wedge W_q \neq W_c\}$ **do**
  forall the $d \in N_q^D$ **do**
   Init operator o;
   $o.insert = \emptyset$;
   $o.delete = construct\_streamlet(W_q, d)$;
   if $sim(d, m(d)) \geq \Delta_D$ **then**
    $o.insert = construct\_data\_streamlet(W_c, d)$;
   $operators = operators \cup \{o\}$;
  forall the $\{d \in N_c^D | \nexists d' \in N_q^D : (m(d') = d \vee sim(d', d) \geq \Delta_D)\}$ **do**
   Init operator o;
   $o.delete = \emptyset$;
   $o.insert = construct\_streamlet(W_c, d)$;
   $operators = operators \cup \{o\}$;
return $operators$

Algorithm 1: Learning algorithm of workflow adaptation operators

Although, the operators describe how to transform the set of data nodes N_q^D to N_c^D the workflow streamlet also contains information about how to exchange delete, or insert tasks or control-flow nodes. This is because identical data nodes that are mapped can possibly be processed differently, i.e., different task nodes are used in their streamlets to process this data.

5   Workflow Adaptation Using Adaptation Operators

We now present the adaptation procedure in more detail. After the retrieval of a most similar workflow W the user might want to adapt it according to his or her preferences.

5.1   Change Request

Following the retrieval, a change request C is defined by specifying sets of tasks or data nodes that should be added C_{add} or removed C_{del} from workflow W. The change request can be either acquired manually from the user after the workflow is presented or it can be automatically derived based on the difference between the query and the retrieved case. As the tasks and data nodes are taxonomically ordered (see Sect. 2.3), the change request can also be defined by a higher level concept of the taxonomy in order to define a more general change of the workflow. For example, a change request specified as "DELETE meat" ensures that the adapted recipe is a vegetarian dish. We define the change request fulfilment $\mathcal{F}(C, W) \rightarrow [0, 1]$ for a change request C and a workflow $W = (N, E)$

as the number of desired nodes contained plus the number of undesired nodes not contained in relation to the size òf the change request:

$$\mathcal{F}(C, W) = \frac{|N \cap C_{add}| + |C_{del} \setminus N|}{|C_{add}| + |C_{del}|} \tag{3}$$

5.2   Adaptation Procedure

The goal of the adaptation procedure is to maximize the value of $\mathcal{F}(C, W)$. Therefore, it uses a kind of a hill climbing local search algorithm in order to optimize the change request fulfillment $\mathcal{F}(C, W)$. Basically, the general idea of the adaptation is that for each streamlet W_d in the retrieved workflow an applicable operator o is searched and applied to the workflow W, if its application increases the change request fulfillment. This leads to a chain of adaptation operators. During this search process, the change request, i.e., the set C_{add} is updated according to the applied operator. This ensures that nodes which were already inserted are not inserted again by subsequent operators.

Prior to the search, a partial order of data nodes is constructed w.r.t. their usage in the workflow. More precisely, data nodes are ordered with respect to which data is used first in the control-flow of the workflow, i.e., as input of a task. During adaptation this partial order is traversed starting with the data node used first in the workflow. For each streamlet W_d in the retrieved workflow W at most one applicable adaptation exchange or delete operator o is selected which maximizes $\mathcal{F}(C, o(W))$. Further, delete operators must not remove a desired node. Operators with the highest similarity between the streamlet W_d of W and the deletion streamlet o_D are preferred during selection, aiming at selecting the best possible operator. Afterwards, insert operators are applied to further improve change request fulfillment. Therefore, we select insert operators whose insertion streamlet o_I contains at least one desired but not an undesired node. Insert operators whose head node of o_I is already in W are disregarded. The adaptation process terminates, if no further insert operator can be applied which improves the change request fulfillment.

6   Evaluation

The described approach on operator-based workflow adaptation has been implemented as component of the CAKE framework [1]. To demonstrate its usefulness, the approach is experimentally evaluated to analyze whether the workflows can be improved regarding the change request fulfillment (Hypothesis H1) and to validate whether the adapted workflows are of an acceptable quality (Hypothesis H2). As only workflow operators are applied leading to a consistent workflow, the resulting workflows are consistent, which was checked and confirmed in order to validate the correctness of the implementation.

H1. The operator-based workflow adaptation considerably improves the change request fulfillment of a workflow.

H2. The operator-based workflow adaptation does not significantly reduce the quality of workflows.

6.1 Evaluation Setup

We manually constructed 60 pasta recipe workflows from the textual recipe descriptions on www.studentrecipes.com with an average size of 25 nodes and 64 edges [23]. Altogether, they contain 162 different ingredients and 67 tasks. For ingredients and tasks, a taxonomy was manually constructed. The extracted workflows, contained AND, XOR, as well as LOOP structures. The repository of 60 workflows was split into two data sets: One repository containing 10 arbitrary workflows (referred to as query workflows) and a case base containing the remaining 50 workflows. For the learning algorithm, we set the parameter $\Delta_W = 0$ as the case base only contained pasta workflows and thus only similar recipes. Further, we chose the threshold $\Delta_D = 0.5$ in order to only create exchange operators if the corresponding head data nodes have been mapped with a similarity of at least 0.5. In total 9416 workflow adaptation operators (1504 exchange, 4460 insert, 3452 delete operators) were learned from the workflows in the case base. For each query workflow QW_i we retrieved the most similar workflow CBW_i from the case base (referred to as case base workflow) and automatically generated a change request for CBW_i by determining the set of nodes to be added and deleted in order to arrive at QW_i. A change request "DELETE prawns", for example, means that the case base workflow CBW_i uses prawns while the query workflow QW_i does not[1]. We executed the proposed operator-based adaptation method for each of the 10 workflows CBW_i using the corresponding change request. Thus, 10 adapted workflows are computed. During adaptation we chose a similarity threshold between the deletion streamlet and streamlet in the workflow Δ_S as 0.5 in order to only apply operators if at least half of the workflow elements are identical. In average, 8.9 operators (1.6 exchange, 2.8 insert, 4.5 delete operators) had been applied in order to adapt a workflow.

6.2 Experimental Evaluation and Results

To verify hypothesis H1 we computed the average change request fulfillment of the 10 adapted workflows which stands at 69,4 %. As the change request was rather large (in average 22,6 nodes) and not any combination can be represented by the automatically learned adaptation operators a change request of 100 % is not to be expected. Hence, Hypothesis H1 is confirmed.

To evaluate Hypothesis H2 a blinded experiment was performed involving 5 human experts. The experts rated the quality of the 10 case base worklows (workflow before adaptation) and the 10 corresponding adapted workflows. These 20 workflows were presented in random order, without any additional information. Thus, the experts did not know whether the workflow was an original workflow from the case base or an adapted workflow. The experts were asked to assess the quality of each workflow based on 3 rating items on a 5 point Lickert scale (from 1=very low to 5=very high). The rating items comprised the culinary quality

---

[1] The change request only contained ingredients and preparation steps present in the workflows from the case base and no ingredients that are used as mixtures of multiple ingredients (e.g., vegetable mix).

Table 1. Item rating assessment

	Better case base workflows	Better adapted workflows	Equal
Correctness of preparation	23	11	16
Culinary quality	21	12	17
Plausibility of preparation order	19	13	18
Aggregated quality	27	16	7

of the recipe, the correctness of the preparation (e.g. slice milk would violate the correctness), and the plausibility of the preparation order. Additionally, we computed an aggregated quality using the three rating items.

The ratings from the 5 experts of all 10 workflow pairs were compared, leading to 50 ratings. We define that one item was rated better for a workflow if it was scored with a higher value than the corresponding item of the compared workflow. Based on this, we conclude that a workflow has a higher aggregated quality, if more of its items were rated better than those of the compared workflow.

The results for each rating item in isolation and for the aggregated quality assessment are given in Table 1. It shows the number of workflows for which the case base workflow or the adapted workflow is better, as well as the number of workflows which were equally rated. In 23 out of 50 rated workflow pairs, the adapted workflow was rated of higher or equal quality (concerning the aggregated quality), whereas 27 case base workflows were rated higher. Thus, in 46 % of the assessments, the adaptation produced workflows with at least the same quality compared to the corresponding workflow from the case base. Additionally, Table 2 illustrates the average rating difference on the items of all 50 workflow pairs. In total, the items of each case base workflow are rated 1.1 higher than those of the adapted workflow, which means the single items were rated about 0.37 times better than the corresponding item of the adapted workflow. Thus, the experts rated the items and hence the quality of the case base workflows only slightly higher. This has also been proved by a paired t-test on the aggregated quality, which showed that the quality difference between the case base workflows and the adapted workflows is statistically not significant ($p = 0.19$). Altogether, Hypothesis H2 is confirmed.

Further, we asked the experts to give textual explanations in case of bad quality ratings. We identified three major reasons for quality degradations caused by the adaptation process and we sketched first ideas to overcome them.

1. It must be ensured that some components should occur only once (e.g. a sauce). This could be achieved by a few general operators that have to be defined manually.
2. After the removal of an input data object from a task the name of the output produced by the task might have to be changed (e.g. removing meat from a mixture doesn't produce a meat mixture anymore). To identify the best

Table 2. Average differences on item ratings

Correctness of preparation	0.46
Culinary quality	0.44
Plausibility of preparation order	0.20
Average per item	0.37
Average per workflow	1.1

suitable output name, outputs from similar tasks with similar input data could be employed.

3. The application of insert operators may also insert new data objects, e.g. ingredients that usually have to be processed before they are used (e.g. bolognese sauce). However, this information was not always included (e.g. a packet sauce could be used instead). To remedy this shortcoming, insert operators processing the desired data can be searched and included into the workflow.

7   Conclusions and Related Work

We presented a novel approach to operator-based adaptation of workflows, including a new representation for workflow adaptation operators, an algorithm for learning such operators, and a search-based process for applying the operators in order to address a specified workflow change request.

The major challenge of adaptation in case-based reasoning is the acquisition bottleneck of adaptation knowledge. Hence, various approaches for learning and applying of adaptation knowledge for cases represented as attribute-values have been proposed [4,8,9,14,16]. Only litte work is published that addresses this problem for more complex case representations (e.g., [15]), or for POCBR in particular. In POCBR, related work was presented by Minor et al. [18]. They propose a workflow adaptation approach which transform workflows by applying a single adaption case that can be acquired automatically [19]. In our own previous work, we introduced a compositional workflow adaptation method [22] which identifies subcomponents of a workflow, called streams. In contrast to this, the operators proposed here represent smaller subcomponents with a higher granularity. Further, the operator-based adaptation not only exchanges similar components, but inserts, deletes or exchanges different components of the workflow. Moreover, in this paper we propose to learn operators from pairs of similar workflows, while in compositional adaptation each single case is decomposed into streams. Dufour-Lussier et al. [6] presented a compositional adaptation approach for processes, requiring additional adaptation knowledge.

Our evaluation showed that the presented approach is promising for the purpose of workflow adaptation. However, the expert evaluation revealed some shortcomings on the quality of the adapted cases. Hence, we sketched ways how to overcome these which we will expore in the future. Future work will also investigate generalization [23] to improve the applicability of the learned adaptation operators. Moreover, an extension by operators exchanging a data node with multiple data nodes and vice versa is planned. We will also investigate methods to control the retention of learned adaptation operators, as already proposed by Jalali and Leake [11] for adaptation rules. Further, we plan to perform comparisons of the different adaptation approaches we already proposed [17, 22, 23] and other related work (e.g., [18]). Finally, we aim at integrating them into a more comprehensive formal framework.

Acknowledgements. This work was funded by the German Research Foundation (DFG), project number BE 1373/3-1.

References

1. Bergmann, R., Gessinger, S., Görg, S., Müller, G.: The collaborative agile knowledge engine cake. In: Proceedings of the 18th International Conference on Supporting Group Work, pp. 281–284, ACM (2014)
2. Bergmann, R., Gil, Y.: Similarity assessment and efficient retrieval of semantic workflows. Inf. Syst. **40**, 115–127 (2014)
3. Bergmann, R., Wilke, W.: Towards a new formal model of transformational adaptation in case-based reasoning. In: Prade, H. (ed.) 13th European Conference on Artificial Intelligence (ECAI 1998), pp. 53–57. John Wiley & Sons (1998)
4. Craw, S., Jarmulak, J., Rowe, R.: Learning and applying case-based adaptation knowledge. In: Aha, D.W., Watson, I. (eds.) ICCBR 2001. LNCS (LNAI), vol. 2080, pp. 131–145. Springer, Heidelberg (2001)
5. Davenport, T.: Process Innovation: Reengineering Work Through Information Technology. Harvard Business Review Press, Boston (2013)
6. Dufour-Lussier, V., Lieber, J., Nauer, E., Toussaint, Y.: Text adaptation using formal concept analysis. In: Bichindaritz, I., Montani, S. (eds.) ICCBR 2010. LNCS, vol. 6176, pp. 96–110. Springer, Heidelberg (2010)
7. Dumas, M., van der Aalst, W., ter Hofstede, A.: Process-aware Information Systems: Bridging People and Software Through Process Technology. Wiley, Hoboken (2005)
8. Fuchs, B., Lieber, J., Mille, A., Napoli, A.: Differential adaptation: an operational approach to adaptation for solving numerical problems with CBR. Knowl. Based Syst. **68**, 103–114 (2014)
9. Hanney, K., Keane, M.T.: Learning adaptation rules from a case-base. In: Smith, I.F.C., Faltings, B. (eds.) EWCBR 1996. LNCS, vol. 1168, pp. 179–192. Springer, Heidelberg (1996)
10. Hung, P., Chiu, D.: Developing workflow-based information integration (WII) with exception support in a web services environment. In: Proceedings of the 37th Annual Hawaii International Conference on System Sciences 2004, p. 10 (2004)
11. Jalali, V., Leake, D.: On retention of adaptation rules. In: Lamontagne, L., Plaza, E. (eds.) ICCBR 2014. LNCS, vol. 8765, pp. 200–214. Springer, Heidelberg (2014)
12. Kapetanakis, S., Petridis, M., Knight, B., Ma, J., Bacon, L.: A case based reasoning approach for the monitoring of business workflows. In: Bichindaritz, I., Montani, S. (eds.) ICCBR 2010. LNCS, vol. 6176, pp. 390–405. Springer, Heidelberg (2010)
13. Leake, D.B., Kendall-Morwick, J.: Towards case-based support for e-Science workflow generation by mining provenance. In: Althoff, K.-D., Bergmann, R., Minor,

M., Hanft, A. (eds.) ECCBR 2008. LNCS (LNAI), vol. 5239, pp. 269–283. Springer, Heidelberg (2008)

14. Li, H., Li, X., Hu, D., Hao, T., Wenyin, L., Chen, X.: Adaptation rule learning for case-based reasoning. Concurr. Comput. Pract. Exp. **21**(5), 673–689 (2009)

15. Lieber, J., Napoli, A.: Using classification in case-based planning. In: ECAI, pp. 132–136, Citeseer (1996)

16. McSherry, D.: Demand-driven discovery of adaptation knowledge. In: Dean, T. (ed.) IJCAI, pp. 222–227, Morgan Kaufmann (1999)

17. Minor, M., Bergmann, R., Görg, S.: Case-based adaptation of workflows. Inf. Syst. **40**, 142–152 (2014)

18. Minor, M., Bergmann, R., Görg, S., Walter, K.: Towards case-based adaptation of workflows. In: Bichindaritz, I., Montani, S. (eds.) ICCBR 2010. LNCS, vol. 6176, pp. 421–435. Springer, Heidelberg (2010)

19. Minor, M., Görg, S.: Acquiring adaptation cases for scientific workflows. In: Ram, A., Wiratunga, N. (eds.) ICCBR 2011. LNCS, vol. 6880, pp. 166–180. Springer, Heidelberg (2011)

20. Minor, M., Montani, S., Recio-Garcia, J.A.: Process-oriented case-based reasoning. Inf. Syst. **40**, 103–105 (2014)

21. Montani, S., Leonardi, G., Lo Vetere, M.: Case retrieval and clustering for business process monitoring. In: Proceedings of the ICCBR 2011 Workshops, pp. 77–86 (2011)

22. Müller, G., Bergmann, R.: Workflow streams: a means for compositional adaptation in process-oriented CBR. In: Lamontagne, L., Plaza, E. (eds.) ICCBR 2014. LNCS, vol. 8765, pp. 315–329. Springer, Heidelberg (2014)

23. Müller, G., Bergmann, R.: Generalization of workflows in process-oriented case-based reasoning. In: 28th FLAIRS Conference, AAAI, Hollywood (Florida), USA (2015)

24. Schumacher, P., Minor, M., Walter, K., Bergmann, R.: Extraction of procedural knowledge from the web. In: Workshop Proceedings WWW 2012, Lyon, France (2012)

25. Taylor, I.J., Deelman, E., Gannon, D.B.: Workflows for e-Science. Springer, London (2007)

26. Weber, B., Wild, W., Feige, U.: CBRFlow: enabling adaptive workflow management through conversational case-based reasoning. In: Funk, P., González Calero, P.A. (eds.) ECCBR 2004. LNCS (LNAI), vol. 3155, pp. 434–448. Springer, Heidelberg (2004)

27. Wilke, W., Bergmann, R.: Techniques and knowledge used for adaptation during case-based problem solving. In: del Pobil, A.P., Mira, J., Ali, M. (eds.) IEA-1998-AIE. LNCS, vol. 1416, pp. 497–506. Springer, Heidelberg (1998)

Fault Diagnosis via Fusion of Information from a Case Stream

Tomas Olsson[1,2]([✉]), Ning Xiong[1], Elisabeth Källström[3],
Anders Holst[2], and Peter Funk[1]

[1] School of Innovation, Design, and Engineering,
Mälardalen University, Västerås, Sweden
{tomas.olsson,ning.xiong,peter.funk}@mdh.se
[2] SICS Swedish ICT, Isafjordsgatan 22, Box 1263, 164 29 Kista, Sweden
{tomas.olsson,anders.holst}@sics.se
[3] Volvo Construction Equipment, 631 85 Eskilstuna, Sweden
elisabeth.kallstrom@volvo.com

Abstract. This paper presents a novel approach to fault diagnosis applied to a stream of cases. The approach uses a combination of case-based reasoning and information fusion to do classification. The approach consists of two steps. First, we perform local anomaly detection on-board a machine to identify anomalous individual cases. Then, we monitor the stream of anomalous cases using a stream anomaly detector based on a sliding window approach. When the stream anomaly detector identifies an anomalous window, the anomalous cases in the window are classified using a CBR classifier. Thereafter, the individual classifications are aggregated into a composite case with a single prediction using a information fusion method. We compare three information fusion approaches: simple majority vote, weighted majority vote and Dempster-Shafer fusion. As baseline for comparison, we use the classification of the last identified anomalous case in the window as the aggregated prediction.

Keywords: Case-based reasoning · Information fusion · Anomaly detection · Fault diagnosis

1 Introduction

Fault diagnosis is about detecting and identifying problems in machines ideally before they lead to a system failure [1,2]. The type of faults we consider in this paper is faults of subcomponents that do not immediately impact the overall function of the system, but can be detected by monitoring the system using sensors. For instance, a minor oil leakage would not immediately affect the function of a gear switch or an engine, but would lead to gradual performance degeneration. Case-based reasoning (CBR) has since the beginning of the field been used for fault diagnosis in various ways [3,4] and also in combination with other machine learning and signal processing approaches [5–7]. Traditionally in

© Springer International Publishing Switzerland 2015
E. Hüllermeier and M. Minor (Eds.): ICCBR 2015, LNAI 9343, pp. 275–289, 2015.
DOI: 10.1007/978-3-319-24586-7_19

CBR, and other related fields, the focus has been on classifying individual cases, although there are attempts to make collective classification using the surrounding as additional source of information [8].

In the new era of big data, we are no longer looking at individual cases but at streams of data that are produced on the fly, many times in a speed that cannot be managed by traditional methods [9,10]. Previously, data streams were mainly stemming from large on-line services such as Google Inc. and Facebook, but currently, traditional industrial companies are also investigating new approaches on how to collect and analyse data generated by their industrial equipment and machines [11,12].

In a previous paper, we presented a framework for remote fault diagnosis of heavy-duty machines, where faults were detected on-board the machines using an anomaly detector and then diagnosed off-board with a CBR approach [13]. However, the diagnosis was only done for individual cases. In another paper, we have presented a probabilistic approach to aggregating individual anomalies in order to assess the anomalousness of a group of cases [12]. In the current work, we propose a new approach to fault diagnosis that builds on top of and combines the above two approaches, where we are not only looking at individual cases, but at a stream of cases.

In this paper, we assume that we are monitoring a stream of events with related segments of sensor signals, such as a series of gear switches as in our previous paper [13]. Then, the signal segments constitute the individual cases that a local anomaly detector classifies as normal or anomalous. Further, we assume that there can be misclassifications, false positives, so that the individual cases are not reliable as the only source for a diagnosis. For instance, individual gear switch segments can appear anomalous also without the present of a true fault. However, if the number of anomalies among the most recent cases increases and becomes larger than normal, it can be used as an indication of a true fault. Therefore, we also assume that there is a method that confidently can assess whether the most recent cases as a group are indeed anomalous [12]. This anomalous group of cases is thereafter considered a composite case. Then, the problem we address is to identify the type of fault that is behind the anomaly. For this purpose, we apply information fusion approaches to fuse the evidence from the individual cases in the composite case in order to infer the most likely reasons for the anomaly.

The paper is organised as follows. Section 2 gives a motivating overview of the heavy-duty machine diagnosis problem where this approach is meant to be applied. In Sect. 3, we present our diagnostic framework and the extension to analysing a stream of cases proposed in this paper. In Sect. 4, the proposed approach is evaluated on three publicly available data sets used for simulating streams of data. In Sect. 5, we describe related work. Section 6 ends the paper with a summary of the work and some conclusions and future work.

2   Background

Improving uptime is important in the heavy-duty construction equipment companies in order to avoid machine failures. Especially important is to avoid major failures that lead to the machine standing still until the fault is corrected. Then, the customer looses money while the machine stands still and may also spend money on repairs if the warranty period is already ended. This altogether leads to customer dissatisfaction. The heavy-duty construction equipment companies also spend a lot of money on warranty costs, which tend to negatively affect the overall profit of the company.

The uptime issues discussed above may be addressed by implementing an intelligent on-board diagnosis in the heavy-duty machine. In this way early faults can be addressed well in advance before a major failure occurs. Further, the current on-board diagnosis systems use simple rules and maps to carry out diagnosis and as such, most failures are not easy to diagnose as a result of too many fault codes generated when there is a failure. Thus, the Engineers and Technicians may have to spend substantial time to identify the failure.

The transmission and axle are parts of such components whose failures may result in costly downtime. Thus, monitoring the components in both the transmission and the axle on-board via an intelligent diagnostic method will result in improved uptime in the heavy-duty machine.

Signals from the transmission that may be of interest to monitor are quantities that provide information about the torque converter, gears, clutches, bearings, the oil quality in the transmission, and of course the vibration levels. For the axle, it could be of interest to closely monitor the planetary gears, oil quality in the axle, the vibration levels, the bearings, and differentials. Thus, monitoring the behaviour of relevant quantities in the transmissions and axles via the proposed fault diagnosis framework on-board the heavy-duty construction equipment may enhance the on-board diagnosis to prevent machine failure.

3   Diagnostic Framework

In Fig. 1, we show the original diagnostic framework, where there are three components: the first two are on-board the machine and the third is off-board the machine [13]. The first component extracts features from the signals on-board the machine. Then, the second component assesses whether the extracted features are anomalous or not. The third component performs case-based diagnosis by estimating the severity of the fault. The original framework only supported individual case diagnosis.

Figure 2 shows the extended framework where we added a fourth component that performs anomaly detection over a stream of cases. Thus, only the most recent cases that together indicate an anomaly are now considered for diagnosis as a composite case. The stream anomaly detector can be located both on-board the machine or off-board the machine. If located on-board the machine, we would only send data when an anomaly occurs, so that the communication could be

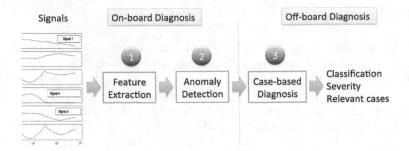

Fig. 1. The original on-board and off-board fault diagnosis system.

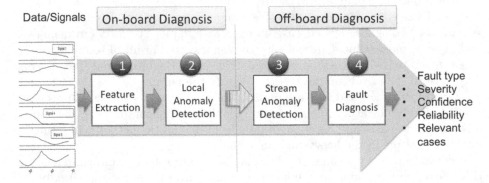

Fig. 2. The extended fault diagnosis framework.

minimised. However, if it is run off-board the machine we can also consider the individual anomalies from many other machines. In addition, with the proposed approach, we can use both real-valued anomaly score (higher value for larger anomalies) and binary-valued anomaly scores (0 for normal or 1 for anomalous cases). For off-board analysis, binary-valued anomaly scores could also minimise the communication, but would lead to a less sensitive detector than for real valued anomaly scores. Nevertheless, in this paper, we only look at a single machine and we simplify the setup by only using binary anomaly scores.

We use the same (local) anomaly detector as in the previous paper. However, a difference will be that we need a data stream instead of isolated individual cases. Thus, in the next section, we will summarise the anomaly detection methods presented in the previous papers [12,13]. Since fault detection of the individual cases is not in focus of this paper, we will use the simplest setup tried in the previous paper for local anomaly detection leaving out the Gaussian mixture model from [13]. The section after that presents the new approach to fusing the most recent individual cases into an aggregated prediction.

3.1   Local Anomaly Detection and Stream Anomaly Detection

As local anomaly detector, we again use logistic regression trained on both normal and anomalous cases where we use weight of importance for each case inverse to the class frequency in order to manage imbalanced data sets [14].

Logistic regression is a probabilistic binary classifier that outputs predictions as probabilities [15]. Logistic regression is defined as follows for $c \in [0, 1]$ and a feature vector x:

$$p(c = 1|x) = \frac{1}{1 + \exp(-\omega^T x)} \qquad p(c = 0|x) = \frac{\exp(-\omega^T x)}{1 + \exp(-\omega^T x)}$$

where ω is a weight vector with $K+1$ weights assuming that x has $K+1$ features including an extra feature that is 1 for all cases. Then, the case is classified as $c = 1$ if $\omega^T x \geq 0$ and $c = 0$ otherwise. The ω is found by using $L1$ regularisation and maximum likelihood estimation optimised for accuracy.

In addition, we assume that there is a fully functioning stream anomaly detector called PRAAG (Probabilistic Anomaly Aggregation) that uses the approach presented in [12], but applied to a sliding window with the most recent cases in a stream. In short, PRAAG works as follows. Assume that we have a large sliding window with the N last cases. Then, we have a smaller sliding window with the m $(< N)$ most recent cases. In addition, for all cases we have an individual anomaly score that can be used to rank the cases from least anomalous to most anomalous. Assuming that the anomaly scores are independent and identically distributed for the normal cases, we can then compute how unlikely the ranking of the cases is in the most recent window with respect to the ranking of cases in the large window. We will not go into the details of using PRAAG for analysing streaming data. However, for the simple case with binary anomalies scores (0 or 1), PRAAG simply computes the probability of observing the number of ones or more in the small window compared to the fraction of ones in the large window as defined below.

Definition 1. *Let P be the number of ones in the large window, then the probability of observing k number of ones in the small window of size m by chance alone is Binomially distributed with $p = \frac{P}{N}$ (the fraction of ones in the large window):*

$$bin(k; p, m) = \binom{m}{k} p^k (1-p)^{m-k} \tag{1}$$

Then, as a measure of anomalousness, we compute the probability of observing k or more number of ones:

$$\bar{A}(k; p, m) = 1 - \sum_{y=0}^{k-1} bin(y; p, m) \tag{2}$$

As an aggregated anomaly score, we use the negative logarithm of (2). Then, we assume that only windows less probable than 1 per 1.000.000 are anomalies, that is, $\bar{A}(k; p, m) < 10^{-6}$. This results in a very low false detection rate.

3.2   Stream Classification with Case Fusion

In this section, we assume that an anomalous window of the most recent cases (the small window) has been detected. The anomalous window constitutes a composite case of the individual cases. Then, the problem is how to identify the fault type of the composite case without relying only on a single anomalous case. Alternatively, a ranked list of the most likely faults can be created that would also work when there are multiple faults. For this purpose, we will give the problem a generic description to which several methods can be applied.

The composite case is a series of $i \in \{1, 2, \ldots, m\}$ tuples containing a feature vector x_i representing a case and an anomaly score a_i. The $a_i \in \{0, 1\}$ is computed using the anomaly detector implemented by logistic regression, where 1 means that the case is anomalous and 0 means it is non-anomalous. Then, we use a method to assign to each case a weight distribution $p_i(y) \in [0, 1]$ over possible fault classes $y \in S$, such that $\sum_{y \in S} p_i(y) = 1$, where the weight measures our belief or confidence in each class y. If $p_i(y) = 1$, we are fully confident of the classification while $p_i(y) = 0$ means that we lack confidence in the classification. A composite case is then a series of tuples $E = \{(x_i, a_i, p_i)\}_{i=1}^{m}$.

We estimate $p_i(y)$ using a k-nearest neighbour (kNN) classifier approach using the Manhattan distance metric for each case x_i of the composite case. The attributes are weighted using the maximum information coefficient [15]. Thus, we also assume we have a case library of a set of individual cases labeled with a specific fault. As estimate of $p_i(y)$, we use the fraction of nearest neighbours labeled with the same class y. This is a similar setup as we used in [13]. However, in principle, any prediction method can be used in our approach.

For fusing the weight distributions p_i of the composite case E into an aggregated prediction, we use three different approaches: (1) the simple majority vote, (2) the weighted majority vote [16], and (3) Dempster-Shafer theory based fusion [17]. As lower bound of the performance, we use the predicted class of the last detected anomalous individual case in the composite case. In the following, p_E is the fused weight distribution, $\hat{y}$ is the fused classification of the composite case computed as the class with maximum weight and S is the set of all classes.

Simple Majority Vote Fusion. The first fusion approach is the simple majority vote among the individual cases with $a_i > 0$. That is:

$$\hat{y} = \arg\max_{c \in S} \sum_{i=1, a_i>0}^{m} I(c = \arg\max_{y} p_i(y)) \tag{3}$$

where $I(\cdot)$ is the indicator function that is 1 when the argument is true or 0 if the argument is false, and the fused weights are $p_E(c) = \sum_{i=1, a_i>0}^{m} I(\ldots), \forall c \in S$.

Weighted Majority Vote Fusion. The second fusion approach is the weighted majority vote among the individual cases with $a_i > 0$. That is:

$$\hat{y} = \arg\max_{c \in S} \sum_{i=1, a_i>0}^{m} p_i(c) \tag{4}$$

where the fused weights is the sum of the individual weights so that $p_{\boldsymbol{E}}(c) = \sum_{i=1,a_i>0}^{m} p_i(c), \forall c \in S$.

Dempster-Schafer Theory Fusion. The Dempster-Schafer theory (D-S theory) is one of the powerful tools to inference with information under uncertainty [17,18]. This approach is especially useful when a case in the window is classified into multiple classes with different degrees of confidence.

Obviously the frame of discernment in this context is the set S of all possible classes. The information that case i in the composite case is classified into multiple possible classes can be interpreted as the following basic probability values:

$$m(c, i) = p_i(c) \text{ for any } c \in S \tag{5}$$

According to D-S theory, the sum of degrees of belief on all subsets is equal to one. Now we have probabilities (degrees of belief) on single faults, whose summation can be less than one in the general case. In principle, by definition of the belief structure, we can distribute the remaining amount on any other subsets. But, in our context, without any other information, it would be a reasonable practice to assign the remaining belief (probability mass) to the subset containing all classes. Thus we can write

$$m(S, i) = 1 - \sum_{\forall c} m(c, i) = 1 - \sum_{\forall c} p_i(c) \tag{6}$$

Equations (5) and (6) can be considered as a basic probability assignment function, which is induced according to the information of case i as evidence.

Next we attempt to combine the basic probability assignment functions for individual cases into an overall assessment using the evidence combination rule. Denote $\boldsymbol{E}_t \subseteq \boldsymbol{E}$ as the ordered set of cases in the composite case containing the individual case $1, 2, \ldots, t$, so that $\boldsymbol{E}_m = \boldsymbol{E}$.

Let $m(c, \boldsymbol{E}_t)$ be the basic probability value to which the hypothesis that the aggregated class c is supported by the evidences in $\boldsymbol{E}_t$ (which also is the fusion weight for this method). By $m(S, \boldsymbol{E}_t)$ we denote the remaining probability mass unassigned to single class after the t first cases in $\boldsymbol{E}$ have been combined. The algorithm to fuse case information according to the evidence combination rule can be formulated in a recursive form as follows:

$$\begin{aligned} m(c, \boldsymbol{E}_{t+1}) &\propto m(c, \boldsymbol{E}_t)m(c, t+1) + m(c, \boldsymbol{E}_t)m(S, t+1) + \\ &\qquad m(S, \boldsymbol{E}_t)m(c, t+1) \\ m(S, \boldsymbol{E}_{t+1}) &\propto m(S, \boldsymbol{E}_t)m(S, t+1) \end{aligned} \tag{7}$$

where we left out a normalising factor (indicated by $\propto$)

$$K_{t+1} = 1 - \sum_{\substack{c_1, c_2 \in C \\ c_1 \neq c_2}} m(c_1, \boldsymbol{E}_t)m(c_2, t+1) \; t = 1, 2, \ldots, m-1 \tag{8}$$

We divide the expressions with the normalising factor to make the sum of the basic probability values induced by the evidences equal to one. To start with the above recursive form, we have $m(c, \boldsymbol{E}_1) = m(c, 1)$ and $m(S, \boldsymbol{E}_1) = m(S, 1)$. The final outcomes of this combination procedure are $m(c, \boldsymbol{E})$ and $m(S, \boldsymbol{E})$, which correspond to the basic probability values after incorporating all cases in the window as evidences. It bears noting that the probability mass $m(c, \boldsymbol{E})$ also represents the degree of belief in class c after considering all individual cases. Hence the combined degrees of belief are directly given by

$$\beta_c = m(c, \boldsymbol{E}) \quad \forall c \in S \tag{9}$$

$$\beta_S = m(S, \boldsymbol{E}) = 1 - \sum_{\forall c} \beta_c \tag{10}$$

where β_S refers to the degree of belief unassigned to any single class after all individual cases have been incorporated. Further $\beta_c + \beta_S$ gives the upper bound of the likelihood for the hypothesis that the aggregated class is c. The lower bound of the likelihood for c as the aggregated class is reflected by the belief degree β_c. In other words, we obtain the interval $[\beta_c, \beta_c + \beta_S]$ as the estimate of the probability for class c by using the combination rule in the D-S theory. For simplicity, the fused weights are also defined as the probability mass $p_{\boldsymbol{E}}(c) = m(c, \boldsymbol{E}), \forall c \in S$.

4   Experiments

The proposed approach was evaluated using a combination of public data sets and simulations. We have used three data sets with multiple classes and more than one thousand cases from the UC Irvine Machine Learning Repository [19]. The first data set is the Steel plate faults data set with 1941 cases and 7 types of steel plate faults [20]. The second data set is the Landsat data set with 6435 cases and 6 classes of soil types. The last data set is the Shuttle data set with 58000 cases and 7 classes.[1] For each data set, we have selected one or more of the original classes to be the normal class, and the remaining classes were used as faults. Table 1 shows a summary of the data sets and which original classes were used as normal class and fault classes respectively. The numbers do not always sum up to the total number since only a subset of the classes was used.

Table 1. The data sets with normal and fault cases with corresponding sizes.

Data set	Normal class	Fault classes	#normal cases	#fault cases
Steel plate	3, 6, 7	1, 2, 4, 5	1466	475
Landsat	1, 3, 7	2, 4, 5	4399	2036
Shuttle	1	4, 5	45586	12170

The simulations are run according to Algorithm 1. We run the simulations 500 times for each class: the normal class and all fault classes. A simulation run

---

[1] Thanks to NASA for allowing us to use the Shuttle data sets.

consists of 999 simulation steps where, in each step, a case is randomly drawn from the normal class. Then, at step 1000 a fault is injected by sampling every following step with 50 % probability from a fault class. However, we iterate over which classes are sampled so that in case of a data set with 3 classes, every third run uses one of the classes as the currently sampled class. This iteration also includes sampling from the normal class so we can catch false detections. The simulation continues additional 1000 steps before a new run is started. Notice that we do not store any anomaly scores as long as the last m computed PRAAG scores are less then 10^{-2} and thus, only store truly non-anomalous scores. However, we do not ignore any cases before the large window is filled, that is, after 500 simulation steps. In all runs, we have used a window size of $N = 500$ for the large window and a size of $m = 50$ for the small window.

At the beginning of each run, we randomly split the data into a training set and a testing set. Since, we want to show that the fusion improves the classification performance for a less than perfect anomaly detector, we only train the anomaly detector on a small subset of the data. In addition, we want to simulate a stream of data and then, we want to be able to sample from a relatively large collection of test data. So, for the steel plate data set, we used 90 % for testing and 10 % for training, and for the other data sets, we used 97 % for testing and 3 % for training. It is also ensured that the fault classes are proportionally distributed between the partitions. Then, we train the anomaly detector to classify cases into normal and anomalous cases, while the kNN is trained to classify cases into the different fault classes without using the normal class. Logistic regression and the kNN were fine-tuned using 5-fold cross validation, optimised for accuracy.

As performance measure we use the accuracy of diagnosing the detected anomalous sliding windows. Accuracy for the information fusion approaches is defined as the fraction of the simulation runs detected by PRAAG that also was correctly classified. Thus, even if PRAAG only detected 70 % of the simulation runs with faults, we can have 100 % accuracy. In addition, although PRAAG is not in focus of this paper, we also report the precision, recall, the mean detection delay and number of false detections for the PRAAG algorithm as a point of reference, including the combined accuracy of the fusion approaches and PRAAG. In the latter case, the accuracy is computed as the fraction of correctly classified simulation runs, where PRAAG detects the normal class. Recall is the fraction of the simulation runs with faults that where accurately identified as anomalous. Precision is the fraction of the simulation runs identified as anomalous that actually had a fault. The detection delay is computed as the number of simulation steps from that a fault was injected until a stream is identified as anomalous. We exclude undetected windows from the computation.

Table 2 shows the accuracy for the different information fusion approaches. As can be seen, the baseline approach performs worst for all data sets. The remaining approaches perform equally well on the Landsat and Shuttle data with almost perfect accuracy (they are not exactly 100 %). However, for the Steel plate data set (the smallest data set), the weighted majority vote is best,

---

Algorithm 1. Pseudocode of the simulation algorithm for classification fusion.

$N \leftarrow 500$ is the number of cases in the large window
$m \leftarrow 50$ is the number of cases in the small window
$CLASSES = [Normal, Fault1, Fault2, \ldots]$
$MaxNumOfRuns \leftarrow 500 \times length(CLASSES)$
for simulation run $\in \{1, 2, 3, \ldots, MaxNumOfRuns\}$ **do**
    Randomly split data into a small **training set** X_0, y_0 and a large **testing set**
    X_1, y_1 (fault classes are stratified)
    Train *Anomaly Detector* to classify normal and anomalous cases using X_0, y_0
    Train *kNN algorithm* to classify faults *only* using the fault cases in X_0,y_0
    for simulation step $\in \{1, \ldots, 2000\}$ **do**
        $currentClass \leftarrow$ next class in CLASSES {Repeatedly iterate over classes}
        if $currentClass = Normal$ **OR** $step < 1000$ **OR** $random() < 0.5$ **then**
            Sample x from X_1 of class Normal
        else
            Sample x from X_1 of the $currentClass$
        end if
        Compute anomaly score $a \in \{0, 1\}$ for case x using *AnomalyDetector*
        Add a to small window (**Remove** oldest value)
        Add a to large window (**Remove** oldest value)
        $k \leftarrow$ number of ones in the small window
        $p \leftarrow$ fraction of ones in the large window
        if $\bar{A}(k; p, m) < 10^{-6}$ **then** {Check anomalousness of small window}
            Classify the small window as $\hat{y}$ using a fusion approach from Sect. 3.2:
            A. Compute p_i for each case x_i with $a_i = 1$ in the small window using $kNNs$
            B. Apply one of the fusion approaches to all p_i to estimate $\hat{y}$
        end if
        if $step > N$ **AND**
        $\bar{A}(k; p, m) < 0.01$ for all of the m most recent computations **then**
        {Keep only truly non-anomalous cases}
        Remove last a from the large window and **restore** old value
        end if
    end for
end for

---

followed closely by the D-S fusion approach, while the simple majority vote is the worst, but not far behind.

Table 3 shows the performance of the PRAAG algorithm, where we notice that the recall is not perfect, although there are very few simulation runs with false detections (only 1 for the shuttle data) and almost perfect precision. The bad recall is reflected in a lower total accuracy as shown in Table 4.

The mean fraction large and small in Table 3 measures the mean fraction of ones (individual anomalous) in each window over all simulation runs at time of detection. With a perfect local anomaly detector, the mean fraction would be 0 (no anomalies) and not larger than 0.5 (50 % probability of a fault class) for the large and small window respectively. Yet, for the Steel plate data set, the large window has about 30 % fraction of ones, which is large, indicating that the

Table 2. Accuracy of the information fusion approaches for detected faults.

Fusion approach	Steel plate	Landsat	Shuttle
Last detected case	0.65	0.87	0.97
Simple majority vote	0.87	1.00	1.00
Weighted majority vote	0.91	1.00	1.00
Dempster-Shafer fusion	0.90	1.00	1.00

Table 3. Metrics showing the performance of the PRAAG algorithm.

Metric	Steel plate	Landsat	Shuttle
Precision	1.00	1.00	0.999
Recall	0.7065	0.667	0.999
#false detections	0	0	1
Mean delay	190 ± 363	26.7 ± 17.8	33.0 ± 26.8
Mean fraction, large	0.297	0.056	0.084
Mean fraction, small	0.639	0.276	0.336

local anomaly detector produces many false positives. This is the reason that the PRAAG recall is low in this case. For the two other data sets, the fraction in the large window is much smaller. Then, the reason of the bad recall in case of the Landsat data set can be seen by looking at the recall for each individual fault (not shown in a table), where the fault class 4 has zero recall while the two remaining fault classes have 100 % recall. Apparently, the local anomaly detector is not able to detect the fault class 4. The mean detection delay also reflects that it is harder to detect the Steel plate data faults, since it takes about 5 or more times longer for the detector to identify the Steel plate faults than the others. So, the results of both the fusion approaches and the PRAAG algorithm indicate that the Steel plate data set is harder to predict than the others, which is probably at least partially due to the smaller number of training cases.

Table 4. Total accuracy of the information fusion approaches for all classes.

Fusion approach	Steel plate	Landsat	Shuttle
Last detected case	0.57	0.69	0.98
Simple majority vote	0.69	0.75	1.00
Weighted majority vote	0.71	0.75	1.00
Dempster-Shafer fusion	0.71	0.75	1.00

The simulation algorithm was implemented in Python using the minepy library for computing the maximal information coefficient [21], the scikit-learn machine learning library for logistic regression and the kNN algorithm [14] and the pyds library for implementing the D-S fusion [22].

5   Related Work

Simple majority vote and weighted majority vote have been used for classifier fusion in many applications [16,23,24]. The same is true for Dempster-Shafer based classifier fusion [25,26]. Yet, our work differs in that we consider not fusing the output from several classifiers, but the outcome of classification for several individual cases. However, it would be easy to also combine the proposed approach with the use of several different classifiers as additional sources for classifications, and thereby, make the diagnosis more robust. Our approach also resembles multi-sensor fusion applications that also can use similar approaches [26,27]. The classification of the individual cases can be considered as information from several sensors.

CBR has been shown to be useful tools for monitoring and diagnostics in various industrial application domains. Two CBR systems were developed in [28] to handle the diagnosis issues with cars, employing direct signals and extracted segment features from signals respectively. Bach et al. [29] used service reports as inputs to CBR to diagnose vehicles with problems. Another system called Cassiopee [30] relies on the combination of CBR and decision trees for dealing with troubleshooting problems of airplane engines. Earlier work in CBR and information fusion has been performed by Olsson et al. [31], exploring the combination of CBR and information fusion for fault diagnosis in industrial robots.

More recently, Yousuf and Cheetham [32] built a hierarchical CBR system to diagnose the failure of turbines with multiple root causes. The overall system consists of reasoning in two levels. The lower level is supported by a set of basic reasoners (implemented as rule-based or case-based reasoning), which have the task to produce the confidence values for each of the five most common root causes. At the upper level, another case-based reasoner is utilised for combining the individual confidences in order to determine the most possible root cause.

6   Conclusions and Future Work

In this paper, we have proposed an approach for fault diagnosis of a stream of cases using a combination of statistical anomaly detection using logistic regression and probabilistic anomaly aggregation (PRAAG), CBR-based diagnosis and fusion of classified cases in a most recent sliding window. Then, we have compared three approaches to information fusion: simple and weighted majority vote and Dempster-Shafer fusion. As baseline, we used the classification of the last detected anomalous individual case. All information fusion approaches were better than the baseline but with comparable performance, but in case of the Steel plate data set, the weighed majority vote was slightly better.

The proposed approach is generic in that any local anomaly detector or diagnostic classifier can be used in the same approach. We can also replace

PRAAG with another detector that can be used for identifying change in the output of the local anomaly detector, for instance, ADWIN [33] or EDDM [34]. So, future work includes investigating the PRAAG algorithm for stream anomaly detection in comparison to other related approaches. Another research direction is to combine the result of multiple classifiers into the prediction and in that way, create an ensemble learner [35].

Further, we do not currently have any examples of composite cases, only individual cases. However, when the system is taken into use, we can start collecting instances of composite cases and store them in an additional case base. In order to support retrieval of composite cases, we will consider a similar approach as we presented in [36]. In [36], cases were retrieved with respect to the similarity of the predictive probability distribution over the classes. In the current work, if normalised, the fused weight distributions produced by the fusion methods can be interpreted as probabilities over the fault classes. Thus, for instance, the distance between two composite cases $E^{(i)}, E^{(j)}$ using the approximate log-prob metric proposed in [36] is then $d(E^{(i)}, E^{(j)}) = \sum_{c \in S} |\log(p_{E^{(i)}}(c)) - \log(p_{E^{(j)}}((c))|$ assuming that $\sum_{c \in S} p_{E^{(i)}} = \sum_{c \in S} p_{E^{(j)}} = 1$. Another approach to representing composite cases would be to store histograms for each feature of the individual cases and use a suitable similarity measure for comparing histograms [37].

Finally, the possibility of multiple faults should also be investigated by for instance looking at multi-label algorithms where the individual cases are labeled with more than one fault or extend the fusion methods to use the conflicting classifications among the individual cases to infer multiple faults.

Acknowledgements. This work has been partially supported by the FP7 EU Large scale Integrating Project SMART VORTEX co-financed by the European Union [38], the Swedish Knowledge Foundation (KK-stiftelsen) [39] through ITS-EASY Research School and Swedish Governmental Agency for Innovation Systems (VINNOVA) grant no 10020, grant no 2012- 01277 and JU grant no 100266.

References

1. Isermann, R.: Fault-Diagnosis Systems: An Introduction from Fault Detection to Fault Tolerance. Springer, Heidelberg (2006)
2. Bengtsson, M., Olsson, E., Funk, P., Jackson, M.: Technical design of condition based maintenance systems - a case study using sound analysis and case-based reasoning. In: 8th International Conference of Maintenance and Realiability, Knoxville, USA (2004)
3. Kockskamper, K., Traphoner, R., Wernicke, W., Faupcl, B.: Knowledge acquisition in the domain of cnc'machining centers: the moltke approach. In: EKAW 1989: Third European Workshop on Knowledge Acquisition for Knowledge-Based Systems, Paris, July 1989, p. 180. AFCET (1989)
4. Althoff, K., Maurer, F., Wess, S., Traphöner, R.: Moltke: an integrated workbench for fault diagnosis in engineering systems. In: Proceedings of the EXPERSYS 1992, Paris (1992)

5. Auriol, E., Crowder, R., McKendrick, R., Rowe, R., Knudsen, T.: Integrating case-based reasoning and hypermedia documentation: an application for the diagnosis of a welding robot at odense steel shipyard. Eng. Appl. Artif. Intell. **12**(6), 691–703 (1999)
6. Yang, B., Han, T., Kim, Y.: Integration of art-kohonen neural network and case-based reasoning for intelligent fault diagnosis. Expert Syst. Appl. **26**(3), 387–395 (2004)
7. Chougule, R., Rajpathak, D., Bandyopadhyay, P.: An integrated framework for effective service and repair in the automotive domain: an application of association mining and case-based-reasoning. Comput. Ind. **62**(7), 742–754 (2011)
8. Gupta, K.M., Aha, D.W., Moore, P.: Case-based collective inference for maritime object classification. In: McGinty, L., Wilson, D.C. (eds.) ICCBR 2009. LNCS, vol. 5650, pp. 434–449. Springer, Heidelberg (2009)
9. Domingos, P., Hulten, G.: Catching up with the data: research issues in mining data streams. In: Proceedings of the Workshop on Research Issues in Data Mining and Knowledge Discovery (2001)
10. Gama, J.: A survey on learning from data streams: current and future trends. Prog. Artif. Intell. **1**(1), 45–55 (2012)
11. Johanson, M., Belenki, S., Jalminger, J., Fant, M., Gjertz, M.: Big automotive data: Leveraging large volumes of data for knowledge-driven product development. In: 2014 IEEE International Conference on Big Data, pp. 736–741. IEEE (2014)
12. Olsson, T., Holst, A.: A probabilistic approach to aggregating anomalies for unsu-pervised anomaly detection with industrial applications. In: Proceedings of the Twenty-Eigth International Florida Artificial Intelligence Research Society Con-ference, May 2015
13. Olsson, T., Källström, E., Gillblad, D., Funk, P., Lindström, J., Håkansson, L., Lundin, J., Svensson, M., Larsson, J.: Fault diagnosis of heavy duty machines: automatic transmission clutches. In: Workshop on Synergies between CBR and Data Mining at 22nd International Conference on Case-Based Reasoning, Septem-ber 2014
14. Pedregosa, F., Varoquaux, G., Gramfort, A., Michel, V., Thirion, B., Grisel, O., Blondel, M., Prettenhofer, P., Weiss, R., Dubourg, V., Vanderplas, J., Passos, A., Cournapeau, D., Brucher, M., Perrot, M., Duchesnay, E.: Scikit-learn: machine learning in Python. J. Mach. Learn. Res. **12**, 2825–2830 (2011)
15. Murphy, K.P.: Machine Learning: A Probabilistic Perspective. MIT Press, Cam-bridge (2012)
16. Kittler, J., Alkoot, F.M.: Sum versus vote fusion in multiple classifier systems. IEEE Trans. Pattern Anal. Mach. Intell. **25**(1), 110–115 (2003)
17. Shafer, G.: A Mathematical Theory of Evidence. Princeton University Press, Princeton (1976)
18. Smets, P.: Belief functions. In: Smets, P., et al. (eds.) Non-Standard Logics for automated Reasoning, pp. 253–286. Academic Press, San Diego (1988)
19. Bache, K., Lichman, M.: UCI machine learning repository (2013)
20. Steel Plates Faults Data Set. Source: Semeion, Research Center of Sciences of Communication, Rome, Italy. www.semeion.it: https://archive.ics.uci.edu/ml/datasets/Steel+Plates+Faults. Accessed July 2015
21. Albanese, D., Filosi, M., Visintainer, R., Riccadonna, S., Jurman, G., Furlanello, C.: minerva and minepy: a C engine for the MINE suite and its R, Python and MATLAB wrappers. Bioinformatics **29**(3), 407–408 (2013)
22. pyDS: a python library for performing calculations in the dempster-shafer theory of evidence (2014). https://github.com/reineking/pyds

23. Bauer, E., Kohavi, R.: An empirical comparison of voting classification algorithms: Bagging, boosting, and variants. Mach. Learn. **36**(1–2), 105–139 (1999)
24. Kuncheva, L.I.: A theoretical study on six classifier fusion strategies. IEEE Trans. Pattern Anal. Mach. Intell. **24**(2), 281–286 (2002)
25. Al-Ani, A., Deriche, M.: A new technique for combining multiple classifiers using the Dempster-Shafer theory of evidence. J. Artif. Intell. Res. **17**, 333–361 (2002)
26. Basir, O., Yuan, X.: Engine fault diagnosis based on multi-sensor information fusion using Dempster-Shafer evidence theory. Inf. Fusion **8**(4), 379–386 (2007)
27. Yang, G.Z., Andreu-Perez, J., Hu, X., Thiemjarus, S.: Multi-sensor fusion. In: Yang, G.Z. (ed.) Body Sensor Networks, pp. 301–354. Springer, London (2014)
28. Wen, Z., Crossman, J., Cardillo, J., Murphey, Y.: Case-base reasoning in vehicle fault diagnostics. In: Proceedings of the International Joint Conference on Neural Networks, vol. 4, pp. 2679–2684. IEEE (2003)
29. Bach, K., Althoff, K.-D., Newo, R., Stahl, A.: A case-based reasoning approach for providing machine diagnosis from service reports. In: Ram, A., Wiratunga, N. (eds.) ICCBR 2011. LNCS, vol. 6880, pp. 363–377. Springer, Heidelberg (2011)
30. Heider, R.: Troubleshooting CFM 56–3 engines for the Boeing 737 using CBR and data-mining. In: Smith, I., Faltings, B.V. (eds.) EWCBR 1996. LNCS, vol. 1168, pp. 512–518. Springer, Heidelberg (1996)
31. Olsson, E., Funk, P., Xiong, N.: Fault diagnosis in industry using sensor readings and case-based reasoning. J. Intell. Fuzzy Syst. **15**(1), 41–46 (2004)
32. Yousuf, A., Cheetham, W.: Case-based reasoning for turbine trip diagnostics. In: Agudo, B.D., Watson, I. (eds.) ICCBR 2012. LNCS, vol. 7466, pp. 458–468. Springer, Heidelberg (2012)
33. Bifet, A., Gavaldá, R.: Learning from time-changing data with adaptive windowing. In: Proceedings of the 2007 SIAM International Conference on Data Mining, Society for Industrial and Applied Mathematics, p. 443 (2007)
34. Baena, M., del Campo, J., Fidalgo, R., Bifet, A., Gavaldà, R., Morales, R.: Early drift detection method. In: Fourth International Workshop on Knowledge Discovery from Data Streams (2006)
35. Maclin, R., Opitz, D.: Popular ensemble methods: an empirical study. J. Artifi. Intell. Res. **11**, 169–198 (1999)
36. Olsson, T., Gillblad, D., Funk, P., Xiong, N.: Explaining probabilistic fault diagnosis and classification using case-based reasoning. In: Lamontagne, L., Plaza, E. (eds.) ICCBR 2014. LNCS, vol. 8765, pp. 360–374. Springer, Heidelberg (2014)
37. Cunningham, P.: A taxonomy of similarity mechanisms for case-based reasoning. IEEE Trans. Knowl. Data Eng. **21**(11), 1532–1543 (2009)
38. SMART VORTEX: scalable semantic product data stream management for collaboration and decision making in engineering. http://www.smartvortex.eu/. Accessed July 2015
39. KK-Stiftelse: Swedish Knowledge Foundation: http://www.kks.se. Accessed July 2015

Argument-Based Case Revision
in CBR for Story Generation

Santiago Ontañón[1]([✉]), Enric Plaza[2], and Jichen Zhu[1]

[1] Drexel University, Philadelphia, PA 19104, USA
santi@cs.drexel.edu, jichen.zhu@drexel.edu
[2] IIIA, Artificial Intelligence Research Institute CSIC,
Spanish Council for Scientific Research,
Campus UAB, 08193 Bellaterra, Catalonia, Spain
enric@iiia.csic.es

Abstract. This paper presents a new approach to case revision in case-based reasoning based on the idea of argumentation. Previous work on case reuse has proposed the use of operations such as case amalgamation (or merging), which generate solutions by combining information coming from different cases. Such approaches are often based on exploring the search space of possible combinations looking for a solution that maximizes a certain criteria. We show how Revise can be performed by arguments attacking specific parts of a case produced by Reuse, and how they can guide and prevent repeating pitfalls in future cases. The proposed approach is evaluated in the task of automatic story generation.

1 Introduction

Case-based reasoning systems are based on the hypothesis that "similar problems have similar solutions", and thus new problems are solved by reusing or adapting solutions of past problems. However, how to reuse or adapt past solutions to new problems, and how to revise these solutions are some of the least understood problems in case-based reasoning. There are multiple open problems such as what knowledge is required for adaptation and how to acquire it [20], the relation between solution reuse and case retrieval [17], and solution revision [10]. This paper builds upon previous work on search-based reuse in case-based reasoning, and specifically on approaches based on amalgam or merge operators [3,12], where a solution to a given problem is generating by amalgamating the problem with one or more retrieved cases.

Specifically, in this paper we focus on the following problem: search-based approaches to case reuse employ some sort of search mechanism over the space of solutions trying to either maximize or satisfy some evaluation function, that hopefully captures the quality of the proposed solution. However, in some domains, such as automated story generation [6] (which we used as our application domain), defining an evaluation function that captures the quality of a solution is a very hard problem. For this reason, we propose a new case revision approach that integrates argumentation into the case reuse process. Each time

© Springer International Publishing Switzerland 2015
E. Hüllermeier and M. Minor (Eds.): ICCBR 2015, LNAI 9343, pp. 290–305, 2015.
DOI: 10.1007/978-3-319-24586-7_20

the case reuse process proposes a solution, this is evaluated against a collection of arguments that may attack the solution, forcing the case reuse to search for alternative solutions. We claim that rather than capture how "good" a story is, it is easier to define a collection of arguments that attack certain negative aspects of the story. These arguments can be kept for future episodes, to prevent generating stories that suffer from the same problems.

The remainder of this paper is organized as follows. We first introduce some background on amalgam-based case reuse, and on argumentation. After that we present our motivating domain: automatic story generation. Argument-based revision is then presented, followed by an experimental evaluation. The paper closes with conclusions and directions for future work.

2   Background

Stories (cases) are represented in the formalism of feature terms [1], and case reuse is implemented as an amalgam of two feature terms: a source term and a target term. We will briefly introduce here the basic notions of feature terms and amalgamation; for more a detailed explanation see [13].

Feature terms are defined by their *signature*: $\Sigma = \langle \mathcal{S}, \mathcal{F}, \leq, \mathcal{V} \rangle$. $\mathcal{S}$ is a finite set of sort symbols, including $\bot$ representing the most general sort ("any"), and $\top$ representing the most specific sort ("none"). $\leq$ is an order relation inducing a single inheritance hierarchy in $\mathcal{S}$, where $s \leq s'$ means s is more general than or equal to s', for any $s, s' \in \mathcal{S}$ ("any" is more general than any s which, in turn, is more general than "none"). $\mathcal{F}$ is a set of feature symbols, and $\mathcal{V}$ is a set of variable names. We define a feature term ψ as: $\psi :: = X{:}s\ [f_1 \doteq \Psi_1, \cdots, f_n \doteq \Psi_n]$, where ψ points to the *root* variable X (that we will note as $root(\psi)$) of sort s; $X \in \mathcal{V}$, $s \in \mathcal{S}$, $f_i \in \mathcal{F}$, and Ψ_i is either a variable $Y \in \mathcal{V}$, or a set of variables $\{X_1, ..., X_m\}$. The set of variables present in a term ψ is noted $Var(\psi)$; for instance, the term shown in Fig. 2 has 18 variables, one for each node.

The basic relation over feature terms is *subsumption* ($\sqsubseteq$), i.e. given two terms ψ_1 and ψ_2 we say $\psi_1 \sqsubseteq \psi_2$ (ψ_1 subsumes ψ_2) when ψ_1 is a generalization of ψ_2, or dually ψ_2 is a specialization of ψ_1. Subsumption generates a total mapping $m\colon Var(\psi_1) \rightarrow Var(\psi_2)$ satisfying certain conditions such as $m(root(\psi_1)) = root(\psi_2)$, or if $X.f = Y$ then $m(X).f = m(Y)$ (formal definition in [13]).

The *unification* of two terms ψ_1 and ψ_2, $\psi_1 \sqcup \psi_2$, is the most general term subsumed by both and the dual notion of *antiunification* of two terms ψ_1 and ψ_2, $\psi_1 \sqcap \psi_2$, is the most specific term that subsumes both. There might be more than one antiunification and more than one unification for two given terms. Also, although an antiunification always exist, two terms might not unify. After this summary, we are able to define the *amalgam* of two terms.

2.1   Amalgam-Based Case Reuse

An *amalgam* of two terms is a new term that contains *parts from these two terms*. For instance, an amalgam of 'a red French sedan' and 'a blue German

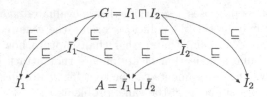

Fig. 1. A diagram of an amalgam A from inputs I_1 and I_2 where $A = \bar{I}_1 \sqcup \bar{I}_2$.

minivan' is 'a red German sedan'; clearly there are always multiple possibilities for amalgams, like 'a blue French minivan'.

In this paper we define an amalgam in a feature term language $\mathcal{L}$ as:

Definition 1 (Amalgam). *A term $A \in \mathcal{L}$ is an amalgam of two inputs I_1 and I_2, with anti-unification $G = I_1 \sqcap I_2$, if there exist two generalizations $\bar{I}_1$ and $\bar{I}_2$ such that (1) $G \sqsubseteq \bar{I}_1 \sqsubseteq I_1$ (2) $G \sqsubseteq \bar{I}_2 \sqsubseteq I_2$, and (3) $A = \bar{I}_1 \sqcup \bar{I}_2$*

When $\bar{I}_1$ and $\bar{I}_2$ have no common specialization then trivially $A = \top$, since their only unifier is "none". For our purpose we will be only interested in non-trivial amalgams (those different from $\top$) of the input pair, which we call their *amalgam space*. This definition is illustrated in Fig. 1, where the anti-unification of the inputs is indicated as G, and the amalgam A is the unification of two concrete generalizations $\bar{I}_1$ and $\bar{I}_2$ of the inputs; for short we call $\bar{I}_1$ and $\bar{I}_2$ the *transfers* of amalgam A. Usually we are interested only on maximal amalgams of two input terms, i.e., those amalgams that contain maximal parts of their inputs that can be unified. Formally, an amalgam A of inputs I_1 and I_2 is maximal if there is no other non-trivial amalgam A' of inputs I_1 and I_2 such that $A \sqsubset A'$.

In our system, amalgamation is used in the Reuse process to create a new story, by combining a story from the case base with a target specifying desired aspects of the story to be generated.

2.2   Argumentation

Computational argumentation, in the abstract framework, consists of a set of nodes (called arguments, intuitively understood as formulas) and an attack relation among pairs of nodes. An abstract argumentation framework $AF = \langle Q, R \rangle$ is composed by a finite set of arguments Q and an attack relation R among the arguments [4]. For instance, an attack relation written $\alpha \twoheadrightarrow \beta$ means that argument α is attacking argument β. In our previous work [14] on learning by communication we integrated inductive concept learning with a concrete argumentation model in the A-MAIL framework (where A-MAIL stands for argumentation-based multiagent inductive learning). In particular, a case e in the case base could serve to attack an argument as a counter-example $e \twoheadrightarrow \beta$. Here A-MAIL is a concrete argumentation framework, not abstract like Dung's, and one main difference is that while Dung's assume a finite, known set of arguments we assume an open-ended set of arguments. As we shall see, using arguments to revise cases during the CBR cycle is essentially an open-ended process, since more often than not the knowledge (here in the form of arguments) used to revise cases is external to the CBR system.

Another difference is that during the Revise process, the case is assumed to be a concrete, instantiated formula — while in the A-MAIL framework examples and counterexamples were instantiated, and the other arguments were assumed to be general formulas. The usual definition of an attack $\alpha \rightarrow \beta$ is that α concludes the opposite of β and $\beta \sqsubset \alpha$ (α is a specialization of β). Section 4 introduces the role of arguments in the Revise process of the CBR cycle.

3   Automatic Story Generation

Compared with the established narrative forms such as prose fiction computer-generated stories are still in their early stage. Despite the recent progress in the area, these stories are still fairly rudimental in terms of both the depth of meanings and the range of their varieties.

Automatic story generation is an interdisciplinary topic focusing on devising models for algorithmically producing narrative content and/or discourse. Story generation is an important area for interactive digital entertainment and cultural production. Built on the age-old tradition of storytelling, algorithmically generated stories can be used in a wide variety of domains such as computer games, training and education. In addition, research in story generation may shed light into the broader phenomena of computational creativity [6].

Different techniques have been studied in story generation, the most common of which is automated planning. Salient examples of planning-based story generation systems include Tale-Spin [11], Universe [9] and Fabulist [16]. By contrast, computational analogy algorithms have not been sufficiently explored in the domain of story generation. An alternative approach is that of using case-based or analogy-based approaches. Examples of this alternative approach are MISTREL [19], MEXICA [15] or the work of Gervás et al. [7], which used case-based approaches, or SAM [21], which uses computational analogy.

Specifically, in this paper we focus on a case-based approach, and address the following problem: given a partially specified story (target), and a collection of fully specified stories (case-base), how can we generate a new story by reusing one of the cases in the case base (source)? This is an important problem in story-generation since it would allow for a significant amount of authorial control over the output of the story generator (controlling the target story), while providing a fully automated way to suggest completed stories based on the target.

4   Argument-Based Revision

The Revision process in the CBR cycle introduces knowledge that is external to the CBR system to evaluate and/or improve the outcome of the Retrieve and Reuse processes. The situation is similar to supervised Machine Learning where an external source (called "oracle") gives new information to the learning system on its output; this feedback is used by the learning system to perform credit and blame assignment on its learnt structures, modify them accordingly, and increase performance over time (i.e. learn from interacting with the supervisor). Now, different forms of interacting with the oracle define different modalities of learning. The most common modality in supervised learning is when, in classification tasks, a system predicts as solution a class for an instance, and the oracle either accepts it as correct or, if not, provides the correct class for that instance. This modality is common in CBR systems in classification tasks, where the oracle "revises" the predicted class when the system is wrong and provides the correct class (thus, the revised case is formed and can be retained in the case base with the correct solution). However, oracles can provide different information: e.g. an oracle can provide a yes/no feedback to the system given an instance, but does not provide the correct answer. In semi-supervised learning approaches, such as reinforcement learning, the oracle provides a numerical value that estimates how good is the solution provided by the system.

For CBR systems in more complex tasks than classification, the Revise phase usually assumes an external oracle (that might be a human expert or a domain model) that can provide a revised solution that is correct, so the system can learn. Other approaches, like "critics" in the CHEF system, are able to detect failures on a recipe and apply repair strategies (e.g. add or remove steps in the recipe [8]). This approach is based on analyzing the failure of the plan being executed (in the real world or a simulated world).

The approach we take is to consider the interaction between system and oracle as a restricted form of dialogue, in which the system provides a tentative solution and the oracle's feedback is an argument attacking the parts of the solution that, according to that oracle, are wrong or unsatisfactory. This is particularly interesting in creative domains, such as storytelling, in which what is a "wrong" output (as in classification) or a "failure" (as in executing a plan), is rather difficult to determine. In the long run, our goal would be to have a dialogue between system and oracle on the features that are positive or negative in a particular story being generated. For the purposes of this paper, we focus on the more simple scenario where (1) the CBR system presents a solution (a story), (2) the oracle's feedback is one or several arguments attacking specific aspects or parts of the story, and (3) the system incorporates arguments (as well as the ones provided by the oracle in previous cycles) and generates a new solution (story) that is coherent with most of the previous oracle's argument. We will presently define the notions of "argument" and generation of new stories coherent with a set of arguments.

4.1   Arguments and Attacks

Computational argumentation usually defines an attack relation $\alpha \twoheadrightarrow \beta$ between two logical statements α and β. However, our situation is slightly different, in that we have a story represented as a feature term ψ that is a large and complex structure, and an argument that will attack not the whole story but a particular

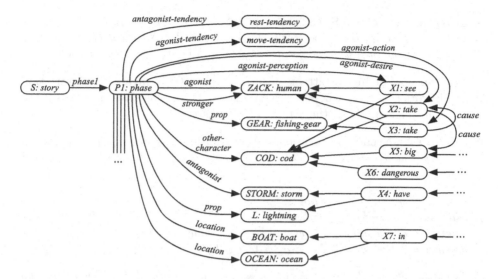

Fig. 2. An example target story used in our experiments (corresponding to the target in S/T3) represented as a feature term. This describes a situation where a human named *Zack*, is in a boat in the middle of the ocean. Zack is taking his fishing gear because he wants to fish a very large cod he has seen. At the same time, there is a storm with lightnings.

part of it (in our formalism, a sub-term of ψ). We will discuss the form of arguments first, and later the attack relation between an argument and an aspect or part of a story.

Definition 2 (Argument). *An argument is a pair* (π, α), *where* π *is a term and* α *is a logical formula over terms with conjunction, disjunction and negation (for terms* $\phi, \phi' \in \mathcal{L}$), *specifically,* α *may have one of the following forms:* ϕ, $\neg\phi$, $\phi \vee \phi'$, *and* $\phi \wedge \phi'$.

Intuitively, an argument (π, α) states that if a story ψ satisfies π (e.g., there is a dragon as antagonist), then the story must also satisfy α (e.g., only a magical weapon can kill the dragon). When the story ψ satisfies π but not α then we say that the argument *attacks* ψ. Moreover, the attacking argument is retained by the system, so in subsequent iterations it would prefer stories that satisfy the argument to those that do not. Therefore, arguments cannot be understood as constraints, but rather as preferences (sometimes called *soft constraints*).

In order to define attacks on stories, represented as feature terms, we need to introduce notation to define subterms of a term. Let $Var(\psi)$ denote the set of variables in term ψ; for instance, the term shown in Fig. 2 has 18 variables, one for each node. Given a variable $X \in Var(\psi)$, the subterm ψ_X is the term with root in variable X, intuitively the subgraph reachable from node X. For instance, in Fig. 2 the variable X_1 has a subterm formed by the root X_1 and the features that go to *Zack* and *cod*, or if we take the variable P_1 then the subterm is the graph shown in Fig. 2 describing Phase 1 of the story.

Definition 3 (Pattern Satisfaction). *Let π and ψ be terms in $\mathcal{L}$, hereby called* pattern *and* description *respectively. Given a variable $X \in Var(\psi)$, we say a description ψ satisfies a pattern π through X if $\pi \sqsubseteq \psi_X$, we write $\pi \sqsubseteq_X \psi$.*

We can now define an attack of an argument (π, α) against a description ψ; the intuition is that whenever ψ satisfies π then, if ψ does not satisfy the patterns in the formula α, the argument attacks ψ.

Definition 4 (Attack). *An argument (π, α) attacks a description ψ, written $(\pi, \alpha) \rightarrowtail \psi$ whenever $\exists X \in Var(\psi)$ such that $\pi \sqsubseteq_X \psi$ and one of the following holds:*

1. *when $\alpha = \phi$ and $\phi \not\sqsubseteq_X \psi$ holds,*
2. *when $\alpha = \neg\phi$ and $\phi \sqsubseteq_X \psi$ holds,*
3. *when $\alpha = \phi \vee \phi'$ and $\phi \not\sqsubseteq_X \psi$ and $\phi' \not\sqsubseteq_X \psi$ hold,*
4. *when $\alpha = \phi \wedge \phi'$ and $\phi \not\sqsubseteq_X \psi$ or $\phi' \not\sqsubseteq_X \psi$ hold.*

Each subsumption relation $A \sqsubseteq_X B$ generates a mapping between $Var(A)$ and $Var(B)$ (as defined in Sect. 2). The mapping generated when testing subsumption of α must respect the mapping generated for π. Moreover, if π subsumes ψ for more than one mapping, each one of these mappings constitutes a different attack.

For the purposes of this paper we do not use nested logical connectives; our arguments use only simple negation, conjunction and disjunctions. The definition of attack is the converse of satisfaction with respect to the formula α; thus satisfying $\phi \vee \phi'$ means that either one (ϕ or ϕ') is satisfied in ψ, and then there is no attack. Because of this, for $\alpha = \phi \vee \phi'$ accomplishing an attack, it means that neither ϕ nor ϕ' are satisfied in ψ. We will use the notation $|a \rightarrowtail \psi|$ to denote the number of variables in $Var(\psi)$ that are attacked by the argument $a = (\pi, \alpha)$.

For the particular case where an argument wants to express that the formula α has to be satisfied regardless of any precondition π, we use the notation $(\bot, \alpha)$.

Finally, notice that a new argument (π, α) is retained by the system, and will be used to generate new stories where the ones that satisfy (are not attacked by) argument (π, α) are preferred. More generally, given a set of known arguments $Args$, the system will generate stories by exploring the space of amalgams and preferring those that satisfy (are not attacked by) more arguments in $Args$.

4.2 Argument-Based Revision Algorithm

This section presents a specific Argument-based Revision Algorithm (ARA) that combines the ideas presented above. Specifically, the algorithm we propose performs a greedy search over the amalgam space, starting with the most general amalgam possible (the anti unification of the two input terms I_1 and I_2), and it specializes it iteratively, employing arguments to determine which of the possible specializations is the most promising to pursue.

Algorithm 1. ARA(I_1, I_2, f, $Args$)

```
1: t = 0, A_0 = A* = I_1 ⊓ I_2, Ī_1^0 = Ī_2^0 = A_0
2: loop
3:     t = t + 1
4:     R_1^t = specializations(Ī_1^{t-1})
5:     R_2^t = specializations(Ī_2^{t-1})
6:     C = {I ⊓ Ī_2^{t-1}|I ∈ R_1^t} ∪ {Ī_1^{t-1} ⊓ I|I ∈ R_2^t}
7:     if C = ∅ then
8:         return A*
9:     else
10:        A_t = argmax_{A∈C} evaluation(A, f, Args)
11:        if evaluation(A_t, f, Args) > evaluation(A*, f, Args) then A* = A_t
12:        Ī_1^t = A_t ⊓ I_1, Ī_2^t = A_t ⊓ I_2
13:    end if
14: end loop
```

ARA is shown in Algorithm 1, and works as follows. Given two terms I_1 and I_2 (which in our case represent the target case, and the retrieved case), an evaluation function f and a set of arguments $Args$:

1. Step 1 initializes the current amalgam, A_0, the currently best amalgam A^* and the current two transfer terms $\bar{I}_1^0$ and $\bar{I}_2^0$ to be equal to the antiunification of the two input terms (the most general amalgam possible).
2. Then, at each iteration t, first ARA finds the set of possible specializations of the current two transfers (this is done using a *refinement operator* over feature terms [13]).
3. Line 6 computes all the possible next amalgams, resulting from unifying the next specializations with the previous transfer terms.
4. Lines 10–11 select the best amalgam, and line 12 updates the transfer terms for the next iteration. The way the best amalgam is determined is where arguments come into play. Each argument a in our framework is assigned a weight w_a. The weight of an argument represents how serious is the issue that this argument tries to prevent[1]. Each amalgam is then assessed as follows:

$$evaluation(A, f, Args) = f(A) - \sum_{a \in Args} w_a \times |a \twoheadrightarrow A|$$

where:

- f is an evaluation function that provides a basic score for an amalgam. For example, f could encode things like "larger amalgams are preferable" by giving higher scores to amalgams with a larger number of variables. In our experiments, we used the function: $f(A) = |A| - k \times |Var(A)|$, where $|A|$ is the size of the term A (number of times we need to specialize $\perp$

---

[1] In the experiments shown later we use a hand-fixed weight equal for all arguments; determining individual weights is discussed in future work.

using the refinement operator to reach A), and captures the size of the story, and $k = 4$ in our experiments. A larger story means that we have been able to transfer more information from the source and target in the amalgam, so a general goal is to maximize $|A|$. The number of variables in A is $|Var(A)|$. When unifying the two transfers, we would like variables of one transfer to be mapped to variables of the other. If this is not the case, the number of variables in the resulting amalgam grows. Thus, minimizing the number of variables in the amalgam has the effect of maximizing the number of variables from the source that are mapped to the target.

- The final term subtracts the weight of each attacked argument multiplied by the number of times the argument attacks some subterm with root X of the story A.

The effect of the ARA algorithm is to find amalgams that strike a balance between maximizing the evaluation function f, and minimizing the number of attacking arguments. The next section describes the generation of stories using amalgams and arguments in an experimental scenario.

5   Experimental Evaluation

In order to evaluate our argument-based revision approach, we prepared an experimental setup that bypasses case retrieval altogether. Thus, we prepared four source/target pairs: S/T1, S/T2, S/T3 and S/T4. In order to compare our results with previous work, we translated the source/target pairs used in our previous work [21] to the feature term formalism used in the approach presented in this paper[2]. Below we provide details on the stories used for evaluation, and then we provide empirical results illustrating the performance of our approach.

5.1   Dataset

As mentioned above, we represent stories using *feature terms* [1]. Specifically, a story is a term composed of a sequence of *phases*, where a phase represents a given instant in a story. Each phase contains a set of characters, locations, props, actions and relations. We separated characters into three groups: the *agonist* (protagonist or main character), the *antagonist* (the main opposing force), and *other characters*. Entities that are not characters are classified into *locations* and *props*. For each of these characters and objects, we specify a collection of properties such as: their relations to other characters, whether they are performing any action, or their desires, likes and possessions.

Additionally, the high-level structure of each phase is captured by annotating who is the agonist, the antagonist, and their force relation (inspired in the cognitive linguistics framework of *force dynamics* [18]). This allows us to represent,

---

[2] The specific source/target pairs used for this experiment can be downloaded from https://sites.google.com/site/santiagoontanonvillar/software.

Table 1. Properties of the source-target (S/T) pairs used in our study.

	Source		Target			
	Phases	Fefinements	Phases	Refinements	Surface similarity	Structural similarity
S/T 1	4	203	1	73	Low	Low
S/T 2	4	203	1	62	Low	High
S/T 3	2	90	1	79	High	Low
S/T 4	3	134	2	89	High	High

in a compact way, the high-level structure of a story. Figure 2 shows an example story used in our dataset (the target in S/T3).

As mentioned above, we translated the four source/target pairs used in [21] to feature terms. These four pairs were selected because they represent a variety of scenarios based on how similar the target is to the source. We distinguish two types of similarity between stories: *surface* similarity and *structural* similarity. The former refers to whether two stories contain similar concepts (e.g., both contain a boat and a fish), while the latter refers to whether they have a similar structure (e.g., both refer to a story where the main character overcame a difficulty and succeeded). Surface similarity can be measured by the percentage of keywords shared between two stories, and structural similarity is measured by how much the force dynamics structures representing the stories match.

Table 1 shows some statistics of the stories used in our evaluation. For each story, we show the size in number of phases, and number of refinements (this is the number of times we have to apply the specialization refinement operator to ⊥ to obtain the given story). This shows that target stories in our dataset tend to be smaller than the source ones (expected, since they are only partially specified). Additionally, we show which source/target pairs are similar in terms of surface and structural similarity.

5.2 Experimental Setup

We evaluated our approach in the following way. First, we run our Reuse approach with an empty set of arguments for each of the four source/target pairs. This gets us a baseline against with which to compare. Then, we iteratively generated arguments to address all the issues we observed in the solution generated for S/T1. This resulted in a total of 15 arguments. Using those arguments, we then report on the performance of the system in generating stories again for all the four source/target pairs. We did not generate arguments for the other source/target pairs purposefully, in order to assess the extent to which arguments generated for one story can be used to improve performance in other stories. To compare results, we report the value of the evaluation function f for the solution found, and the number of attacks that the resulting story receives with the 15 arguments we generated. The number of attacks should be seen as a proxy for the number of syntactic or semantic mistakes that the generated stories contain

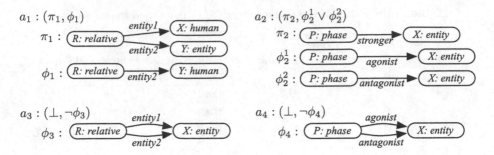

Fig. 3. Four example arguments generated during Revise and used in our experimental evaluation.

(where a "syntactic mistake" would be something like having a character that is never specified as agonist, antagonist or other-character, and a "semantic mistake" would be something like driving a boat on land). Additionally, we report subjective impressions on the generated stories. Finally, we compare the results obtained with those obtained in our previous work with the same source/target pairs, but with a different algorithm [21].

Figure 3 shows four of the 15 arguments we generated. Specifically, those four arguments capture the following:

- a_1: If a human in the story has a relative, it must also be human.
- a_2: If an entity is marked as "stronger" in a phase (a force dynamics annotations), that entity must be the agonist or the antagonist.
- a_3: An entity cannot be a relative (parent, son, sibling) of itself.
- a_4: An entity cannot be the agonist and the antagonist at the same time.

Notice that those four example arguments are basically capturing things that could be specified using a stricter ontology. However, having them as arguments, allows us to be flexible, and allowing violations in a story in some cases. For example, violating a_1 would allow for fantasy stories where a human has a parent being a magical being; violating a_4 would allow for "split-self" stories, where the main character is both the agonist and the antagonist of the story. Other arguments capture common sense (e.g., a boat cannot be driven on land), or story aesthetics (e.g., the location of a story should not change from phase to phase), but they are not hard constraints, and there are specific stories that do not comply to one of them. The weight w_a of all the arguments was set to 10.

5.3   Results

The left-hand side Table 2 shows the amount of time taken and number of amalgams explored generating stories using amalgams, but without argument-based revision. Automatically quantifying the quality of stories generated using automatic story generation is an open challenge, and thus, in this paper we report

Table 2. *Score* achieved (result of using the evaluation function f in the final story), number of *attacks* that the final story receives from the set of 15 arguments we used, *time* taken, and number of amalgams *explored* in the four story pairs used in our experiments.

	Not using arguments				Using arguments			
	Score	Attacks	Time	Explored	Score	Attacks	Time	Explored
S/T1	224	9	43 s	5435	207	3	83 s	5785
S/T2	219	3	232 s	4015	210	2	185 s	4081
S/T3	157	2	7 s	1393	144	2	8 s	1463
S/T4	230	6	21 s	2056	215	1	13 s	1909

the number of attacks received by the stories using the 15 arguments we generated as a proxy for the number of errors those stories contain (although this is by no means a reflection of their literary quality, it reflects their coherence). We will also list a collection of issues or interesting details of each of the generating stories we observed.

- S/T1: In the resulting story, our system made a series of semantic mistakes: created a story where the father of one of the characters is a butterfly, and when the sister of the main character passed away, the main character threw her into the toilet (since in the source story, a pet fish died and was flushed down the toilet); and a collection of syntactic mistakes: incorrect force dynamics structure, listed the sister of the main character as a prop, and used a location in an incorrect place of the story structure.
- S/T2: The resulting story is almost syntactically correct since the stories share strong structural similarity, however, given that they talk about very disparate things (they share very little surface similarity), the story is rather surreal. In the resulting story, there is a fish inside of a flower in a backyard of the main character (who wants to play a game there); the fish later dies.
- S/T3: In this case, the resulting story is perfectly valid: the main character wants to fish a giant cod, but he ends up not being able to, since the fishing gear breaks while pulling the cod out of the ocean. It only contains a couple of syntactic mistakes.
- S/T4: Source and target here are significant similarity (in one the main character drives a car up a mountain, and in the other he drives a motor-boat in a bay). The resulting story is almost correct, except for a few semantic mistakes: first, in the generated story, a motorboat is driven up a mountain (which clearly cannot be done), and also the bottom of the mountain is "in the main character" (which doesn't make sense), also there appear to be two motorboats instead of one. The rest of the story is coherent: after driving for a while, the main character realizes he did not fill the tank, and needs to turn around.

The right-hand side of Table 2 shows that, obviously, when incorporating the arguments into the search process, the resulting stories receive fewer attacks, since the search process was directed towards parts of the amalgam space containing stories with fewer attacks. For example the output for S/T1 received only 3 attacks (while the same 15 arguments would generate 9 attacks against the story generated without taking them into account). Moreover, even if the 15 arguments were generated with S/T1 in mind, other stories also receive attacks, and thus benefit from these 15 arguments. Also, as the table shows, the time taken and the number of amalgams explored vary from the case when we do not use argument-based revision, but do not significantly increase. Looking closely at the generated stories, we observed the following:

- S/T1: After argument-based revision, all of the semantic mistakes in this story disappeared, since we provided arguments to address each of them. Only two small syntactic problems persisted (a prop and a location were used without being defined in their appropriate place in the feature term). Notice that this illustrates both the strengths and weaknesses of our approach: on the one hand, it is easy to provide arguments that prevent semantic or syntactic mistakes, but on the other hand, given that stories are generated by amalgamating information from source and target, it is not possible to force the resulting story to have something that was not present in the source nor target just by using arguments.
- S/T2: The only aspects that were improved in this story are the syntactic ones, concerning some minor force dynamic structures. The overall story is the same as without arguments.
- S/T3: No changes were observed in this story.
- S/T4: Almost all the syntactic and semantic errors were eliminated in this story: the "bottom of the mountain" is now "in the island", and not "in the main character", and there is a single motorboat instead of two. The only semantic error that remains is the fact that a motorboat cannot be driven up a mountain (since we had no argument to address this, as all arguments were generated just to attack the issues of S/T1).

Comparing the results obtained using argument-based revision with the results obtained using the same four story pairs by the SAM algorithm [21], we observed the following. First, our approach is able to transfer much more information from the source case to the target case. SAM is based on computational analogy (it uses the SME algorithm [5] internally), and only transfers elements of the source that are related in some way (via analogical mapping) to the target. The amalgam-based approach naturally achieves the same result, but can also transfer information that is not mapped directly to the transfer. For example, SAM was barely able to transfer any information at all for S/T2, whereas our approach generates a full story consisting of four phases. Another example is S/T4 where SAM generated a story that did not have the semantic mistake of driving a motorboat up a mountain (since the mountain was not transferred to the final story), but had a different semantic mistake: driving a motorboat *after* it had ran out of fuel. Additionally, being able to use arguments to guide the search process provides our new approach a natural way to guide the generation process toward regions of the amalgam space that contain better stories.

6   Related Work

We already discussed the use of critics in CBR planning systems [8] and how it relates to our approach. A related topic is that of critiquing-based recommenders. The main goal in recommender systems is to acquire, via user feedback, a better model of the user preferences: "Critiquing systems help users incrementally build their preference models and refine them as they see more options" [2]. Probably user-initiated critiquing is the more similar to our approach, where a user is presented with product features that can be selected as candidates to be changed. Our approach is different, allowing richer feedback using arguments. These arguments, in the form of pairs (condition, soft-constraint), are acquired by the system to improve on its own task, in this domain generating stories, not for user personalization. Moreover, the arguments are integrated as a driving force into the search process of generating amalgams of stories during Reuse.

7   Discussion and Future Work

This paper has presented an approach to case revision based on arguments. The main idea is to generate a collection of arguments that *attack* specific aspects of a given solution that we want to prevent. These arguments are kept by the system to prevent the same aspects from appearing in future solutions (albeit as *soft constraints* only). The approach was incorporated into a search-based case reuse framework and evaluated in a story generation task.

Our results indicate that by generating a small collection of arguments, our approach was able to generate stories of higher quality, and that the same arguments generated to attack a specific story were successfully used to increase the quality of a separate set of stories.

Future work includes allowing the system to defend itself against attacking arguments, by generating counter-arguments based on stories in the case base, or even arguments that support a specific aspect of a story, rather than attack it, moving closer to a full-fledged argumentation model, such as in [14]. However, we'd need a larger case base of stories for achieving a richer dialogue. Moreover, providing some support in deciding the weights w_a used for strengthening the arguments remains as future work. Eventually, we would like to model the notion of an *audience* to which to generated story is generated. Value-based argumentation offers this possibility, by associating arguments to values and modeling an audience as a partially ordered set of values. The order among values may help in determining the weights w_a for their associated arguments.

Acknowledgements. This research was partially supported by projects Colnvent (FET-Open grant 611553) and NASAID (CSIC Intramural 201550E022).

References

1. Carpenter, B.: The Logic of Typed Feature Structures. Cambridge Tracts in Theoretical Computer Science, vol. 32. Cambridge University Press, Cambridge (1992)
2. Chen, L., Pu, P.: Critiquing-based recommenders: survey and emerging trends. User Model. User-Adap. Interac. **22**(1–2), 125–150 (2012)
3. Cojan, J., Lieber, J.: Belief revision-based case-based reasoning. In: Proceedings of the ECAI-2012 Workshop SAMAI: Similarity and Analogy-based Methods in AI, pp. 33–39 (2012)
4. Dung, P.M.: On the acceptability of arguments and its fundamental role in nonmonotonic reasoning, logic programming and n-person games. Artif. Intell. **77**(2), 321–357 (1995)
5. Falkenhainer, B., Forbus, K.D., Gentner, D.: The structure-mapping engine: Algorithm and examples. Artif. intell. **41**(1), 1–63 (1989)
6. Gervás, P.: Computational approaches to storytelling and creativity. AI Mag. **30**(3), 49–62 (2009)
7. Gervás, P., Díaz-Agudo, B., Peinado, F., Hervás, R.: Story plot generation based on CBR. J. Knowl.-Based Syst. **18**(4–5), 235–242 (2005)
8. Hammond, K.: Explaining and repairing plans that fail. Artificial Intelligence **45**, 173–228 (1990)
9. Lebowitz, M.: Creating characters in a story-telling universe. Poetics **13**, 171–194 (1984)
10. de Mántaras, L., López, R., McSherry, D., Bridge, D., Leake, D., Smyth, B., Craw, S., Faltings, B., Maher, M.L., Cox, M.T., Forbus, K., Keane, M.T., Aamodt, A., Watson, I.D.: Retrieval, reuse, revision and retention in case-based reasoning. Knowl. Eng. Rev. **20**(03), 215–240 (2005)
11. Meehan, J.: The Metanovel: writing stories by computer. Ph.D. thesis, Yale University (1976)
12. Ontañón, S., Plaza, E.: Amalgams: a formal approach for combining multiple case solutions. In: Bichindaritz, I., Montani, S. (eds.) ICCBR 2010. LNCS, vol. 6176, pp. 257–271. Springer, Heidelberg (2010)
13. Ontañón, S., Plaza, E.: Similarity measures over refinement graphs. Mach. Learn. **87**(1), 57–92 (2012)
14. Ontañón, S., Plaza, E.: Coordinated inductive learning using argumentation-based communication. Auton. Agents Multi-Agent Syst. **29**(2), 266–304 (2015). http://dx.doi.org/10.1007/s10458-014-9256-2
15. Pérez y Pérez, R., Sharples, M.: Mexica: A computer model of a cognitive account of creative writing. J. Exp. Theor. Artif. Intell. **13**(2), 119–139 (2001)
16. Riedl, M.: Narrative generation: balancing plot and character. Ph.D. thesis, North Carolina State University (2004)
17. Smyth, B., Keane, M.T.: Adaptation-guided retrieval: questioning the similarity assumption in reasoning. Artif. Intell. **102**(2), 249–293 (1998)
18. Talmy, L.: Force dynamics in language and cognition. Cogn. Sci. **12**(1), 49–100 (1988)
19. Turner, S.R.: A model of creativity. In: The Creative Process: A Computer Model of Storytelling and Creativity. Lawrence Erlbaum Associates, Hillsdale (1994)

20. Wilke, W., Bergmann, R.: Techniques and knowledge used for adaptation during case-based problem solving. In: Mira, J., Moonis, A., de Pobil, A.P. (eds.) IEA/AIE 1998. LNCS, vol. 1416, pp. 497–506. Springer, Heidelberg (1998)
21. Zhu, J., Ontañón, S.: Shall I compare thee to another story? - an empirical study of analogy-based story generation. IEEE Trans. Comput. Intell. AI Games 6(2), 216–227 (2014)

CBR Model for Predicting a Building's Electricity Use: On-Line Implementation in the Absence of Historical Data

Radu Platon[✉], Jacques Martel, and Kaiser Zoghlami

Natural Resources Canada, CanmetENERGY, Varennes, Canada
{radu.platon,jacques.martel,kaiser.zoghlami}@nrcan.gc.ca

Abstract. This paper presents the development and on-line implementation of a case-based reasoning (CBR) model that predicts the hourly electricity consumption of an institutional building. Building operation measurements and measured and forecast weather information are used to predict the electricity use for the next 6 h. The model's ability to efficiently deal with an initial absence of historical data and continuously learn as more data becomes available was tested by emptying the database holding historical data prior to the on-line implementation. The prediction accuracy was monitored for almost 4 months. The results show significant improvement as more data becomes available: the initial error, 1 h following the on-line implementation is close to 44 %, it decreases by almost half after 16 h, and reaches 12.8 % at the end of the monitored period. This shows the applicability of a CBR predictive model for new and retrofit buildings where historical data is not available.

Keywords: Case-based reasoning · Building · Electricity consumption · Prediction · Historical data · Continuous learning

1 Introduction

Buildings represent the largest energy consuming sector in the world, being responsible for more than one-third of total energy consumption and for significant greenhouse gas emissions [1]. In North America (U.S. and Canada) alone, institutional and commercial buildings account for 40 % of the total energy consumption [2].

The prediction of energy use in buildings plays a significant role in achieving an efficient operation and reducing the energy load and the environmental impact. The prediction of the electricity use enables the optimization of the operation of building systems in order to reduce consumption, peak demand and costs. It is critical for planning and optimizing the operation of thermal energy storage devices linked to electromechanical heating, ventilation and air conditioning (HVAC) systems, and it can identify periods of excessive consumption in order to improve control strategies.

This paper presents the application of case-based reasoning (CBR) model for predicting the electricity consumption of buildings in the absence of historical operational data. Since CBR models use specific knowledge collected on previously encountered situations to solve new situations, historical data is critical for achieving accurate

© Springer International Publishing Switzerland 2015
E. Hüllermeier and M. Minor (Eds.): ICCBR 2015, LNAI 9343, pp. 306–319, 2015.
DOI: 10.1007/978-3-319-24586-7_21

results. However, in the case of new buildings, or retrofit buildings (having undergone major modifications to the point that previous data is no longer representative of the current operation), historical data is not available. In this situation, a model able to learn and improve its predictive accuracy as more data becomes available is required. The objective of the study described in this paper and the approach used are presented next; a literature review of building energy prediction follows. The building data and modelling input selection are presented in Sect. 2. The CBR model is presented in Sect. 3, and its on-line implementation in the absence of historical and its predictive performance are presented in Sect. 4. Section 5 contains directions for future work, and the conclusion is presented in Sect. 6.

1.1 Objective and Approach

Sound prediction is required for all buildings, not only those having large amounts of historical operational data available. A CBR model that predicts ahead of time the hourly electricity consumption of a Canadian institutional building was developed and implemented on-line. In order to simulate the case of a building where historical data is not available, the database storing measurements was emptied. The objective of this study was to test the model's ability to efficiently deal with an initial absence of historical data and continuously learn as data becomes available. The prediction horizon was set to 6 h, and the prediction accuracy was monitored for a period of almost 4 months.

Hourly measurements of the following variables were available:

- variables related to the operation of building systems – chiller, boilers and air handling units
- building electricity consumption
- weather information – current and forecast values of outside air temperature and relative humidity
- indoor conditions – air temperature

The modelling inputs were selected amongst all available variables based on their statistical importance, as determined by a Principal Component Analysis (PCA); this procedure is explained in the Sect. 2.1 – Selection of Modelling Inputs.

Building operating modes corresponding to office working and non-working hours were identified and a CBR model corresponding to each operating mode was used in order to better represent the electricity use profiles.

1.2 Literature Survey

Early reports on predicting building energy consumption include the Great Energy Predictor Shootout competitions organized by American Society of Heating, Refrigerating and Air-Conditioning Engineers (ASHRAE): artificial neural networks (ANN) models were among the most accurate for predicting hourly building electricity use, chilled water load and heating water load [3], and for estimating hourly energy baselines [4]. A review of different methods used for predicting building energy consumption, including physical and thermodynamic principles, statistical methods and ANN models, is presented in [5]. ANN models were reported to deliver accurate predictions for: building energy loads based on characteristics such as the thickness of the insulation [6]; hourly, daily and monthly energy loads [7–10]; electrical load based on building end-uses [11]. Adaptive ANN models, re-trained periodically with new building operational data to handle changes in operating conditions, were shown to successfully estimate energy use [12]. A comparison between a detailed building energy simulation and an ANN model is presented in [13].

Some CBR building environment applications can be found in the published literature: a decision support model for selecting buildings most suitable to achieve energy savings [14], data selection for developing a building energy load model [15] and a thermal comfort evaluation method that uses knowledge based on past experiences [16]. However, very few examples of CBR being used directly for predicting the energy of a building are reported: a CBR model predicting the electricity use is currently implemented on-line at a Natural Resources Canada (NRCan) building located near Montreal [17]; an improved version of this model, using readily available operational data, was developed for a NRCan building located in Calgary [18].

2 Building Data

A CBR model was implemented on-line at an institutional Canadian facility located in Calgary (Alberta, Canada). The building has a total floor surface area of almost 17,000 m^2 and houses mainly office and storage spaces. The HVAC equipment consists of 5 dual-duct air handling units (AHU) served by a one chiller and 3 natural gas boilers. The building heating load dictates the number of boilers functioning at a given time, and the boiler outlet measurement represents the combined output of all functioning boilers. There is only one AHU that recirculates indoor air, the other four AHUs supply 100 % fresh air.

Hourly measurements of the following variables related to the building operation were available:

- boiler outlet temperature (°C) and flowrate (l/s)
- chiller outlet temperature (°C) and flowrate (l/s)
- hot duct supply air temperatures for 4 AHUs (°C)
- cold duct supply air temperatures for 5 AHUs (°C)
- supply air flowrates for 4 AHUs (%) – the air flow rates were indicated by the speed at which the fans are operated, in terms of the motor variable frequency drive percentage
- return air fan flowrate for 1 AHU (%)
- indoor air temperatures of the east and west wings (°C)
- building electricity consumption (kWh)

Weather information – current outside air temperature and relative humidity, as well as their forecast values for the next 6 h – was also available. The weather forecast is obtained from an Environment Canada website to which the CBR model connects automatically on an hourly basis to ensure that the latest available forecast information is used in the prediction. In all, 25 variables were available.

2.1   Selection of Modelling Inputs

The modelling inputs were selected using a PCA-based procedure. PCA is a statistical modelling technique that identifies correlations between variables and summarizes the dataset using linear combinations of the variables – the principal components. Each component contains interrelated variables and accounts for a certain level of the overall data variability. The components are extracted in decreasing order of importance: the first component reflects the greatest source of variance in the original data, and each succeeding component accounts for as much of the remaining variability as possible. The PCA-based method was carried out on all available variables – except the building electricity consumption, which is the variable to be predicted – and the relevant variables were selected according to their statistical significance, in terms of explained overall variability in the original dataset. Optimal input selection using PCA was extensively reported in the literature, such as for models of the electricity consumption in residential dwellings [19] and in office buildings [20].

Following the PCA analysis, 10 of the available variables were deemed relevant and used as modelling inputs. Weather conditions – outside air temperature and relative humidity – as well as variables related to heating and cooling plant operation, and AHU air temperatures were deemed relevant. Measurements of the indoor air temperature were not deemed relevant, most probably due to the fact that the inside air temperature does not vary significantly; for example, it remains relatively constant during working hours to ensure occupant comfort. A detailed description of the procedure used to select the modelling inputs can be found in [18].

Forecast weather information was also added as input to the model. The modelling inputs are presented in Table 1.

Table 1. Modellig inputs

Variable and measuring units
IN 1: Measured outside air temperature (°C)
IN 2: Measured outside air relative humidity (%)
IN 3: Forecast outside air temperature (°C)
IN 4: Forecast outside air relative humidity (%)
IN 5: Air handling unit 2 supply hot air temperature unit (°C)
IN 6: Air handling unit 3 supply hot air temperature (°C)
IN 7: West wing air handling unit supply cold air temperature (°C)
IN 8: Air handling unit 4 supply cold air temperature unit (°C)
IN 9: Chiller water outlet temperature (°C)
IN 10: Chiller water outlet flow rate (l/s)
IN 11: Boiler outlet water temperature (°C)
IN 12: Boiler outlet water flow rate (l/s)

2.2 Working and Non-working Hours

Two building operating modes were identified according to the electricity use during working hours and during non-working hours. The building consumes approximately 80 % more electricity during working hours – 7 AM to 5 PM – than during non-working hours. The hourly building electricity consumption in 2013 and 2014 for both working and non-working hours operating modes is shown in Fig. 1 – please note that the gap observed from March 29th to May 1st 2013 corresponds to a period of time when measurements were not available.

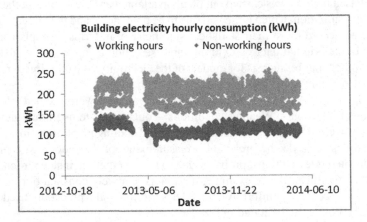

Fig. 1. Building electricity consumption during working and non-working hours

To appropriately represent the building electricity use profile, a CBR model corresponding to each operating mode was used. However, the analysis presented in this paper concentrates only on the model developed using working-hours measurements.

3 CBR Model

The CBR model uses hourly operational measurements and weather information to predict the electricity consumption of the building for the next 6 h. The CBR prediction of building electricity use presented in this study is based on the concept that the current trend of the building electricity consumption can be approximated using past trends occurring at similar operation and weather conditions.

The present conditions for which the prediction is calculated are stored in the present case, while past cases contain similar past conditions. The electricity consumption corresponding to the present case is predicted using consumptions stored in similar past cases. The present case contains current and past hourly measurements of all the input variables, except weather forecast; three measurements are stored: one taken at the present hour, and two corresponding to two prior hours. The forecast weather values – outside air temperature and relative humidity – for the 6 h following the current hour are also stored in the present case.

Selection of similar past cases starts with the indexing process, where past measurements taken at conditions similar to the present conditions are identified. The identification of past similar measurements is carried out based on the following criteria:

- building operating mode – working or non-working hours
- time of measurement – past measurements taken at the same hour of the day as the present measurements, as well as measurements taken one hour before and one hour after: for example, if the prediction is to be carried out today at 2 PM, past measurements taken at 1, 2 and 3 PM are selected
- outside air temperature – past measurements corresponding to outside air temperatures within a ± 2°C interval with respect to the present temperature value

Only measurements satisfying the above criteria are selected during the indexing process, and they become current hours of past cases.

The procedure for arranging the information in past cases is identical to that used for the present case, with the exception that the values of past building electricity consumptions following the current hour are included, since these past values will be used to predict the electricity use of the present case.

Next, for each input variable of a case, the variable similarity between the present case and its corresponding measurements from past cases is calculated using the Euclidean distance between present and past measurements. A factor that decreases linearly as the measurement time is further away from the current hour is used, such that recent measurements are more significant than older ones. This distance is calculated as follows:

$$d\left(X, X'\right) = \frac{\sqrt{\sum_{i=1}^{n} w_i \left(x_i - x'_i\right)^2}}{\sum_{i=1}^{n} w_i} \tag{1}$$

where n represents the number of measurements, $d(X,X')$ is the distance between the measurements of a variable from the present case to a past case, w_i stands for time-decreasing factor of measurement at time i and x_i and x'_i represent the measurement at time i of the variables from the present and past cases, respectively.

Next, the variable similarity is calculated using this distance and minimum and maximum thresholds – d_{min} and d_{max}, respectively. If the distance between two variables is less than d_{min}, the variables are considered to be perfectly similar – similarity of 1. If the distance is superior to d_{max} the variables are considered not at all similar – similarity of 0. The d_{min} and d_{max} values are usually selected based on expert knowledge of the building operation. The variable similarity VS varies linearly between 0 and 1:

$$VS = \begin{cases} 1, & d\left(X, X'\right) \le d_{min} \\ \dfrac{d_{max} - d\left(X, X'\right)}{d_{max} - d_{min}}, & d_{min} < d\left(X, X'\right) < d_{max} \\ 0, & d_{max} \le dd\left(X, X'\right) \end{cases} \tag{2}$$

Next, the similarity between the present case and each past case is calculated using the variable similarities previously calculated, along with weights associated to each input variable. These weights represent the significance of the input variable in predicting the building electricity use. Having M input variables, the case similarity CS is calculated as the weighted average of the input variables' similarities:

$$CS = \frac{\sum_{j=1}^{M} VS_j W_j}{\sum_{j=1}^{M} W_j} \tag{3}$$

where VS_j and w_j represent the variable similarity and the weight of the jth variable, respectively. The weight of each variable was set to 1, since it was assumed all variables have the same impact on the building electricity use.

Next, the building electricity consumption corresponding to the present case is calculated. From all the past cases identified previously, only those having a similarity value superior to a predetermined threshold are used for prediction. This threshold was set to 0.8, meaning that past cases with a similarity value less than 0.8 are not considered similar enough with the present case and they will not be used for prediction. The building electricity consumption of the present case P_t is predicted as a weighted average of the consumptions from each similar past case:

$$P_t = \frac{\sum_{k=1}^{K} CS_k Y_{k,t}}{\sum_{k=1}^{K} CS_k} \tag{4}$$

where K is number of similar cases, CS_k is the similarity of kth case and $Y_{k,t}$ represents the electricity consumption of the kth case at time t. One similar past case is sufficient to carry out the prediction.

An example of a present and past case is shown in Fig. 2:

- the modelling inputs are the variables presented in Table 1 – represented by IN1 to IN12
- the modelling output is the building electricity consumption, denoted by OUT
- the current hour is 10 AM for both the present and past case
- the outside air temperature (IN1) of the present case is 10°C, and that of the past case is 11°C (the difference is within the ± 2°C interval used during the indexing process)
- the forecast values of the outside air temperature and humidity (IN3 and IN4) are available for the 6 h following the current hour
- the electricity consumption values are available in the past case, and they will be used to predict the consumption of the present case, along with the case similarity value determined using variable similarities (VS1 to VS12)

Present case

Date & time	IN 1	IN 2	IN 3	IN 4	IN 5	IN 6	IN 7	IN 8	IN 9	IN 10	IN 11	IN 12	OUT
t6 2015-03-06 14:00			10	32.92									?
t5 2015-03-06 15:00			9	39.92									?
t4 2015-03-06 16:00			8	39.32									?
t3 2015-03-09 07:00			9.3	36.84									?
t2 2015-03-09 08:00			10	34.62									?
t1 2015-03-09 09:00			11	32.14									?
t0 2015-03-09 10:00	10	34.36	12	29.29	29.49	28.76	15.58	24.89	30.85	-0.05	67.33	76.12	203.12
t-1 2015-03-09 11:00	10	34.36	12	31.39	29.28	28.36	10.11	21.66	30.85	-0.05	67.33	76.12	203.51
t-2 2015-03-09 12:00	14	24.47	12	31.60	28.26	28.36	9.71	22.47	30.85	-0.05	66.31	73.05	197.06

VS 1   VS 2   VS 3   VS 4   VS 5   VS 6   VS 7   VS 8   VS 9   VS 10   VS 11   VS 12

Past case

Date & time	IN 1	IN 2	IN 3	IN 4	IN 5	IN 6	IN 7	IN 8	IN 9	IN 10	IN 11	IN 12	OUT
t6 2015-03-05 14:00			9	26.36									*215.00*
t5 2015-03-05 15:00			9.7	28.50									*209.31*
t4 2015-03-05 16:00			11	26.66									*166.38*
t3 2015-03-06 07:00			6.4	38.53									*166.56*
t2 2015-03-06 08:00			5.4	40.97									*195.55*
t1 2015-03-06 09:00			7.6	34.94									*217.54*
t0 2015-03-06 10:00	11	25.53	9.6	30.52	30.09	30.40	17.31	28.28	31.18	-0.05	67.99	76.59	215.09
t-1 2015-03-06 11:00	12	23.89	13	23.15	29.56	28.80	16.27	23.22	31.85	-0.05	67.33	76.59	216.25
t-2 2015-03-06 12:00	12	23.89	12	22.74	29.22	28.46	10.36	21.40	31.85	-0.05	66.65	76.59	196.71

Fig. 2. Example of a present and past case

4   On-Line Model Implementation and Performance

In order to simulate the case of a building where historical data is not available, the database storing measurements was emptied before the on-line implementation of the model. The model's ability to learn and improve its predictive accuracy as more data becomes available was monitored.

The model was implemented on-line on January 19th 2015, and the hourly predictions calculated until April 7th 2015 were analyzed in this study. The predictive performance was calculated in terms of the Coefficient of Variation of the Root Mean Square Error – CV(RMSE) – calculated as the square root of the average of the squares of the error for each observation, normalized to the mean of the measured building electricity consumption.

One hour following the on-line implementation, 1 similar past case is used to carry out the prediction; the error is high, close to 44 %. However, as more data becomes available, the case library grows and the error sharply drops: 16 h following the on-line implementation, the error decreases by half, being close to 22 %. The error continues to decrease, reaching 12.8 % at the end of the monitored period – 468 h following the on-line implementation. The error evolution is shown in Fig. 3, and summarized, with respect to the number of hours following the on-line implementation, in Table 2.

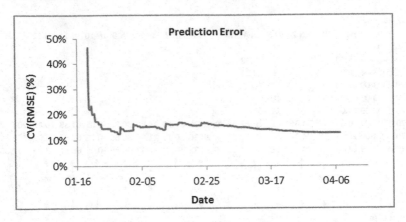

Fig. 3. Predictive error evolution

Table 2. Predictive error evolution summary

Date (2015)	Number of hours after on-line implementation	Error
Jan. 19th	1	44.42 %
Jan. 20th	16	22.39 %
Jan. 23rd	39	15.98 %
March 9th	309	14.99 %
March 17th	375	13.97 %
March 27th	452	12.98 %
April 7th	468	12.79 %

To be noted is the fact that despite having almost 4 months' worth of data, only 468 hourly predictions are generated. This is normal, since this analysis covers only a building operating mode corresponding to working hours (7 AM to 5 PM) and there are fewer working-hours than non-working hours in this period.

The error evolution plot presented in Fig. 3 shows small variations in the error values: the error has a decreasing trend, and suddenly slightly increases before decreasing again. This is most probably explained by the inherent nature of building measurements that can vary due to different reasons, such as equipment start-up or shut-down or erroneous readings from measuring instruments. In case of such variations, the CBR model might experience difficulties in finding similar past cases. However, the predictive accuracy clearly shows significant improvement as more data becomes available.

The CBR model performs very well, having an error of 12.8 % error at the end of the monitored period, which is well within the recommended ASHRAE limits – 30 % for hourly predictions [21]. This is a very accurate result, considering off-line tests showed that the error calculated with a case library containing one year's worth of past data was approximately 13 % [18].

The prediction is carried out for a 6-hour horizon, and situations might occur when, at a given hour, similar past cases could not be formed since the current conditions were not similar enough to those present in the historical dataset. In these situations, the prediction of the building electricity use is not calculated, but predicted values from the previous hour are retained. Out of 468 hourly observations recorded during the monitored period, 57 such situations occurred. The predictive error of these observations is significantly greater than that of the observations corresponding to situations where the prediction was calculated each hour: 23.9 % vs. 10.5 %, respectively. This highlights the importance of having predictions each hour in order to maintain accurate results. These results are presented in Table 3.

Table 3. Predictive error with respect to the presence of hourly predictions

Observations	Number	Error
With and without hourly predictions	468	12.79 %
Without hourly predictions only	57	23.86 %
With hourly predictions only	411	10.54 %

5 Future Work

5.1 Hybrid Predictive Approach

The results obtained in this study show that the CBR model has significant capabilities to deal with an initial absence of historical data and continuously learn and improve as more data becomes available. This makes it suitable for new and retrofit buildings where historical data is not available.

However, when using one year's worth of past data for model development, an ANN model outperformed this CBR model, having an error almost 50 % lower [18]. Although being more accurate, ANN models have weak extrapolation capabilities outside the dataspace used for modelling and they need to be re-calculated when presented with measurements that are not representative of the general trend of the data used for modelling; moreover, these models require a relatively large amount of data to achieve high predictive accuracy. CBR models are able to continuously learn from new data automatically, without having to be re-calculated.

A hybrid approach, combing the strengths of the CBR and ANN modelling methods will be investigated:

- initially, when a small volume of historical or no data at all is available, a CBR model would carry out the predictions and continuously learn and improve as more data becomes available
- later on, as sufficient historical data is available, an ANN will be used to further improve the predictive accuracy

The size of the dataset required to develop an accurate ANN model will be an important part of this investigation.

5.2   CBR Model Weights

In this study, the weights of the inputs used in the CBR model were all set to 1, as it was considered that all the input variables have an equal impact on the prediction and no a-priori knowledge of these variables is available. However, not all the variables might have the same impact on the energy consumption of a building: for example, weather conditions – temperature and relative humidity – typically play a major role in the energy use.

The PCA analysis indicated that some variables are more responsible than others for the overall variability of the dataset; therefore, the weights of the inputs in the CBR model can be selected to reflect the variable significance as determined by the PCA procedure. However, the same variable has different coefficients in each of the principal components (PCs) of the PCA model, and each PC is responsible for explaining different levels of overall data variability. For example, the outside air relative humidity has the following coefficients in the first 3 PCs:

- 0.41 in the first PC, which explains 60.4 % of the data variability
- 0.89 in the second PC, which explains 24.3 % of the data variability
- 0.19 the third PC, which explains 8.4 % of the data variability

A procedure for determining the weights of the inputs in the CBR model as a function of the variable coefficients in each PC as well as the PC's contribution to explaining the overall data variability will be investigated.

An approach for optimizing the weights, either through a mathematical procedure or by using expert knowledge of the building operation, should also be investigated.

5.3   On-Line Data Preprocessing

The quality of the data is critical to obtaining accurate predictions. Measurements not representative of the general trend of the data used for model development might occur due to a variety of factors, such as sudden changes from a normal operation regime, measuring instrument malfunctions, database problems, and equipment start-up or shut-down.

The model might experience predictive difficulties in the presence of such data, and we consider that it is better not to calculate the prediction in this situation, than to calculate an erroneous value that might lead to wrong decisions regarding the operation of the building. As an example, deviations from the general trend of the boiler water outlet temperatures are shown in Fig. 4.

The development of an automatic procedure to identify on-line deviations will be investigated.

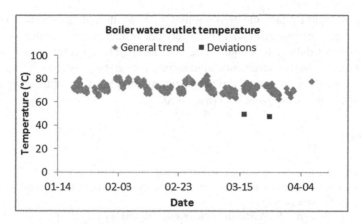

Fig. 4. Boiler water temperature (general trend and deviations)

6 Conclusion

This paper presented the application of a CBR model for predicting the hourly electricity consumption of an institutional building located in Calgary, Canada. The model uses measurements related to the building operation, as well as measured and forecast weather information to predict the building electricity consumption for the next 6 h. The model's ability to efficiently handle an initial absence of historical data and continuously learn and improve as more data becomes available was tested: prior to on-line model implementation, the database holding historical data was emptied. The predictive accuracy was monitored for a period of almost 4 months.

Building operation modes corresponding to working and non-working hours were identified. In order to appropriately represent the building electricity use profiles, a predictive model corresponding to each operating mode was used; as the building consumes significantly more electricity during working hours than during non-working hours, the study presented in this paper concentrates on the working-hours model.

Results show that the predictive accuracy significantly improves as more data becomes available:

- the initial error was 44.4 %, but then it decreased to 22.4 % after only 16 h following the model on-line implementation
- 39 h following the implementation, the error was below 16 %
- after almost 4 months, the error was 12.8 %

The error is well within the recommended ASHRAE limits for hourly predictions, and represents a very accurate result, since off-line tests showed that the CBR model had a predictive accuracy of approximately 13 % when one year's worth of past data was available.

The results indicate that the CBR predictive model is highly applicable for new and retrofit buildings, where historical data is not available. A hybrid approach that combines the strengths of a CBR model – being able to achieve a good predictive performance with a small amount of data and continuously learn as more data becomes available – with the strengths of an artificial neural network model – high predictive accuracy provided sufficient data is available – were discussed. The need for automatic on-line detection of erroneous measurements, or measurements not representative of the general trend of the data used for model development, and the need for optimizing the input weights of the CBR model were also discussed.

References

1. Transition to Sustainable Buildings. International Energy Agency, Paris (2013)
2. North American Intelligent Buildings Roadmap. Continental Automated Buildings Association, Ottawa (2011)
3. Kreider, J.F., Haberl, J.S: Predicting hourly building energy use: the great energy predictor shootout – overview and discussion of results. In: Proceedings of the ASHRAE Annual Meeting, June 25–29 1994, pp. 1104–1118, Florida (1994)
4. Haberl, J.S., Thamilseran, S.: Great energy predictor shootout II measuring retrofit savings. ASHRAE J. **40**, 49–56 (1998)
5. Zhao, H.-X., Magoules, F.: A review on the prediction of building energy consumption. Renew. Sustain. Energy Rev. **16**, 3586–3592 (2012)
6. Ekici, B.B., Aksoy, U.T.: Prediction of building energy consumption by using artificial neural networks. Adv. Eng. Softw. **40**, 356–362 (2009)
7. Gonzalez, P.A., Zamarreno, J.M.: Prediction of hourly energy consumption in buildings based on a feedback artificial neural network. Energy Build. **37**, 595–601 (2005)
8. Karatasou, S., Santamouris, M., Geros, V.: Modeling and predicting building's energy use with artificial neural networks: methods and results. Energy Build. **38**, 949–958 (2006)
9. Ucenic, C., Atsalakis, G.: A neuro-fuzzy approach to forecast the electricity demand. In: Proceedings of the 2006 IASME/WSEAS International Conference on Energy & Environmental Systems, pp. 299–304, Chalkida, Greece (2006)
10. Escrivá-Escrivá, G., Roldán-Blay, C., Álvarez-Bel, C.: Electrical consumption forecast using actual data of building end-use decomposition. Energy Build. **82**, 73–81 (2014)
11. Escrivá-Escrivá, G., Álvarez-Bel, C., Roldán-Blay, C., Alcázar-Ortega, M.: New artificial neural network prediction method for electrical consumption forecasting based on building end-uses. Energy Build. **43**, 3112–3119 (2011)
12. Yang, J., Rivard, H., Zmeureanu, R.: On-line building energy prediction using adaptive artificial neural networks. Energy Build. **37**, 1250–1259 (2005)
13. Neto, A.H., Fiorelli, F.A.S.: Comparison between detailed model simulation and artificial neural network for forecasting building energy consumption. Energy Build. **40**, 2169–2176 (2008)
14. Hong, T., Koo, C., Jeong, K.: A decision support model for reducing electric energy consumption in elementary school facilities. Appl. Energy **95**, 253–266 (2012)
15. Breekweg, M.R.B., Gruber, P., Ahmed, O.: Development of a generalized neural network model to detect faults in building energy performance – Part I. In: ASHRAE Transactions, Atlanta (2000)
16. Kumar, S., Mahdavib, A.: Integrating thermal comfort field data analysis in a case-based building simulation environment. Build. Environ. **36**, 711–720 (2001)
17. Monfet, D., Corsi, M., Choiniere, D., Arkhipova, E.: Development of an energy prediction tool for commercial buildings using case-based reasoning. Energy Build. **81**, 152–160 (2014)

18. Platon, R., Dehkordi, V.R., Martel, J.: Hourly prediction of a building's electricity consumption using case-based reasoning, artificial neural networks and principal component analysis. Energy Build. **92**, 10–18 (2015)
19. Ndiayea, D., Gabriel, K.: Principal component analysis of the electricity consumption in residential dwellings. Energy Build. **43**, 446–453 (2011)
20. Lam, J.C., Wan, K.W., Cheung, K.L., Yang, L.: Principal component analysis of electricity use in office buildings. Energy Build. **40**, 828–836 (2008)
21. ASHRAE Guideline 14: Measurement of energy and demand savings. In: ASHRAE, Atlanta (2002)

Modelling Hierarchical Relationships in Group Recommender Systems

Lara Quijano-Sánchez[✉], Juan A. Recio-García, and Belen Díaz-Agudo

Universidad Complutense de Madrid, Madrid, Spain
{lara.quijano,jareciog,belend}@fdi.ucm.es

Abstract. Group recommender systems have become systems of great interest in the CBR community. In previous papers we have described and validated a social recommendation model that solves different group recommendation challenges using knowledge from social networks. In this paper we have run across two identified limitations of our model, *unprofiled* users and "hierarchical relations" within a group, and have proposed and validated CBR solutions for them.

1 Introduction

Recommender systems are tools that provide suggestions to users. These suggestions are aimed at supporting their users in various decision-making processes, such as what items to buy, what music to listen, or places to visit [19]. Development of recommender systems is a multi-disciplinary effort which involves experts from various fields such as Artificial Intelligence, Case-Based Reasoning (CBR), Personalization, Human Computer Interaction, etc. CBR has a long history of contributing to recommender systems [1,2,11]. Most simply, we can build a case-based recommender system where the cases represent the items (e.g. products) and the CBR application recommends cases that are similar to the user's partially-described preferences. More interestingly, the cases in the case base can instead describe recommendation experiences [4], can alleviate the cold-start problem [15], or can replay previous group behaviour [14].

Up till now, our main line of research has focused on the generation of a set of recommendations that satisfy a group of users, with potentially competing interests. To do so, we have reviewed different ways of combining peoples' personal preferences and proposed an approach that takes into account the social behaviour within a group. Our approach, named *Social Recommendation Model* (*SRM*), defines a set of recommendation methods that include the analysis and use of several social factors such as the *personality* of group members in conflict situations or the *trust* between them [16,18]. Besides, in order to verify our *SRM* we have developed a movie group recommender application, *HappyMovie*, in the social network Facebook that helps us to automatically retrieve the social factors in our *SRM* and implements our proposal [17].

In the course of developing our *SRM*, we have uncovered two main limitations related to the different behaviour patterns users have when involved in

E. Hüllermeier and M. Minor (Eds.): ICCBR 2015, LNAI 9343, pp. 320–335, 2015.
DOI: 10.1007/978-3-319-24586-7_22

decision-making processes with different group configurations. In this paper we have run across two identified limitations: (1) Related to *HappyMovie*'s usability and therefore to *SRM*'s applicability. Our models' main premise is to improve group recommendations through the usage of knowledge stored in social networks. For this reason, our implementation of *SRM*, *HappyMovie*, is embedded in Facebook. This limits our model to provide recommendations to users that belong to this social network, leaving users such as children, elderly or in general people outside this network *unprofiled*. Besides, people inside Facebook mostly belong to a limited age range that mainly represents groups of friends. This leaves for example family outings out of the scope of our system. (2) It is a fact that users inside different social environments behave differently [3] and have different hierarchical relations. For example, a parent won't behave equally when going to the movies with friends than when taking her/his children. This fact is due to the hierarchical relations that emerge in any group decision-making process. In our *SRM*, we have modelled users social behaviour through the computation of social factors. However, these factors are computed as a fixed value for each user (or pair of users). Hence, we fail to capture the different hierarchical relations each user has across the different groups s/he belongs to.

In order to increase *HappyMovie*'s usability and our model's applicability (limitation (1)), we have defined "prototypical" users that represent those members that can't, don't or won't have a profile inside a social network. Preferences and social knowledge for these *unprofiled* users in the group are retrieved and reused, in a certain group context, from a case base CB_{usr} of prototypical users and a case base CB_{grp} of groups' social behaviour. Regarding limitation (2), in this paper we propose the use of an additional social factor related to "hierarchical relations" within a group: *dominance* factor $d_{u,u'}$. This factor can be static (i.e. family relationships) or situation-dependent (i.e. someone decides the movie when it is her/his birthday). In order to better model users' different behaviour patterns across different group configurations, we retrieve and reuse a group's "hierarchical relations" from an extension of our social behaviour case base CB_{grp}. We believe that it is natural to use CBR approaches because, in leisure domains, similar events recur: the same group (perhaps with some small variations) repeats activities together; and some age, gender and hierarchy distributions will tend to recur too (e.g. two adults with two children, or several friends in the same age range).

The paper runs as follows. Section 2 summarises our previous research, that is, our *Social Recommendation Model (SRM)* [16,18]; Sect. 3 presents some identified limitations of our *SRM* and some solutions in a case base fashion to them; Sect. 4 presents a case study where we verify the improvement of group recommendation results when including our case base approaches and Sect. 5 concludes and presents some ideas for future work.

2   Social Recommendation Model

The task of making group recommendations is quite challenging as it has to present a set of interesting products not only to a single person but to a group of people whose concerns are not always compatible. Our approach considers that the real satisfaction of a group regarding a group recommendation cannot be accurately estimated using the simple aggregation of its members' individual preferences. Thus, considering people as social entities that relate with each other allows the better estimation of their individual satisfaction regarding the result of the recommendation and, therefore, improves the global group satisfaction. To address this issue, we have focused our line of work on modelling groups social dynamics by capturing human affective and social processes [16–18], a task which is novel to group recommenders research. To do so, we introduce two main social factors: *personality* and *trust*.

In order to model how easily influenced a user is in a decision-making process we use a personality test (the TKI test [22]) to obtain different user profiles (personality values $p_u{}^1$) that interact differently in conflict situations. According to Thomas-Kilmann's study [22] users that present a low personality value ($p_u <$ 0.4) are considered *cooperative*, which reflects highly tolerant people, meaning that even if the selected item is not the one of their choice, it is good enough for them if the group selects it. On the other hand users that present a high personality value ($p_u > 0.6$) are considered *assertive*, which reflects more selfish people, meaning that other people's choices do not satisfy them.

The second factor that we introduce, *trust*, measures closeness between group members as in how much users can mutually influence each other. Its computation consists of a novel technique of eliciting information from users connected through Social Networks. This trust factor represents the tie strength between users. We have studied several social factors that affect users in order to calculate the tie strength between them. The analyzed factors (which follow the literature regarding tie strength elicitation [7]) are: number of common friends, intensity of the relationship, duration of the relationship and pictures in common. The concrete process followed to compute the trust factor[2], $t_{u,u'}$, in our approach is fully detailed in [17].

Our *SRM* proposes a variation of traditional preference aggregation approaches [13], where before aggregating users individual predicted ratings ($\hat{r}_{u,i}$) we modify them with users' personality and trust [16,18]. Hence, our *SRM* can be defined as the set of methods that follow Eq. 1:

$$\hat{r}_{G_a,i} = \bigsqcup_{\forall u,u' \in G_a \wedge u \neq u'} SocialFunction(\,\hat{r}_{[u|u'],i}\,,\,p_{[u|u']}\,,\,t_{u,u'}\,,\,sf(u,u')\,) \ (1)$$

---

[1] p_u represents user u's predominant behavior according to her/his TKI evaluation [22]. It fits within a range of (0,1], 0 being the reflection of a very cooperative person and 1 the reflection of a very assertive one. This value is computed through a compulsory personality test in *HappyMovie* as detailed in [17].

[2] This factor fits within a range of (0,1], 0 being the reflection of a person someone is not close to and 1 the reflection of a person someone is really close to.

where $\hat{r}_{G_a,i}$ is the estimated rating for a given item i and active group G_a. $\sqcup$ represents any possible aggregation function to be used[3]. p_u represents users personality, $[u|u']$ reflects that this factor is applied in the *SocialFunction* for users u or u', $t_{u,u'}$ represents users tie strength relationships and, sf, is a set of different social factors $\{fs_1, .., fs_n\}$ that can be included or not, depending on whether we want to further apply more social factors or not.

We have devised several recommendation methods based on Eq. 1. For example, the Delegation-Based Recommendation method (*DBR*) [16] or the Influence-Based Recommendation method (*IBR*) [18]. The other main contribution of our work has been the instantiation of our model in a real-life scenario, the social network Facebook. As a result we have built *HappyMovie* [17], which provides a group recommendation for a group of people who wish to go to the cinema together.

3 Limitations of Our SRM

With our *SRM* we have been able to improve the recommendation outcome of traditional preference aggregation approaches [16–18]. The key factor in the success of our *SRM* is the inclusion of social factors. These social factors define each person (our users involved in the recommendation process) as a potentially influenced component of a social community (or group) determined by the environment. In our *SRM* (and therefore in *HappyMovie*), each social factor is computed just once, for each user or pair of users, and stays fixed throughout all possible group recommendations and configurations. Besides, groups of users are formed through the creation of events in Facebook. This fact limited our model to the premise that the groups to be recommended were formed by Facebook users, that is, mostly groups of friends performing joint activities. A completely different situation would be family group recommendation processes, where some of the members (elder or kids) might not have Facebook accounts. In order to solve this limitation (limitation (1)) there is a need to extend *HappyMovie*'s usability. Besides, in family outings, age difference (elder, kids) considerably varies the possible activities, the different priorities that must be taken into account when trying to satisfy the different group members and the different hierarchies that users have inside a group. However, our *SRM* fails to model the different weights and priorities that users inside a group have depending on their group configuration. In order to solve this limitation (limitation (2)) we study how the different social factors (*personality* and *trust*) vary inside different group configurations.

After this discussion of the limitations of our model, we believe that a necessary step in the improvement of our *SRM* should be the analysis of group behaviour according to group hierarchies and the increase of *HappyMovie*'s usability (and hence of *SRM*'s applicability) by being able to include in the system's group representation those users that are *unprofiled* inside the social network.

---

[3] There are several techniques for individual preferences aggregation [12], being *least misery* (where the minimum is taken), *most pleasure* (where the maximum is taken) and *average satisfaction* (where the average is taken) the most common ones.

We believe that these problems can be ideally solved using a CBR approach. Next, we present our solution to each of the identified limitations. To do so we will introduce two complementary CBR systems, CB_{usr} and CB_{grp}, that solve limitation (1) by inferring *unprofiled* users' missing information[4] (Sects. 3.1 and 3.2) and an extension of CB_{grp} that solves limitation (2) (Sect. 3.3) by including "hierarchical relations" in SRM.

3.1   Representation of *unprofiled* users

When studying the problem of including in the group configuration users that do not have Facebook profiles (limitation (1)), a simple solution could be having a case base CB_{usr} of real users where each case $C_{usr} \in CB_{usr}$ models a different user that can be used as a prototype for *unprofiled* users. In order to retrieve from CB_{usr} the case C_{usr}, that represents v, the most similar user to an *unprofiled* user u that needs to be added to an active group G_a requesting a recommendation, we must acknowledge the fact the user u itself has no access to the group recommender application (because s/he does not have a social network account). Therefore, it must be another user $u' \in G_a$ that includes her/him in the group and provides the information about u. Note that a person's knowledge about others is limited, specially if related to sensible information. Therefore, we only request demographic information such as the age or gender of the *unprofiled* user. Next, we need to infer the missing information that our SRM needs for the recommendation process. This information is retrieved from CB_{usr} to model the *unprofiled* user and consists of users' personality (p_u) and preferences ($r_{u,i}$). Hence, each case needs to have the following structure: $C_{usr} = \langle D_{usr}, S_{usr} \rangle$, where the description part is defined as $D_{usr} = \langle id_c, u.age, u.gender \rangle$ and the solution part is defined as $S_{usr} = \langle p_u, r_{u,i} \rangle$. In them:

- id_c is a case identification number, used to distinguish the case from others, but otherwise not used by the CBR system CB_{usr}.
- The *problem description* D_{usr} part of the case comprises the demographic information of each user: user u's age ($u.age$) and user u's gender ($u.gender$);
- The *solution* S_{usr} part of the case contains the social information of each user: user u's personality (p_u) and user u's preferences ($r_{u,i}$);

Due to the little information we request as part of D_{usr}[5], a lot of users fit in the description part of the case. This forces us to create different prefixed models that serve as general solutions. To do so, we divide users in CB_{usr} by

---

[4] That is: users' personality (p_u) and individual preferences ($r_{u,i}$) that are obtained through tests in *HappyMovie* and users' trust with each other ($t_{u,u'}$) that is automatically computed through user's personal information stored in Facebook profiles (see [17] for *HappyMovie*'s functionality details).

[5] This is done for obvious reasons: (1) user u', inscribing in the group the *unprofiled* user u, is not able to provide the system with concrete values representing user's u personality and preferences. (2) We believe it will not be adequate or practical/possible to ask at this point to user u to start answering the needed tests.

gender and age ranges of <12, $[12–18)$, $[18–30)$, $[30–40)$, $[40–50)$ and $>50^6$. Then, when a solution S_{usr} is needed we compute for *unprofiled* user u's corresponding age range and gender a prototype user as the average of the personality and preferences tests results of the users in CB_{usr} that fit description D_{usr} (providing us with p_u and the set of $r_{u,i}$, $i \in$ *HappyMovie*'s preferences test catalogue of movies). Users in CB_{usr} are *HappyMovie*'s users and children (that have answered the tests outside the application) of friends, colleagues and family. CB_{usr} has the following age ranges and number of users:

Age range	Number of males	Number of females
<12	10*	10*
[12–18)	11*	18*
[18–30)	112	41
[30–40)	7*	9*
[40–50)	4*	5*
>50	3	4

Note that the prototypes keep refining the more users our system has. For example prototypes of cells with an * in the user distribution table have been updated after the experiment carried out in Sect. 4. A trend we have found when processing users answers to the given tests is that the standard deviation in users answers for both female and male users grew bigger the older the users were (with an increase of ~0.116). This translates into the conclusion that children have less deviation on their preferences and they all have quite similar preferences on movies. Adults have more variation on their ratings, more experience, more defined tastes and different personal preferences. Besides, another observed fact is that, even if the preferences test should reflect individual preferences, parents have a clear tendency of filling the test taking into account their children's preferences, foreseeing that the whole family will go to the cinema together.

Note that, at this point, the system has all the personal information needed by SRM about users in groups, both the ones that do have a profile in Facebook and the ones that do not. This information includes users' personality p_u and users' individual preferences $r_{u,i}$. However, there is one important factor missing in our model's *unprofiled* users, trust factor $(t_{u,u'})$. As we have said before our SRM automatically computes $t_{u,u'}$ by extracting social information from users profiles in the social network (see Sect. 2). However, now that we have introduced "prototypical" users, the trust factor between *unprofiled* users and: *profiled* users and/or other *unprofiled* users is missing. Next, we explain how we compute it and how we later extend this approach to help us better model users different behaviour patterns inside different group configurations.

---

[6] These ranges could be of course extended and have been selected as representations of the main stages of life.

3.2   Inferring Interpersonal Trust for Unprofiled Users

To estimate the trust of an *unprofiled* user we have developed a second CBR system, CB_{grp}, that stores the information about the social relationships among group members. Cases inside CB_{grp} are defined as follows: $C_{grp} = \langle D_{grp}, S_{grp} \rangle$, where the description part is defined as $D_{grp} = \langle id_c, U \rangle$ and the solution part is defined as $S_{grp} = \langle V, T \rangle$. In them:

- id_c is a case identification number, used to distinguish the case from others, but otherwise not used by the CBR system CB_{grp};
- The *problem description* D_{grp} part of the case comprises the composition of each group. Group compositions are defined as the set of users U containing users' demographical and social information, that is both the information used and retrieved from the approach explained above, the CB_{usr} case base. Therefore for each user u in the group description U we have her/his gender $u.gender$, age range $u.age$, personality p_u and preferences $r_{u,i}$;
- The *solution* S_{grp} of the case contains a graph representing the trust among users: Users in the group ($V = \{v_1, ..., v_n\}$) are the vertices of the graph and the trust among users is represented as a list of edges $T = \{\langle v, v', t_{v,v'} \rangle\} \; \forall v, v' \in V$;

We use C_{grp}, that represents the most similar group to G_a in our case base CB_{grp}, to retrieve the trust factor information that may be missing if G_a has "prototypical" users (see Fig. 1). For each G_a we retrieve the case $C_{grp} \in CB_{grp}$ that is most similar according to the group composition. In order to retrieve the most similar group we have designed an algorithm that pairs each user u in G_a with the one that plays the most similar role in the retrieved group G_c. This algorithm is a variance of the ones described in [14] and [15]. We have altered the original definitions for two reasons: (1) In [14] the goal was to find the pairing that maximized total similarity, meaning that a person in G_a might not be paired with the person who is most similar in G_c, it just optimized total similarity. In [15], *gsim* definition did pair each user with its most similar user, however, the mapping was not bijective, meaning we did not prevent two or more people from G_a being associated with the same user G_c. As in this paper we do

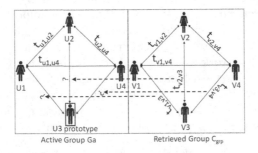

Fig. 1. Group retrieval: retrieve the missing trust values from the most similar user.

wanted to pair each user with its most similar user and also needed the mapping to be bijective (because we want to later infer the "hierarchical relations" within that group, hence we need each user to paired with the one that played the most similar role, Sect. 3.3), we have used [15]'s *gsim* definition but using [14]'s bijective mapping. (2) In [15] the group G_a had the complete graph of trust values in the problem description part of the case, this means that $t_{u,v}$ was used as part of the case definition and not as part of the solution and therefore it was used in the user similarity computation *usim*. Here we have reused *usim* without taking the $t_{u,v}$ similarity into account. Next we describe our similarity metric.

We denote the group similarity by *gsim*. It is a form of graph similarity where users are nodes, which we provide (D_{grp}); trust relationships and roles inside the group are weighted edges which we retrieve (S_{grp}). In our definition of group similarity, we pair each user from the active group G_a with exactly one user from the group in the case G_c and vice versa. In other words, we will be finding a *bijection* from G_a to G_c. This raises a problem when comparing groups of different sizes, where a bijection is not possible. In such situations, we could simply say that $gsim(G_a, G_c) = 0$. However, we did not want to do this. It might force the system to retrieve unsuitable cases. Consider a case base that just happens to contain many families of four (two adults, two children), no families of five, but many parties of five friends. If the active group is a family of five (two adults, three children), it is surely not appropriate to prevent retrieval of families of four and only retrieve parties of five friends. To enable comparisons, this is the point, prior to computing similarity, that we insert additional virtual users into either G_a or G_c, whichever is the smaller, in order to make the groups the same size.

Now, we can define the group similarity measure. Consider any pair of equal-sized groups, G and G' and a bijection, $f : G \mapsto G'$. We will map members of G to G', and so for any $u \in G$, we can compute the similarity, *usim*, to his/her most similar partner $v \in G'$, obtaining a tuple $< u, v >$. We will do this for each user and his/her partner, and take the average:

$$gsim(G, G', f) \hat{=} \frac{\sum_{u \in G} usim(u, G, f(v^*, G'))}{|G|} \tag{2}$$

where

$$v^* \hat{=} \arg \max_{v \in G'} usim(u, G, v, G') \tag{3}$$

The definition of *gsim* (Eq. 2) uses *usim*, the similarity between a person u in one group G and a person v in another group G', which we have not yet defined. We make use of their ratings, age, gender and personality values. Specifically, we combine local similarities into a global similarity. The local similarities are as follows. For the users' ratings, we use the Pearson correlation [9] ($\rho_{[0,1]}$). For gender, we use an equality metric ($eq(x, y)$) and for ages and personalities, we use the range-normalized difference ($rn\_diff_{attr}(x, y)$):

$$eq(x, y) \hat{=} \begin{cases} 1 & \text{if } x = y \\ 0 & \text{otherwise} \end{cases} \quad rn\_diff_{attr}(x, y) \hat{=} 1 - \frac{|x - y|}{range_{attr}}$$

Finally, the global similarity, *usim*, is simply an average of $\rho_{[0,1]}$, eq_{gender}, rn_diff_{age} and $rn\_diff_{p_u}$. However, we have the problem of virtual users, who do not have ages, genders, personalities, or ratings. If either user is a virtual user, we simply take *usim* to be the mid-point of the similarity range. Empirically, this is 0.6. This means that there is neither an advantage nor a disadvantage to being matched with a virtual user and, since everyone must be paired with someone, this seems appropriate.

While this completes the definition of $gsim(G, G', f)$, it assumes that we give it a particular bijection, f, which pairs members of G with members of G'. But, for the similarity, we want to consider *every* such bijection and settle on the *best one*, the one that gives the best alignment between the group members (their ages, genders, personalities, ratings). We must compute *usim* for each bijection. Let $\mathcal{B}(A, B)$ denote all bijections between equal-sized sets A and B. For example, if A is $\{a, b, c\}$ and B is $\{x, y, z\}$, then one bijection is $\{a \mapsto x, b \mapsto y, c \mapsto z\}$, another is $\{a \mapsto y, b \mapsto x, c \mapsto z\}$, and so on. Our definition of the similarity of group G and G' is based on finding the bijection, out of all the possible bijections, that maximizes the similarity between users, *usim*, Eq. 3.

$$gsim(G, G') \hat{=} \max_{f \in \mathcal{B}(G,G')} gsim(G, G', f) \qquad (4)$$

Think of this as finding the pairing that maximizes similarity between users so that we get the correct matching between users that play similar roles in groups. Note that as this is a bijection each user u in G_a will be pair to a different user v in G_c.

Now that we have retrieved the most similar group to G_a along with its trust distribution graph T and we have a set of tuples $< u, v >$ that match each user in G_a to its corresponding user in C_{grp} (which is the one that played the most similar role). We assign to each pair of users (u_1, u_2) in G_a that have a missing trust value (t_{u_1,u_2}), the trust value between their corresponding users (v_1, v_2) in C_{grp} according to the set of tuples $< u, v >$, that is t_{v_1,v_2} (see Fig. 1). This way we are able to address the problem of the missing trust values when introducing *unprofiled* users in the group configuration.

3.3   Modelling Hierarchical Relationships

Users inside different social environments behave differently [3]. As we have said before, this is a fact that our *SRM* did not contemplate when computing the different social factors that model users' social behaviour (limitation (2)). There is evidence that people's personality (understood as how cooperative or easily influenced they are) and trust (understood as how much others can influence them) vary depending on their social environment. This evidence can be found in several studies [23] and in several personal statements: "I find that I often *adapt* my personality to certain people. It's like with each individual (co-workers, friends, family) I develop a certain image or personality that *fits* with that person" [21].

There are different social theories that identify and formalize the hierarchical behaviour in a group. For example, the social dominance theory (SDT) is a theory of group relations that focuses on the maintenance and stability of group-based social hierarchies [20]. SDT begins with the empirical observation that social systems have a threefold group-based hierarchy structure: age-based, gender-based and arbitrary set-based. Age-based hierarchies invariably give more power to adults and middle-age people than children and younger adults, and gender-based hierarchies invariably grant more power to one gender over other, but arbitrary-set hierarchies are truly arbitrary and group dependent. This kind of social behaviour is also referred as "power relationships" [8]. Although this term is usually applied to the relation of superiors with respect to subordinates and related to the concept of *authority*, it can be also applied to social groups. According to the social psychologists French and Raven [6], power is that state of affairs which holds in a given relationship, A-B, such that a given influence attempt by A over B makes A's desired change in B more likely. Another theory that tries to explain this social behaviour is Game Theory [5]. Therefore, there are different social theories that define from different points of view the relationships that influence the outcome of a decision within a group. Leaving aside the discussion about the correct term to define this type of relationship, in this paper we introduce this factor in our model named as *dominance* factor. This new social factor, both static (family, friendship, etc.) or situation-dependent (celebrations, events, etc.), is included in our SRM to provide better group recommendations. In order to define this new social factor we have followed Leavitt [10] that suggested that there are certain "hierarchical relations", which are useful in predicting human behaviour:

- *Equality.* This relationship represents that users treat each other as equals. When applied in roles inside groups it can represent children of the same age and adults of the same/similar age, that is brothers/sisters and friends.
- *Dominant.* This relationship represents higher priority and necessities. When applied in roles inside groups it can represent children towards adults, elderly towards others or guests (celebrations, birthdays) towards others.
- *Submissive.* This relationship represents lower priority and necessities. When applied in roles inside groups it can represent adults towards children, others towards elderly or others towards guests (celebrations, birthdays).

Asking users to manually indicate their social hierarchical relations and behaviour inside each group every time they ask for a recommendation or that the group configuration varies is unaffordable. Also, to our knowledge, there are no personality tests or tie strength prediction tests that extract this type of information. Hence, this type of problem suggests addressing it with a CBR solution that allows us to retrieve from the case base CB_{grp} of groups, their social hierarchical patterns and adapt the previously computed social factors of the requesting active group G_a to the patterns retrieved in the most similar case C_{grp}. To do so we can extend our CBR system for trust inference and include the dominance relationship in the representation of the solution: $S_{grp} = \langle V, T, D \rangle$

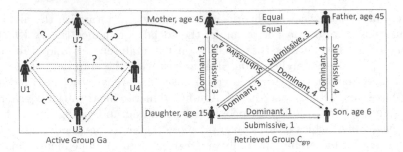

Fig. 2. Group retrieval: Retrieve behaviour pattern roles from the most similar user + example of the social hierarchical patterns in a family composed by two adults of age 45 and two children of age 15 and 6 respectively.

being V and T equally defined as in the previous version of the CBR system CB_{grp} explained in Sect. 3.2 and $D = \{\langle v, v', d_{v,v'}\rangle\}\ \forall v, v' \in V$.

This extension includes the social hierarchical pattern distribution of each group represented as the dominance graph $D = \{\langle v, v', d_{v,v'}\rangle\}$. Hence, $d_{v,v'}$ is a tuple that represents the type of relationship and a degree, $< [equality|dominant|submissive], \delta >$. This *dominance* factor labels the type of hierarchical relationship between two users in a group and modifies their p_v and $t_{v,v'}$ according to it making them now *dynamic* social factors. The degree factor δ *in* $d_{v,v'}$ measures the difference between age ranges. That is, for example in the case of an adult v of age 45 having a *submissive* relationship with a child v' of age 6, the difference in their age range is of 4 and hence this would be the degree value δ in the dominance tuple (see Fig. 2 for a complete graphical example). Note that $d_{v,v'} \neq d_{v,v'}$, but that they are complementary.

In order to compute the *dominance* factor ($d_{u,u'}$), we retrieve it as a label in the edge that connects two users/nodes. This is done in the same way we retrieved the missing $t_{u,u'}$ values for the *unprofiled* users. That is, we retrieve the most similar group $G_c \in CB_g$ to G_a along with its social behaviour pattern distribution D and the set of tuples $< u, v >$ that match each user in G_a to its corresponding user in C_{grp} (which is the one that played the most similar role). We assign to each pair of users (u_1, u_2) in G_a their *dominance* factor d_{u_1,u_2} as the *dominance* value between their corresponding users (v_1, v_2) in G_c according to the set of tuples $< u, v >$, that is d_{v_1,v_2} (see Fig. 2).

Once we have all the social factors for G_a: p_u, $t_{u,u'}$ and $d_{u,u'}$, we update p_u and $t_{u,u}$ according to the label and degree indicated in $d_{u,u'}$ as follows: $p_u = p_u + \alpha\delta$; $p_v = p_v - \alpha\delta$; $t_{u,u'} = t_{u,u'} + \alpha\delta$; $t_{u',u} = t_{u',u} - \alpha\delta$; here the α weight depends on the type of dominance relationship[7]:

---

[7] We have empirically assigned a value of ± 1 to α in order to have a moderate impact of the *dominance* factor.

- If $d_{u,u'} = <\ equality, 0>$ then $\alpha = 0$ and the social factors stay fixed.
- If $d_{u,u'} = <\ Dominant, degree>$. Then $\alpha = 0.1$. Which represents that if user u has a *dominant* behaviour over user u' s/he will be less likely to give in and be therefore more assertive and s/he will be less likely to trust u''s preferences. As we have said that $d_{u,u'}$ is complimentary to $d_{u',u}$ then u' will have a *submissive* behaviour as explained next.
- If $d_{u,u'} = <\ Submissive, degree>$. Then $\alpha = -0.1$. Which represents that if user u has a *submissive* behaviour over user u' s/he will be more likely to give in and be therefore more cooperative and s/he will be more likely to trust u''s preferences.

Summing up, with these two approaches of *including unprofiled users* and *including hierarchical relationships* we are now able to represent groups with members without social network profiles and make better group recommendations to them by better modelling their group social behaviour (limitations (1) and (2)). In our previous experiments [16–18] users inside groups belonged to the same age range, and therefore, we were only able to recommend to groups of friends. Now, we have increased *HappyMovie*'s usability and are able to recommend not only to groups of friends but also to families too. Next, we will prove through a case study that by including our new social factor $d_{u,u'}$ and hence making our previous social factors p_u and $t_{u,u'}$ adaptive to the group composition we improve the results of the recommendations. Note that due to the existing problems in obtaining a real and big enough case base that represents enough types of users and group configurations, we have opted for using prototypical group configurations and social behaviour patterns. Therefore, in order to build our case base CB_g of "prototypical" groups we have asked 10 "prototypical" families with different configurations (size 3 with 2 adults and a kid, size 3 with 1 adult and 2 kids of the same age range, size 3 with 1 adult and 2 kids of different age range, size 4 with 2 adults and 2 kids of the same age range, etc.) to indicate their configuration related to the *trust* and *dominance* social factors. This is done to represent different family outings and established prototypical behaviours that represent both of our graphs, the trust distribution graph T (Fig. 1) and the dominance distribution graph D (Fig. 2) for their different configurations. Also note that for now we have focused on representing family configurations and that we leave for future work representing different "hierarchical relations" in friend outings, like events with birthdays, etc.

4   Case Study

We have evaluated the performance of our new social factor $d_{u,u'}$ that models users' social behaviour patterns inside different group configurations by comparing the performance of our *SRM* with and without this factor. Our goal with this experiment has been to prove that by including the *dominance* factor we are able to better model users social dynamics inside a group. To do so we have focused on the concrete case of recommending to a family that wants to go to the movies. We have gathered 17 real families of sizes 6, 5, 4 and 3 where 30 adults and 33 kids have participated. Note that this set is different from the set that belongs to both our case bases C_{usr} and C_{grp}. Next:

1. For each family we ask the adults to use *HappyMovie* and answer its two compulsory tests, the personality and preferences test. This provides us with the p_u, $r_{u,i}$ and $t_{u,u'}$ factors (note that $t_{u,u'}$ is automatically inferred from users' profiles in Facebook) as well as the demographic information of all users in the group that have a Facebook account.
2. For each family we ask how many, the age and gender of their group members do not have Facebook accounts (this includes mostly the kids). For these members we have used our *including unprofiled users* technique (Sects. 3.1 and 3.2) to infer the missing factors that our *SRM* needs.
3. We provide each family with the current movie listing (20 movies) and ask them to debate between themselves and provide an ordered list of the 3 movies that they would like to go to watch together as a family, G_m.
4. We run our *SRM* with and without the *dominance* factor for two social methods *DBR* and *IBR* [16,18] and also a method with no social factors at all. Therefore, we compare a traditional group recommendation approach (Non Social), two of our original *SRM* methods (*DBR* and *IBR*) and a modification of them by including the *dominance* factor introduced in this paper (*DBR+Dominance* and *IBR+Dominance*). Each recommender presents the top $k' = 3$ movies from the 20 candidates. Then, let R be the set of recommendations made by a particular recommender, we compare R with G_m (the ordered set of movies that the families provided). We compute total *success@n* for $n = 1, 2, 3$, where *success@n* = 1 if $\exists i, i \in R \land i \in G_m$ and is 0 otherwise. For example, when using *success@2*, we score 1 each time there is at least one recommended movie in the top two positions of G_m. We also compute total *precision@n* for $n = 1, 2, 3$, where *precision@n* $\hat{=} |\{i : i \in R \land i \in G_m\}|/n$. For example, if no recommended movie is in the top two positions in G_m, then *precision@2* = 0; if one recommended movie is in the top two positions in E, then *precision@2* = 0.5.

Results (Fig. 3) show that for all the different measures, the inclusion of social factors improves the traditional group recommendation approach (Non Social) and that all the recommenders that also include the *dominance* factor obtain better (or equal in the case of *success@3*) results than our original *SRM* methods without it (*DBR* and *IBR*). We have performed the non-parametric Kruskal-Wallis H Test which does not require the assumption of normality and that measures if two results are statistically different or not and have confirmed that the difference between the Non Social results and the rest of the recommenders is significant. Besides, the difference between our original methods *DBR* and *IBR* and the improved ones with the dominance factor *DBR+Dominance* and *IBR+Dominance* is also significant (for all measures save for the *success@3* where we can easily see that they are very similar). However, the difference between both methods *DBR* and *IBR* with or without the dominance factor has

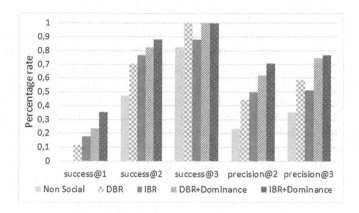

Fig. 3. Analysis of our algorithms for different evaluation metrics.

not been proven to be significative. This experimentation allows us to conclude that we have improved our *SRM* by better modelling users' social dynamics inside different group configurations.

We have performed an additional experiment in order to check the viability of our prototypes. To do so, we have asked the *unprofiled* users in our experiment to fullfil an offline version of *HappyMovie*'s tests, providing us with their real p_u and $r_{u,i}$ values. Next, we have compared these values with the ones used in the prototypes that represented them in the case study (step 2). Results have shown a MAE of 0.98, which means that we rate correctly each users' values with a precision of ± 1. This allows us to conclude that the prototype users can indeed be used in the presence of *unprofiled* users as the error in the estimated values is affordable and does not affect the recommendation results.

5   Conclusions and Future Work

SRM is a validated model that simulates people's decision-making behaviour in groups based on the influence of social relationships between individuals [16, 18]. In this paper we have identified two limitations of our model and have proposed two solutions to them: (1) Related to represent users that have no social network profiles, our first solution is based on prototypes that are retrieved and reused in a CBR manner. (2) Related to the study of "hierarchical relations" inside groups, our second solution extends the use in *SRM* of our previously fixed social factor values to dynamic social factor values that adapt depending on users' different behaviour patterns in different social environments and group compositions. We have modelled the "hierarchical relations" that emerge in any group decision-making process through a new social factor, dominance $d_{u,u'}$. In order to prove the benefits of including $d_{u,u'}$ we have run an experiment using a case study with *HappyMovie* where we have modelled families. Results of these experiments have allowed us to: (1) validate the designed user prototypes, that have been proved to be accurate with a 0.98 MAE error. And hence, increase the population inside our application and its usability. (2) validate the success of including our new social factor $d_{u,u'}$ that improves the results of the previous SRM methods and allows us to better model users social behaviour and dynamics inside groups.

As future work, related to our *dominance* factor and different dynamics inside the same group, we could extend our study of "hierarchical relations" inside SRM to the groups' context dependency. This is, a group of friends may not have the same behaviour one day or another. Whether because it is someone's birthday in which case the preferences of this group member may have more weight or because of the groups' general emotional state, where members could point out that they are not in a good mood and feel like watching a comedy for example.

References

1. Baccigalupo, C., Plaza, E.: Case-based sequential ordering of songs for playlist recommendation. In: Roth-Berghofer, T.R., Göker, M.H., Güvenir, H.A. (eds.) ECCBR 2006. LNCS (LNAI), vol. 4106, pp. 286–300. Springer, Heidelberg (2006)
2. Bridge, D., Göker, M.H., McGinty, L., Smyth, B.: Case-based recommender systems. Knowl. Eng. Rev. **20**(3), 315–320 (2006)
3. Christakis, N.A., Fowler, J.H.: Social contagion theory: examining dynamic social networks and human behavior. CoRR (2011)
4. Dong, R., Schaal, M., O'Mahony, M.P., McCarthy, K., Smyth, B.: Opinionated product recommendation. In: Delany, S.J., Ontañón, S. (eds.) ICCBR 2013. LNCS, vol. 7969, pp. 44–58. Springer, Heidelberg (2013)
5. Dowding, K.M.: Power. Concepts in the Social Sciences. Open University Press, Buckingham (1996)
6. French, J.R.P., Raven, B.: The Bases of Social Power. In: Research and Theory on Group Dynamcis, New York, pp. 607–623 (1959)
7. Gilbert, E., Karahalios, K.: Predicting tie strength with social media. In: CHI 2009, pp. 211–220. ACM, New York (2009)
8. Greiner, L.E., Schein, V.E.: Power and Organization Development: Mobilizing Power to Implement Change. Addison-Wesley OD Series. Addison-Wesley, Reading (1988)
9. Herlocker, J.L.: Understanding and improving automated collaborative filtering systems. Ph.D. thesis, University of Minnesota (2000)
10. Leavitt, H.: Managerial Psychology. University of Chicago Press, Chicago (1972)
11. Lorenzi, F., Ricci, F.: Case-based recommender systems: a unifying view. In: Mobasher, B., Anand, S.S. (eds.) ITWP 2003. LNCS (LNAI), vol. 3169, pp. 89–113. Springer, Heidelberg (2005)
12. Masthoff, J.: Group modeling: selecting a sequence of television items to suit a group of viewers. User Model. User-Adap. Interact. **14**(1), 37–85 (2004)
13. Masthoff, J., Gatt, A.: In pursuit of satisfaction and the prevention of embarrassment: affective state in group recommender systems. User Model. User-Adap. Interact. **16**(3–4), 281–319 (2006)
14. Quijano-Sánchez, L., Bridge, D., Díaz-Agudo, B., Recio-García, J.A.: Case-based aggregation of preferences for group recommenders. In: Agudo, B.D., Watson, I. (eds.) ICCBR 2012. LNCS, vol. 7466, pp. 327–341. Springer, Heidelberg (2012)

15. Quijano-Sánchez, L., Bridge, D., Díaz-Agudo, B., Recio-García, J.A.: A case-based solution to the cold-start problem in group recommenders. In: Agudo, B.D., Watson, I. (eds.) ICCBR 2012. LNCS, vol. 7466, pp. 342–356. Springer, Heidelberg (2012)
16. Quijano-Sánchez, L., Recio-García, J.A., Díaz-Agudo, B.: An architecture for developing group recommender systems enhanced by social elements. Appl. Intell. **40**(4), 732–748 (2014)
17. Quijano-Sánchez, L., Recio-García, J.A., Díaz-Agudo, B.: Development of a group recommender application in a social network. Knowl.-Based Syst. **71**, 72–85 (2014). Special Issue on Knowledge-Bases for Cognitive Infocommmunications, KBS
18. Quijano-Sánchez, L., Recio-García, J.A., Díaz-Agudo, B., Jiménez-Díaz, G.: Social factors in group recommender systems. ACM TIST **4**(1), 8 (2013)
19. Ricci, F., Rokach, L., Shapira, B.: Introduction to recommender systems handbook. In: Ricci, F., et al. (eds.) Recommender Systems Handbook, pp. 1–35. Springer, New York (2011)
20. Sidanius, J., Pratto, F.: Social Dominance: An Intergroup Theory of Social Hierarchy and Oppression. Cambirdge University Press, New York (2001)
21. SocialPhobiaWorld. Behaving differently around different people/groups (2013). http://www.socialphobiaworld.com/behaving-differently-around-different-people-groups-57798/
22. Thomas, K.W., Kilmann, R.H.: Thomas-Kilmann Conflict Mode Instrument. Xicom, Tuxedo (1974)
23. Turniansky, B., Hare, A.P.: Individuals and Groups in Organizations. SAGE Publications, London (1998)

Semi-automatic Knowledge Extraction from Semi-structured and Unstructured Data Within the OMAHA Project

Pascal Reuss[1,2](✉), Klaus-Dieter Althoff[1,2], Wolfram Henkel[3],
Matthias Pfeiffer[3], Oliver Hankel[4], and Roland Pick[4]

[1] German Research Center for Artificial Intelligence, Kaiserslautern, Germany
pascal.reuss@dfki.de
http://www.dfki.de
http://www.uni-hildesheim.de
[2] Intelligent Information Systems Lab, Institute of Computer Science,
University of Hildesheim, Hildesheim, Germany
[3] Airbus, Kreetslag 10, 21129 Hamburg, Germany
[4] Lufthansa Industry Solutions, Norderstedt, Germany

Abstract. This paper describes a workflow for semi-automatic knowledge extraction for case-based diagnosis in the aircraft domain. There are different types of data sources: structured, semi-structured and unstructured source. Because of the high number of data sources available and necessary, a semi-automatic extraction and transformation of the knowledge is required to support the knowledge engineers. This support shall be performed by a part of our multi-agent system for aircraft diagnosis. First we describe our multi-agent system to show the context of the knowledge extraction. Then we describe our idea of the workflow with its single tasks and substeps. At last the current implementation, and evaluation of our system is described.

1 Introduction

This paper describes the concept of a semi-automatic knowledge extraction workflow, which is developed for a distributed decision support system for aircraft diagnosis. The system will be realized as a multi-agent-system. It is based on the SEASALT architecture and includes several case-based agents for various tasks. The knowledge extraction workflow will be realized using several agents within the decision support system. In the next section we give an overview of the OMAHA (Overall Management Architecture For Health Analysis) project, the SEASALT architecture and the application domain. In Sect. 3.1 we describe the instantiation of our decision support system based on SEASALT. Section 3.2 contains the initial concept for the knowledge extraction workflow, while 3.3 describes the current implementation status of the workflow. The Sect. 3.4 shows the evaluation setup and the evaluation results and Sect. 4 contains the related work. Finally, Sect. 5 gives a short summary of the paper and an outlook on future work.

© Springer International Publishing Switzerland 2015
E. Hüllermeier and M. Minor (Eds.): ICCBR 2015, LNAI 9343, pp. 336–350, 2015.
DOI: 10.1007/978-3-319-24586-7_23

2   OMAHA Project

The OMAHA project is supported by the Federal Ministry of Economy and Technology in the context of the fifth civilian aeronautics research program [6]. The high-level goal of the OMAHA project is to develop an integrated over-all architecture for health management of civilian aircraft. The project covers several topics like diagnosis and prognosis of flight control systems, innovative maintenance concepts and effective methods of data processing and transmission. A special challenge of the OMAHA project is to outreach the aircraft and its subsystems and integrating systems and processes in the ground segment like manufacturers, maintenance facilities, and service partners. Several enterprises and academic and industrial research institutes take part in the OMAHA project: the aircraft manufacturer Airbus (Airbus Operations, Airbus Defense & Space, Airbus Group Innovations), the system and equipment manufacturers Diehl Aerospace and Nord-Micro, the aviation software solutions provider Linova and IT service provider Lufthansa Systems as well as the German Research Center for Artificial Intelligence and the German Center for Aviation and Space. In addition, several universities are included as subcontractors.

The OMAHA project has several different sub-projects. Our work focuses on a sub-project to develop a cross-system integrated system health monitoring (ISHM). The main goal is to improve the existing diagnostic approach with a multi-agent system (MAS) with several case-based agents to integrate experience into the diagnostic process and provide more precise diagnoses and maintenance suggestions.

2.1   SEASALT

The SEASALT (Shared Experience using an Agent-based System Architecture Layout) architecture is a domain-independent architecture for extracting, analyzing, sharing, and providing experiences [4]. The architecture is based on the Collaborative Multi-Expert-System approach [1,2] and combines several software engineering and artificial intelligence technologies to identify relevant information, process the experience and provide them via an user interface. The knowledge modularization allows the compilation of comprehensive solutions and offers the ability of reusing partial case information in form of snippets. Figure 1 gives an overview over the SEASALT architecture.

The SEASALT architecture consists of five components: the *knowledge sources*, the *knowledge formalization*, the *knowledge provision*, the *knowledge representation*, and the *individualized knowledge*. The *knowledge sources* component is responsible for extracting knowledge from external knowledge sources like databases or web pages and especially Web 2.0 platforms, like forums and social media platforms. These knowledge sources are analyzed by so-called Collector Agents, which are assigned to specific Topic Agents. The Collector Agents collect all contributions that are relevant for the respective Topic Agent's topic [4]. The *knowledge formalization* component is responsible for formalizing the

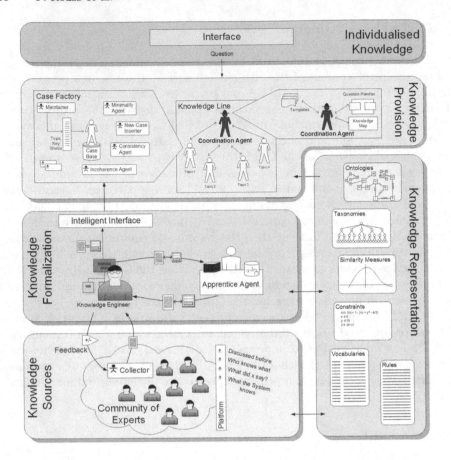

Fig. 1. Overview of the SEASALT architecture

extracted knowledge from the Collector Agents into a modular, structural representation. This formalization is done by a knowledge engineer with the help of a so-called Apprentice Agent. This agent is trained by the knowledge engineer and can reduce the workload for the knowledge engineer [4]. The *knowledge provision* component contains the so called Knowledge Line. The basic idea is a modularization of knowledge analogous to the modularization of software in product lines. The modularization is done among the individual topics that are represented within the knowledge domain. In this component a Coordination Agent is responsible for dividing a given query into several sub queries and pass them to the according Topic Agent. The agent combines the individual solutions to an overall solution, which is presented to the user. The Topic Agents can be any kind of information system or service. If a Topic Agent has a CBR system as knowledge source, the SEASALT architecture provides a Case Factory for the individual case maintenance [3,4]. The *knowledge representation* component contains the underlying knowledge models of the different agents and knowledge sources. The synchronization and matching of the individualized knowledge models improves the knowledge maintenance and the interoperability between the components. The *individualized knowledge* component contains the web-based user interfaces to enter a query and present the solution to the user [4].

2.2   Application Domain

The domain of our application is aircraft fault diagnostic. An aircraft is a highly complex machine and an occurring fault cannot be easily tracked to its root cause. The smallest unit, which can cause a fault, is called Line Replacement Unit (LRU). While a fault can be caused by a single LRU, it also can be caused by the interaction of several LRUs or by the communication line between the LRUs. The data about the fault is in some cases very well structured (e.g., aircraft type, ATA chapter), but in other cases semi-structured (e.g., displayed fault message, references) or unstructured (e.g., fault description, electronic logbook entries, recommendations). These data have to be transformed into vocabulary, similarity measures, and cases.

The application is a first prototype demonstrator with several CBR systems. The systems represent different data sources and subsystems of an aircraft. The data sources are service information letters (SIL) and in-service reports (ISR) and we focus on the subsystems hydraulic and ventilation system. Service information letters contain exceptions to the usual maintenance procedure. These exceptions are described with information like the aircraft type and model, failure code, ATA chapter, displayed message, fault description, recommendations, actual work performed, and references to manuals. In-service reports are failure reports from airlines and contain partially overlapping information with the SIL like aircraft type, ATA chapter, fault description, but contain additional information like starting and landing airport, engine type, and the flight phase in which the fault occurred.

3   Semi-automatic Knowledge Extraction

In this section the instantiation of the SEASALT architecture within the OMAHA project is described. The focus is set on the component *knowledge formalization* to show the idea behind the automatic vocabulary building. The current implementation of the knowledge formalization is described as well as the evaluation of the formalization work flow.

3.1   OMAHA Multi-agent System

For the multi-agent demonstrator we will instantiate every component of the SEASALT architecture. The core components are the *knowledge provision* and the *knowledge formalization*, but the other components will be instantiated, too. The *individualized knowledge* component contains two interfaces for receiving a query and sending the solution. The first interface is a website to send a query to the multi-agent system and to present the retrieved diagnosis. In addition, a user can browse the entire case base, insert new cases or edit existing cases. The second interface communicates with a data warehouse, which contains data about Post Flight Reports (PFR), aircraft configuration data, and operational parameters. A PFR contains the data about the occurred faults during a flight and is the main query for our system. If additional information is required that is not provided by the data warehouse, it can be added via the website. Figure 2 shows the instantiation of the multi-agent system.

The *knowledge provision* component contains all agents for the diagnostic process. We defined several agent classes for the required tasks during the process: interface agent, output agent, composition agent, analyzer agent, coordination agent, solution agent, and topic agent. Each agent class is instantiated through one or more agents. A PFR and additional data is received by the data warehouse agent and/or the webinterface agent. A PFR contains several items that represent occurred faults. The PFR and the additional data are sent to the composition agent, which correlates the additional data with the individual PFR items. The correlated data are sent to the query analyzer agent and the coordination agent in parallel. The query analyzer agent is responsible for checking the correlated data for new concepts, which are not in the vocabulary, and sending a maintenance request to the Case Factory. The Case Factory checks the maintenance request, derives the required maintenance actions and executes the required actions after confirmation from a knowledge engineer. The coordination agent has two main tasks: sending a correlated PFR item to the right solution agent and integrating the returned diagnoses to an overall diagnosis. To determine the right solution agent, the coordination agent uses a so-called Knowledge Map that contains information about the existing solution and topic agents and their dependencies. The Knowledge Map tasks can be outsourced to an additional agent, the knowledge map agent, to provide more parallel processing. The knowledge map agent has access to the general Knowledge Map and to a CBR system that contains individual retrieval paths from past requests. The knowledge map agent uses the CBR system to determine the required topic agents for solving the query from successful past retrieval paths. After determining the required agents, the coordination agents sends the query to the corresponding solution agents. For each aircraft type (e.g., A320, A350, A380, etc.) an own agent team exists to process the query and retrieve a diagnosis. Each agent team consists of several agents: the solution agent receives the query, decomposes it, and sends the query parts to the required topic agents. One topic agent is used to process the configuration data and determine the configuration class of an aircraft. Because the occurrence of many faults depends on the hard- and software configuration of an aircraft, the configuration class can be used to reduce the number of cases in the retrieval process. The other topic agents are distinguished by the content of the case base and the ATA chapters. We derived cases from SIL and ISR for our prototype, but additional data sources are available. The ATA chapter decomposes an aircraft into several subsystems. By distinguishing the CBR systems this way, we get several smaller CBR systems, which have a smaller case structure and are easier to maintain. Each topic agent performs a retrieval on the underlying CBR systems and sends the solutions to the solution agent. The solution agent ranks the individual solutions and sends a ranked list back to the coordination agent and forwarded to the output agent. Each individual solution represents a possible diagnosis for the occurred fault described in the query. Therefore a combination of solutions is not appropriate. All found solutions above a given threshold have to be displayed to the user. The output agent passes the diagnoses to the web interface and the data warehouse.

The *knowledge formalization* component is responsible for transforming the structured, semi-structured, and unstructured data into structured information for the vocabularies, the similarity measures, and the cases itself of the CBR systems. The required maintenance actions for the CBR systems are performed by the Case Factory. For the CBR systems a structural CBR approach was chosen, because almost half of the provided data has the form of attribute value pairs. The other part of the data has to be transformed to be represented as attribute value pairs. The analysis and transformation of the data is done by a so-called case base input analyzer agent. This agent reads the data from different data sources like excel sheets, database result sets, or text documents. Then several information extraction techniques are used to extract keywords and phrases and to find synonyms and hypernyms. In addition, the data is analyzed to find associations within the allowed values of an attribute as well as across different attributes. This way we want to extract Completion rules[1] for query enrichment. The next step in the process is to add the found keywords, their synonyms and phrases to the vocabulary and set an initial similarity between a keyword and its synonyms. Furthermore, taxonomies can be generated or extended using the keywords and their hypernyms. After the vocabulary extension, the cases are generated and stored in the case bases. The last step is the generation or adaptation of the relevance matrices[2] to set or improve the weighting for the problem description attributes. The idea and the top level algorithm of this tool chain and the current implementation status is described in more detail in the following sections.

In the *knowledge sources* component a collector agent is responsible for finding new data in the data warehouse, via web services or in the existing knowledge source of Airbus. New data in the data warehouse could be new configuration data or operational parameters, which have to be integrated into the vocabulary. Web services could be used to update the synonym and hypernym database and from the existing knowledge sources of Airbus new cases can be derived.

The *knowledge representation* component contains the generated vocabulary, the similarity measures and taxonomies, the extracted completion rules, and constraints of the systems to be provided for all agents and CBR systems.

---

[1] Completion rules derive attribute values with a certainty factor if the respective condition is fulfilled (a set of attribute values).

[2] A relevance matrix describes the relevance of available attributes concerning available diagnoses (e.g., [9]).

3.2   Initial Concept for Semi-automatic Knowledge Extraction

There are more than 100.000 documents and data sets with fault descriptions and exceptions within the Airbus data sources. Every document or data set could contain useful information for our case-based diagnosis or even represent a complete case. This amount of data cannot be reasonably analyzed manually, but semi-automatedly with the help of software agents. The result of the analysis and the transformation has to be checked by a knowledge engineer to get feedback. This feedback can be used to improve the analysis and transformation process.

We designed a workflow with ten tasks for processing the data, extracting the knowledge, extending the knowledge containers, and importing cases. Each task consists of several steps. Figure 3 shows the workflow tasks and the associated steps. The input for the workflow is a set of documents with SIL or ISR content and a mapping document. This can be excel sheets, database result sets, or free text documents. The mapping document contains information to which attributes of a case structure the content of the document should be mapped.

The first task in the workflow is the extraction of keywords. Based on the type of the input document, the individual columns and rows or the entire text are processed. This task starts with the steps stopword elimination and stemming of the remaining words. The next step is to replace all abbreviations with the long form of the word. Therefore a list of used abbreviations within the aircraft domain is used to identify abbreviations. The result of this task is a list of keywords extracted from the document.

The second task in the workflow is to find synonyms and hypernyms for each keyword on the list. For the search we use a synonym database from Wordnet extended with technical terms from the avionics domain. For each found synonym and hypernym a search loop for additional synonyms and hypernyms is performed, too. This loop is repeated until no more new synonyms are found. Duplicate synonyms and hypernyms are eliminated and the remaining words are added to the keyword list.

The third task is to identify collocations in addition to the single keywords in the document. While collocations are based on frequently occurring words, the collocation extraction is enhanced by using a vocabulary of technical terms provided by Airbus. This way collocations can be identified even if they occur only a few times, but are relevant to the content. Based on the given technical terms, extracted collocations have a maximum length of five words. All identified collocations are added to a phrase list, while duplicate collocations are removed.

In the next task, all keywords and collocations are added to the vocabulary. The first step is to check the collocations against the keywords, to find combinations of keywords that occurred only as collocation in the given data. The idea is that keywords that do not occur as an individual keyword or as a part of a collocation, but only in the combination of the collocation, will not be added to the vocabulary. This way the growth of the vocabulary can be slowed down.

The fifth task in the workflow contains the setting of initial similarity values between keywords and their synonyms. Due to the fact that words are similar to

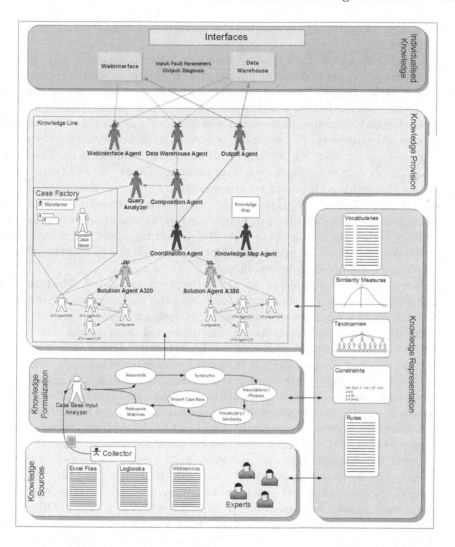

Fig. 2. Instantiation of the SEASALT components within OMAHA

their synonyms, an initial similarity value of 0.8^3 can be assumed between a word and its synonym. The keywords and synonyms are organized in a matrix. Then the found synonyms and hypernyms are used to build taxonomies for similarity assessments. The hypernyms serve as inner nodes, while the keywords and the synonyms are the leaf nodes. Keywords and their synonyms are sibling nodes if they have the same hypernym. Between sibling nodes a similarity of 0.8 can be

---

[3] Assuming, here and the further occurrences, that the similarity measures can take values from the [0;1] interval.

assumed. This way existing taxonomies can be extended or new taxonomies can be generated.

Task six is responsible for finding associations between keywords and phrases within a text or between different columns. The idea is to define completion rules based on these associations. An association between keywords or phrases exists, if the combined occurrence frequency exceeds a given threshold. This threshold defines the minimum occurrence of the combination over all analyzed documents and data sets. For example, a combination between two keywords that occurs in more than 70 percent of all analyzed documents, may be used as a completion rule with an appropriate certainty factor. In addition to the occurrence threshold, a threshold for the minimum number of documents to be analyzed during this task has to be defined. This second threshold is required to avoid the generation of rules by analyzing only few documents, but to generate rules with a high significance. Therefore, the second threshold should be more than 1000 documents or data sets. The higher both thresholds are, the more a generated rule is assumed to be significant.

The seventh task is to generate cases from the given documents. The first step uses the mapping document to map the content of the document to a given case structure. The data from the documents are transformed into values for given attributes to fit the structural approach. The generated cases are not added to a single case base, but assigned to several case bases using a cluster algorithm. The idea behind the clustering strategy is to test the scalability of our approach. The idea is to split the cases based on problem description attributes to get smaller case bases for maintenance. Based on the historical data stored at Airbus, a single case base will contain many thousand cases anyway. Generating an abstract case for each case base, a given query can be compared to the abstract cases and this way a preselection of the required case bases is possible.

We assume a homogenous case structure for all cases generated from the documents. The first case is added to a new case base. For the next case, the similarity to the case in the first case base is computed. If the similarity is below a given threshold, a new case base is created and the new case is added. Otherwise the case is added to the existing case base. Each following case is processed in the same way. The similarity to all cases in the case bases is computed and the new case is added into the case base that contains the case with the highest similarity. If the similarity is below the threshold, a new case base is generated. This step is repeated until all generated cases are added to a case base. While the order of the cases has an impact on the clustering, the dimension of the impact has to be cleared.

Task eight uses sensitivity analysis to determine the weights of the problem description attributes, depending on the content of the cases. This sensitivity analysis is processed for every case base created in the task before. As a result initial relevance matrices are created with the diagnoses as rows and the problem description aka symptoms as columns. These relevance matrices will be used to compute the global similarity during a retrieval.

Task nine contains a consistency check of the vocabulary, similarity measures, and cases by a knowledge engineer to confirm or revise the changes made during the workflow. The feedback from the knowledge engineer is used in task ten to improve the individual tasks and steps within the workflow. The task nine and ten should be processed in periodic intervals and during each workflow execution.

This workflow is designed to be executed beside the CBR cycle as a maintenance workflow. Therefore the before mentioned Case Factory is responsible for the changes to the knowledge containers of a CBR system. This way the workflow is distributed to the knowledge formalization component and the knowledge provision component of the SEASALT architecture. One or more agents in the knowledge formalization component are responsible for the analysis tasks and steps and agents in the Case Factory performing the maintenance actions based on the analysis. But the workflow cannot only be used for maintenance beside the CBR cycle, but also within the CBR cycle. During the retrieval step, a query, especially a natural language query, could be analyzed in the same way as a new case. Therefore a "lighter" version of the workflow could be used, only containing tasks one to six and tasks nine and ten.

3.3 Current Implementation

This section describes the current implementation of our workflow for semi-automated knowledge extraction. We implemented the workflow in Java, because the used CBR tool and the agent framework are Java based, too. Different import mechanisms are implemented to process data from CSV files, text files, and result sets from a database. Because of the different content and data structures of the documents, the data is processed differently for each document type. CSV files and result sets are processed row-wise, while text documents are processed in the whole. The mapping file is written in XML format and contains the information

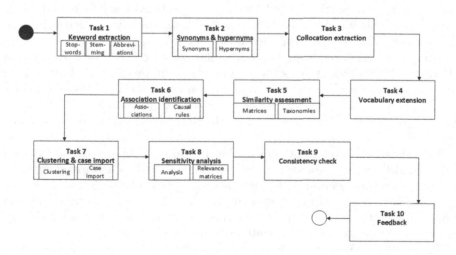

Fig. 3. Workflow for semi-automated knowledge extraction

which column in a CSV file or result set should be mapped to which attribute in the case structure. The following code is an excerpt from the mapping file:

```
<mapping>
<part>problem</part>
<column>AC Type</column>
<attribute>ac_Type</attribute>
</mapping>
```

The keyword extraction is implemented using Apache Lucene and a part-of-speech tagger from the Stanford NLP group. Lucene provides several functions for text analysis, like stopword elimination and stemming and is combined with the Maxent part-of-speech tagger. At first a given input string is tagged with the Maxent tagger and then stopwords are eliminated based on a extended list of English stopwords. This extended list contains all stopwords from the common list of Lucene and some additional words from Airbus' simplified english document. After the elimination of the stopwords, for the remaining words stemming is performed. The result of this step is a list of stemmed keywords. This list is searched for abbreviations based on the Airbus document of used abbreviations in the aircraft domain. All found abbreviations are replaced with the appropriate long word. At last duplicate keywords are removed from the list.

The second task of the workflow is implemented using Wordnet, which provides a large database of synonyms and hypernyms for the English language. For each keyword from the result list of Task 1 the synonyms are determined via Wordnet database and the found synonyms are stored. After searching for synonyms for the given keywords, an additional search is performed based on the found synonyms. This additional search is repeated until the returned synonyms from the Wordnet database contain only already known synonyms. Based on this list of keywords and synonyms, the Wordnet database is requested for hypernyms and for single worded hypernyms a synonym search is performed. The result of this implemented task is a list of keywords with their synonyms and hypernyms in form of a multiple linked list.

In the third task, collocations are identified based on the raw data with the help of the Dragon toolkit. This toolkit provides a phrase extractor based on the frequent occurrence of collocations and a given set of technical terms provided by Airbus. Before using the extractor the abbreviations in the input string are replaced to match the technical terms. The found collocations are stored in a list.

The next task is implemented using the open source tool myCBR. This tool is used to model the case structure, vocabulary, and similarity measures of our CBR systems. It also provides an API to interact with our workflow. This API is used to add all keywords, synonyms, hypernyms, and collocations to the vocabulary of our CBR systems. The mapping information is used to distribute the added words and phrases to the appropriate attributes in the case structure.

The fifth task is only implemented partially at this time. For the added keywords and their synonyms initial similarity values are set in a symmetric similarity matrix. Each keyword has a similarity value of 0.8 to each synonym. This relationship is bidirectional. Additional content-based similarity values have to be assigned manually. The taxonomy creation is not implemented yet.

After extending the vocabulary and setting the similarity values, cases are generated based on the rows of CSV files or database result sets. For each case a retrieval is performed with the problem description of the case as query using the API of myCBR. If the computed similarity is below 80 percent, a new case base is created and the case is added, otherwise the case is added to the case base with the case that has the highest similarity to the query. This process is repeated until all generated cases are added to a case base. If more than one case base has to be considered for adding a case, the case base with the first found case is enlarged.

3.4 Evaluation Setup and Results

This section describes the evaluation setup of the current implementation of our workflow and the diagnosis retrieval. The workflow was used to analyze and process 670 data sets with SIL context and 120 data sets with ISR context. From each data set a case was generated. During the first and third task 872 keywords and 76 collocations were extracted. The second task produced 2862 synonyms and 213 hypernyms. In the first evaluation scenario the raw data and the extracted keywords, synonyms, and hypernyms are compared by maintenance experts from Airbus and Lufthansa. In the second evaluation scenario 50 queries are performed on the system with ten cases as retrieval result. These retrieval results are checked by the maintenance experts from Airbus and Lufthansa Systems for appropriate diagnoses to the given queries.

As a result from the first evaluation scenario the experts rated 628 keywords as correct (ca. 72 percent). From the remaining 244 keywords, 98 keywords are wrongly extracted because of false abbreviation replacement or stemming problems, while 146 keywords are false because of an inappropriate word sense. This means there is an overhead of 27 percent from word sense problems. 62 collocation are rated correctly (82 percent), while 14 collocations are wrong, because of false abbreviation replacement. The synonyms and hypernyms have a similar success rate. 2260 synonyms were rated correct and useful, while 602 synonyms were wrong because of inappropriate word sense. Only 124 hypernyms were rated correct, while the remaining 89 hypernyms are wrong as a consequence of the inappropriate synonym word sense.

The result of the second evaluation scenario is that an average of 78 percent of the retrieved cases have an appropriate diagnosis. For each query this number differs slightly. For some queries all retrieved cases were appropriate, for other queries only a few cases were appropriate. Not only the cases itself were checked, but also the ranking of the cases. An average of 18 percent of the retrieved cases were ranked wrong from an expert point of view.

The evaluation shows that the initial version of our workflow produces good result, but there is still potential for improvement. The results from the workflow are good enough to perform a meaningful retrieval, while the number of correct diagnoses has to be improved. The main problem in both scenarios is the word sense of keywords and synonyms that is in many cases not compatible with the aircraft domain. This problem has to be addressed to identify the useful word senses. Another problem is the missing similarity measures for attribute values, which are not synonyms.

4   Related Work

There is a lot of related work on CBR and information extraction, association rule mining, processing textual data in CBR and text mining. This section contains a selection of related work from these topics. Bach et al. describe in their paper an approach for extraction knowledge from vehicle in-service reports. This approach is also based on the SEASALT architecture like our approach, but uses only automated keyword extraction to process the reports. As an additional step the extracted keywords are classified. Then the extracted keywords are reviewed by experts and inserted manually into the vocabulary [5]. Our approach still has the review process of an expert or knowledge engineer, but aims on a more detailed text processing workflow with phrases, synonyms and hypernyms. We try to create a more automated workflow to populate the vocabulary and initial similarity measures.

In their article about knowledge extraction from web communities, Sauer and Roth-Berghofer describe the KEWo Workbench and the mechanisms provided by this workbench to extract knowledge from semi-structured texts. The KEWo workbench is able to create taxonomies from extracted keywords and phrases based on the relative frequency of the occurrence [11]. In our approach we will generate the taxonomies not from the relative frequency, but from found hypernyms and synonyms from the Wordnet database and useful technical terms from the aircraft domain vocabulary.

Many systems with textual knowledge use the textual CBR approach, like [7,10,12]. The data sources available for our project are mainly structured data, therefore we choose a structural CBR approach. But the most important information about an occurred fault can be found in fault descriptions and logbook entries, which are free text. We decided to use a hybrid approach with the combination of structural CBR and textual CBR techniques, to integrated all available information.

[8] describes an approach for enriching the retrieval using associations. They use the Apriori algorithm to extract relevant cases for correlation between cases. We will use algorithm like Apriori or FP-Growth to extract associations between attribute values in a case. This aims on generating completion rules to enrich a query by setting attribute values automatically based on the completion rules.

5   Summary and Outlook

In this paper we described the idea of a semi-automatic knowledge extraction workflow for a decision support system within the aircraft domain. We give an overview over the decision support system and the tasks and substeps of the workflow. In addition, we show our current implementation of the workflow and the evaluation results, based on the current implementation.

As the evaluation shows there is potential for improvement of the individual tasks of the workflow as well as for the complete workflow. The main problem of the inappropriate word sense, that causes the overhead of the vocabulary and the similarity measures, will be addressed by the extend use of an aircraft domain vocabulary provide by Airbus and Lufthansa Systems. Another idea for solving this problem is to restrict the adding of keywords, based on the relative occurrence frequency. In addition to the enhancement of implemented tasks, the next steps will be the implementation of the tasks for taxonomy creation, the sensitivity analysis and association extraction.

References

1. Althoff, K.D.: Collaborative multi-expert-systems. In: Proceedings of the 16th UK Workshop on Case-Based Reasoning (UKCBR-2012), located at SGAI International Conference on Artificial Intelligence, Cambride, UK, 13 December, pp. 1–1 (2012)
2. Althoff, K.D., Bach, K., Deutsch, J.O., Hanft, A., Mänz, J., Müller, T., Newo, R., Reichle, M., Schaaf, M., Weis, K.H.: Collaborative multi-expert-systems - realizing knowledge-product-lines with case factories and distributed learning systems. In: Baumeister, J., Seipel, D. (eds.) KESE @ KI 2007, Osnabrück, September 2007
3. Althoff, K.D., Reichle, M., Bach, K., Hanft, A., Newo, R.: Agent based maintenance for modularised case bases in collaborative mulit-expert systems. In: Proceedings of the AI2007, 12th UK Workshop on Case-Based Reasoning (2007)
4. Bach, K.: Knowledge acquisition for case-based reasoning systems. Ph.D. thesis, University of Hildesheim (2013). Dr. Hut Verlag Mnchen
5. Bach, K., Althoff, K.-D., Newo, R., Stahl, A.: A case-based reasoning approach for providing machine diagnosis from service reports. In: Ram, A., Wiratunga, N. (eds.) ICCBR 2011. LNCS, vol. 6880, pp. 363–377. Springer, Heidelberg (2011)
6. BMWI: Luftfahrtforschungsprogramms v (2013). www.bmwi.de/BMWi/Redaktion/PDF/B/bekanntmachung-luftfahrtforschungsprogramm-5,property=pdf, bereich=bmwi2012,sprache=de,rwb=true.pdf
7. Ceausu, V., Després, S.: A semantic case-based reasoning framework for text categorization. In: Aberer, K., Choi, K.-S., Noy, N., Allemang, D., Lee, K.-I., Nixon, L.J.B., Golbeck, J., Mika, P., Maynard, D., Mizoguchi, R., Schreiber, G., Cudré-Mauroux, P. (eds.) ASWC 2007 and ISWC 2007. LNCS, vol. 4825, pp. 736–749. Springer, Heidelberg (2007)
8. Mote, A., Ingle, M.: Enriching retrieval process for case based reasoning by using certical association knowledge with correlation. Int. J. Recent Innov. Trends Comput. Commun. 2, 4114–4117 (2015)
9. Richter, M., Wess, S.: Similarity, uncertainty and case-based reasoning in PATDEX. In: Boyer, R.S. (ed.) Automated Reasoning - Essays in Honor of Woody Bledsoe, vol. 1, pp. 249–265. Kluwer Academic Publishers, Dordrecht (1991)
10. Rodrigues, L., Antunes, B., Gomes, P., Santos, A., Carvalho, R.: Using textual CBR for e-learning content categorization and retrieval. In: Proceedings of International Conference on Case-Based Reasoning (2007)

11. Sauer, C.S., Roth-Berghofer, T.: Extracting knowledge from web communities and linked data for case-based reasoning systems. Expert Syst. Spec. Issue Innov. Tech. Appl. Artif. Intell. **31**, 448–456 (2013)
12. Weber, R., Aha, D., Sandhu, N., Munoz-Avila, H.: A textual case-based reasoning framework for knowledge management applications. In: Proceedings of the Ninth German Workshop on Case-Based Reasoning, pp. 244–253 (2001)

Evidence-Driven Retrieval in Textual CBR: Bridging the Gap Between Retrieval and Reuse

Gleb Sizov$^{(\boxtimes)}$, Pinar Öztürk, and Agnar Aamodt

Department of Computer Science, Norwegian University of Science and Technology,
Trondheim, Norway
{sizov,pinar,agnar.aamodt}@idi.ntnu.no

Abstract. The most similar case may not always be the most appropriate one to guide a problem-solving process. It is often important that a retrieved past case can be easily adapted to a target problem. The presented work deals with the retrieval and adaptation in textual case-based reasoning (TCBR) where cases are described textually. In TCBR, it is common to use similarity-based retrieval methods from information retrieval where adaptability of the retrieved cases is not considered. In this paper we introduce a novel case retrieval method called evidence-driven retrieval (EDR). It uses the notion of evidence to determine which parts of the new problem text have been useful in the past solutions and will be used in the adaptation to a new problem. This allows EDR to retrieve cases that are not only similar but also adaptable. We evaluated EDR as part of our TCBR approach that aims to support human experts in root cause analysis of transportation incidents. This approach relies on causal knowledge automatically extracted from incident reports from the Transportation Safety Board of Canada, which are used as textual cases in our experiments. The results for EDR are compared with information retrieval methods traditionally applied in TCBR.

Keywords: Textual CBR · Incident analysis · Causal relations · Adaptation-guided retrieval · Adaptation

1 Introduction

The fundamental assumption in CBR is that similar problems have similar solutions. Therefore, case retrieval in CBR is often based on the similarity between a new problem description and cases in the casebase. Sometimes, however, the most similar case is not the best one to guide the problem-solving process. As argued by Smyth and Keane [19], for many types of applications it is also necessary to consider whether a case can be easily adapted to a target problem. They proposed the adaptation-guided retrieval method that uses adaptation knowledge to retrieve adaptable cases. In TCBR, where cases are described textually, case retrieval is often accomplished using methods from information retrieval (IR) [20]. These methods are similarity-based and do not account for adaptability.

© Springer International Publishing Switzerland 2015
E. Hüllermeier and M. Minor (Eds.): ICCBR 2015, LNAI 9343, pp. 351–365, 2015.
DOI: 10.1007/978-3-319-24586-7_24

In the current paper we propose a novel case retrieval method for TCBR that aims to retrieve adaptable cases, called evidence-driven retrieval (EDR). This method was developed as part of our work on TCBR for automated incident analysis using reports from the Transportation Safety Board of Canada as textual cases. The focus of our previous work was on the representation and adaptation of textual cases while for case retrieval we used a standard IR method [18]. Experimental evaluation revealed that this IR-based retrieval method was the bottleneck for the whole system because many of the retrieved cases, despite reasonable similarity to the new case, could not be adapted to a new problem. This intensified our motivation to develop EDR, which brings together retrieval and reuse of textual cases.

EDR, as well the rest of our TCBR approach, is based on the Text Reasoning Graph (TRG) representation, which we first proposed in [18] and have further improved in the current paper. This representation is designed to capture the so-called reasoning knowledge contained in text, which is automatically extracted from textual reports by our system. Imagine a detective investigating a criminal case where she needs to identify evidences, connect the facts and make conclusions about what might have happened and who is involved. The knowledge used by the detective is often of a relational nature connecting pieces of information together in a coherent reasoning chain. This type of knowledge is essential for complex problems that do not have an immediate answer but require to be analysed in order to be solved. The resultant analysis constitutes the case solution.

EDR uses the notion of *evidence*, which can be defined as a piece of information in the problem description that is instrumental for the analysis. EDR automatically identifies which snippets in the new problem text may be conveying an evidence and assesses their informativeness. The decision of which information in the new problem can be considered as evidence is done in the context of a certain past case. The retrieval mechanism selects the past case that includes most number of evidences with high informativeness. The rationale behind EDR is that information that was important in the analysis in the past can have the same value for the new problem as well. In this way, EDR acts as a feature selection method. Since evidences are starting points in the adaptation process, EDR is biased towards retrieving an adaptable case than merely a similar case.

The rest of the paper is organised as follows. Section 2 explains the incident analysis task. Sections 3 and 4 describe the TRG representation and the procedure for automatic acquisition of it from text. The overview of our TCBR approach is provided in Sect. 5. Section 6 describes EDR together with the adaptation procedure. In Sect. 7 we evaluate EDR and compare it with IR methods. Related work is reviewed in Sect. 8 followed by a discussion of future work and concluding remarks in Sect. 9.

2   Incident Analysis

Complex problems such as diagnosing a patient, investigating an accident or predicting the outcome of a legal case need to be analysed in order to be solved. It is a non-trivial task even for human experts so we are investigating methods to support them in such tasks. The type of the analysis we are aiming for is closely related to *root cause analysis* (RCA) [15] used by human analysts to answer why a certain problem occurred in the first place [16]. A problem is characterised by an undesired outcome such as a failure, accident, defect, dangerous situation, etc. Causes are the events or conditions that lead to the undesired outcome, removal of which would prevent the occurrence of that outcome. RCA goes beyond causes that immediately precede the outcome and aims to identify causal chains reaching the root causes of the problem. These causes are of vital importance for the prevention of the same problems to occur in the future.

We study this analysis task in the transportation domain because incident reports are easily available in this domain. The Transportation Safety Board of Canada (TSBC) provides open access to aviation, marine and rail incident reports between 1990 and the present day[1]. These reports are semi-structured in the sense that they commonly contain sections that can be attributed to one of the parts:

1. *Summary* provides a brief description of an incident.
2. *Factual information* describes the details.
3. *Analysis* documents the reasoning of the analysts
4. *Conclusion* enumerates root causes and contributing factors for the incident.

Most governments impose companies to write such analysis reports. Regardless of that, these reports are important for the companies from a knowledge management perspective; companies find such reports beneficial because they constitute an important source for revision of the companies' safety regulations. Furthermore, they are important for sharing the experiences and reasoning knowledge of the analysts, to be put to use when analysing a new incident. Our overarching goal is to automate parts of this analysis task, and support the report writing process after an incident has been analysed.

3   Representation of Reasoning Knowledge

In our work, we aim to use incident reports as cases for CBR-based incident analysis. Textual representation makes effective case retrieval and adaptation very challenging. One way to overcome this problem, is to convert the text to a semi-structured representation with a well defined semantics. For the analysis task we need a representation that is able to capture the line of reasoning embedded in the analysis text. For this purpose we introduced a representation, coined Text Reasoning Graph (TRG) [18]. To illustrate the idea behind TRG consider the following excerpt from an aviation incident report:

---

[1] Reports from Transportation Safety board of Canada are available from http://www.tsb.gc.ca/eng/rapports-reports.

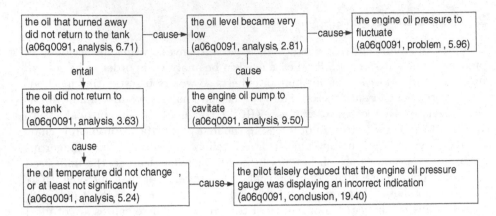

Fig. 1. Example of the text reasoning graph representation with metadata in the following format: (report id, part of the report, informativeness)

The oil that burned away did not return to the tank and, after a short time, the oil level became very low, causing the engine oil pump to cavitate and the engine oil pressure to fluctuate. Furthermore, since the oil did not return to the tank, the oil temperature did not change, or at least not significantly, and the pilot falsely deduced that the engine oil pressure gauge was displaying an incorrect indication.

This excerpt captures reasoning of the expert about the incident, what we call *reasoning knowledge*, which reflects how the analyst put together the pieces of information in order to make a sense out of it. Phrases and sentences in this excerpt are connected through causal relations making the whole excerpt logically coherent. The TRG shown in Fig. 1 makes these relations explicit by collecting them in one graph and adding entailment relations between nodes when it applies. A TRG representation enables automatic inference, e.g. given the TRG in Fig. 1, from the node "the engine oil pressure to fluctuate", through abduction we can infer the explanation "the oil that burned away did not return to the tank", and by deduction the conclusion "the pilot falsely deduced that the engine oil pressure gauge was displaying an incorrect indication".

In addition to a phrase or a sentence, each node in TRG also contains the following metadata:

1. The report id that the node is extracted from.
2. Part of the report containing the same information as the node, e.g. summary, analysis and conclusion.
3. Informativeness of the node, i.e. more specific nodes such as "the oil that burned away did not return to the tank" has higher informativeness than general node such as "the oil did not return to the tank".

4   Acquistion of Reasoning Knowledge from Text

TRG is automatically acquired from text of the incident reports using natural language processing. This process can roughly be divided into two phases: preprocessing and graph construction. In the preprocessing phase, a report in the html format is converted into a structured text annotated with syntactic and semantic information. This information is then used in the graph construction phase to generate a TRG.

Steps in the preprocessing phase are as follows:

1. Extract text and sections from the report in html format.
2. Split the report into summary, analysis and conclusion parts based on section titles, e.g. a section with the title containing the words "findings" or "causes" is assigned to the conclusion part. Similar lexical patterns were constructed for each part.
3. Text of the report is preprocessed with the CoreNLP [13] pipeline that includes tokenization, sentence splitting, part-of-speech tagging, syntactic parsing and co-reference resolution.
4. Causal relations are extracted from text of the report as described in Sect. 4.1.

The graph construction phase includes the following steps:

1. Causal relations are collected in one graph with arguments as nodes and relations as edges. Causal relations are the bare bones of the TRG representation.
2. Nodes that are not arguments of the same causal relation are connected by entailment relations. The longest common paraphrase technique, described in Sect. 4.2, is used for this purpose. This step is necessary to make the graph more connected.
3. Nodes that are paraphrases of each other are merged into one node, preserving the corresponding entailment and causal relations. This makes the graph more compact by eliminating redundant nodes.
4. Nodes with low informativeness are removed from the graph as described in Sect. 4.3, e.g. the phrase "after a short time" does not carry enough concrete information by itself and is considered uninformative.

All the steps in the described process are fully automated, eliminating case acquisition costs. Most of the steps are the same as described in our previous work [18]. Two improvements are the use of co-reference resolution to replace pronouns in text with corresponding references, e.g. "he" is replaced by "pilot", and the use of the longest common paraphrase instead of textual entailment and paraphrase recognition, which facilitates EDR as described in Sect. 6.

4.1   Causal Relation Extraction

Causal relations are used in the generation of the TRG representation. To extract them from text of the report we implemented the pattern matching algorithm proposed by Khoo [8]. Khoo manually constructed 651 patterns and 352 subpatterns for causal relation extraction. The algorithm matches these patterns to sentences in the incident report. If matching succeeds, phrases corresponding to cause and effect arguments are extracted according to the applied pattern, e.g.

pattern: `because of [cause], [effect]`
source: Because of the durability of the coverings, it would be extremely difficult for a survivor with hand or arm injuries to open the survival kit.
cause: the durability of the coverings
effect: it would be extremely difficult for a survivor with hand or arm injuries to open the survival kit.

4.2 Longest Common Paraphrase

Longest common paraphrase (LCP) is the technique we introduce in the current work to connect causal relations into a connected graph. It is also used in the retrieval to identify evidences as described in Sect. 6. LCP identifies the longest pair of phrases in two text fragments that are paraphrases of each other, e.g.

$arg1$ The minimum required radar separation in this airspace was 5 nautical miles laterally or 1000 ft vertically.
$arg2$ No alternate to radar separation minima was in place during the time that communication with the two aircraft was not available.
$lcp1$ minimum required radar separation in this airspace
$lcp2$ radar separation minima.

Two arguments are linked through the longest paraphrase e.g. $arg2 \xrightarrow{entail} lcp1 \xleftarrow{entail} arg2$. To find the longest paraphrases, LCP iterates over all pairs of phrases (S, NP and VP nodes in the syntax tree) starting with the longest ones and stops when paraphrases are found. The paraphrase identification component is based on a semantic text similarity measures. First, it obtains similarity values between each pair of words inside the phrases using the Leacock and Chodorow (LCH) similarity measure [12], which is based on the shortest path between words through WordNet with all senses considered. Then, each word in one phrase is assigned to a similar word in another phrase using the Hungarian algorithm [10]. It makes sure that no two words from one phrase are assigned to the same word in another phrase and the sum of similarities between the assigned words is maximized. Stop words like articles and prepositions are ignored. The text similarity value is obtained by normalizing this sum by the number of words in both phrases. Phrases with a similarity value above 0.7 are considered paraphrases.

4.3 Node Informativeness

Nodes in TRG contain phrases and sentences of various sizes. Informativeness of a phrase measures how much concrete information it contains. In our approach, this measure is used to remove uninformative nodes from TRG as well as to assess the quality of evidences in the retrieval as described in Sect. 6.

Many different approaches exist to measure text or term informativeness [9]. For our task, we use the informativeness measure based on the inverse document frequency (IDF), computed as follows:

$$Info(node) = \sum_{word \in node} \log \frac{|CaseBase|}{|\{case \mid word \in case \wedge case \in CaseBase\}|} \qquad (1)$$

It measures the informativeness of a node as the sum of informativeness of the contained words, which are inversely proportional to their occurrence in the case base. A node is considered informative if its informativeness is higher than 1.0. In addition it should contain at least two, non-stop words and have a direct or indirect connection to a node from the conclusion part of the report.

5   Textual CBR for Incident Analysis

The overview of our TCBR approach for incident analysis is shown in Fig. 2. It follows the classical CBR cycle with retrieve, reuse, revise and retain steps [1]. The case base is a collection of CaseGraphs, where a CaseGraph is a TRG automatically extracted from an incident report which represents an analysis of the incident. Unlike a traditional case representations with separate problem description and solution parts, a CaseGraph contains both of them together in one graph. Nodes in a TRG include the metadata that indicates which part of the report the information in the node is contained in (see example in Fig. 1). Problem nodes correspond to the problem description part of a case and analysis with conclusion nodes to the solution part.

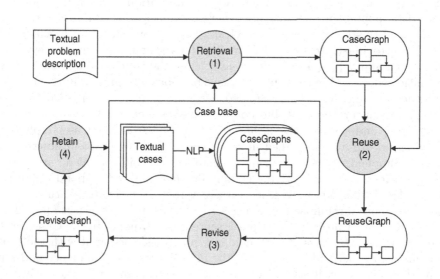

Fig. 2. Textual CBR cycle for incident analysis

In addition to a CaseGraph, our cycle contains a ReuseGaph and a Revise-Graph, that are also TRGs generated at different steps of the CBR cycle. The cycle starts with the textual description of a new problem. In the retrieval step, this description is used to retrieve a CaseGraph from the case base. This Case-Graph is then adapted to the new problem generating a ReuseGraph, which represents the adapted solution. The ReuseGraph is then validated and modified by a human expert in the revise step resulting in the ReviseGraph, which is then stored in the case base for future use.

6 Retrieval and Adaptation

The general idea behind evidence-driven retrieval (EDR) is to assess whether the analysis of a previous problem can be adapted for solving a new problem. As described in Sect. 5, a case in our system is represented by a CaseGraph, which captures an incident analysis by means of causal and entailment relations, while a new problem is described textually. The EDR process matches each CaseGraph in the case base with the new problem description to find the most relevant and adaptable past case. To assess the adaptability, EDR identifies so-called evidences in the new problem text in the context of each past case separately. An evidence is a phrase in the new problem description that carries information that was proven to be instrumental in the analysis of a previous problem. Identification of evidences acts as a method for feature selection where features are selected based on their usefulness in analysing a past problem. In addition to finding evidences, EDR assesses their informativeness so that more informative evidences contribute more to the ranking of cases in the retrieval process.

Evidences also play an important role in adaptation because they serve as the starting points for the analysis generated during the adaptation process. The result of this analysis is a ReuseGraph like the one shown in Fig. 3, which we will use to illustrate how EDR works. This graph contains two evidence nodes: "pilot applied carburettor heat" and "engine abruptly lost all power". The information contained in these phrases have previously been used in the analysis of another incident, e.g. "pilot applied carburettor heat" is entailed by "the pilot applied carburettor heat, but noted engine resulted in a further decrease in engine power and selected the carburettor heat off" from the CaseGraph. The two evidences are also contained in sentences of the new problem description, e.g. "pilot applied carburettor heat" is entailed by "pilot applied carburettor heat and attempted to restart the engine, but heat did not respond" from the problem description. These evidence phrases were identified automatically by the longest common paraphrase method described in Sect. 4.2, which is applied for all combinations of problem description sentences and nodes in the CaseGraph. Then, the informativeness of each evidence phrase is calculated with the IDF-based measure described in Sect. 4.3. The informativeness of evidence nodes in Fig. 3 (shown in parenthesis) is 5.67 and 4.04, resulting in total informativeness of 9.71. This value is used to rank cases in the case base, taking the top one as the retrieved case.

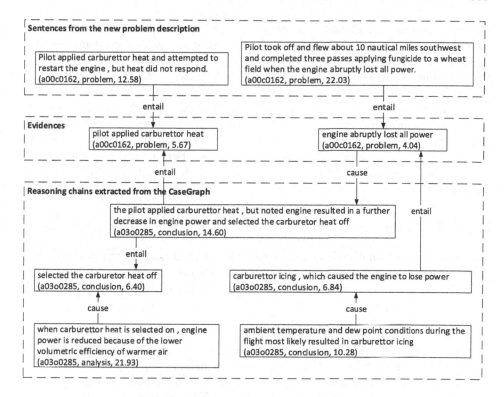

Fig. 3. Part of the ReuseGraph.

Formally, the described process of finding and measuring informativeness of evidences can be formulated as follows:

$$Info_e(S_p, CaseGraph) = \sum_{s \in S_p} \sum_{n \in CaseGraph} Info(LCP(s, n))$$

where S_p is the set of all sentences in the new problem description text, n is a node in the CaseGraph from the case base, $Info$ is the informativeness measure, and LCP is the longest common paraphrase described in Sect. 4.2.

The result of the retrieval process is the CaseGraph with the most informative evidences for the target problem. This CaseGraph together with the evidences is further used in the adaptation process that was first proposed in our previous work [18]. The main idea of this process is to find reasoning chains through the CaseGraph connecting evidences to conclusion nodes. Conclusion nodes are nodes in the CaseGraph that were extracted from the conclusion part of the corresponding report as described in Sect. 2. A reasoning chain is a shortest path through the CaseGraph from an evidence node to a conclusion node. One example of a reasoning chain in Fig. 3 is "engine abruptly lost all power" $\overset{entail}{\longleftarrow}$ "carburettor icing, which caused the engine to lose power" $\overset{cause}{\longleftarrow}$ "ambient temperature and dew point conditions during the flight most likely resulted in carburettor icing". The adaptation process attempts to find a reasoning chain for every evidence and conclusion pair. The shortest path algorithm doesn't consider the direction or weights of edges in the graph. All the reasoning chains are combined in one graph, called ReuseGraph, which is the final result of the adaptation process.

In the described adaptation approach, only one case is reused to solve a target problem meaning that only evidences for the currently retrieved past case are considered. Evidences not used in the retrieved case but possibly used in other past cases may also be important for the new case. For our future work, we are considering a compositional adaptation approach that combines evidences and CaseGraphs from multiple cases.

7   Evaluation

The goal of our evaluation is to investigate the effect of EDR on the performance of the TCBR system for incident analysis. In particular we compare performance of EDR with the bag-of-words and semantic indexing models.

7.1   Dataset and Preprocessing

In our work we use incident investigation reports from the Transportation Safety Board of Canada (TSBC). Compared to other incident report datasets such as The United States Aviation Safety Reporting System (ASRS) dataset used in previous works on incident analysis [2,14,21], TSBC reports are more detailed with extensive analysis sections. A fairly consistent structure of the reports makes it easier to accomplish automatic evaluation. Another advantage of the TSBC dataset is that it contains collections from several different transportation domains which enables us to validate the domain-independent nature of our approach.

Three of TSBC collections were used for evaluation: 922 aviation reports, 375 marine reports, 298 rail reports. Each report is a text document 5–10 pages long that describes an incident and provides textual analysis of the causes and contributing factors as described in Sect. 2. For evaluation, we randomly split each collection into a test and a training set, which contain 20 % and 80 % of the reports, respectively, with no overlap.

7.2   Evaluation Procedure

CBR methods are evaluated based on the quality of the produced solution given a problem description. In our approach, problem description is a textual summary from the test report that briefly describes the incident. Given this summary, the system generates a solution in a form of a ReuseGraph as described in Sect. 2. Four scores are computed in our evaluation: *adaptability, precision, recall* and *f-score*. Adaptability is a binary score indicating whether the adaptation is possible (1) or not (0), which is determined by whether any evidences can be identified in the problem description given the retrieved case.

Adaptability does not evaluate the results of adaptation, which is the role of precision, recall and f-score. Our implementation of these measures is different from IR and is based on the similarity between *conclusion nodes* in the Reuse-Graph and *conclusion sentences* in the test report. These conclusion sentences were written by human experts and enumerate actual causes for the incident, e.g. "During the auto-rotation, the helicopter was not levelled at the time of the landing, which resulted in a hard landing." We use the same text similarity measure as described in Sect. 4.2. Each conclusion node is assigned to a similar conclusion sentence such that no two nodes are assigned to the same sentence and the sum of similarities is maximized. It is the instance of the *assignment problem* solved by the Hungarian algorithm [10]. Given this sum (referred to as *Similarity*), precision, recall and f-score are computed as follows:

$$Similarity = Assignment(ConclusionNodes, ConclusionSentences)$$

$$Precision = Similarity/|ConclusionNodes|$$

$$Recall = Similarity/|ConclusionSentences|$$

$$F\text{-}score = 2 \cdot \frac{Precision \cdot Recall}{Precision + Recall}$$

where precision indicates the correctness of the proposed conclusions and recall shows to what extent the proposed conclusions cover actual conclusions. As in IR, it is important to consider these measures together, which is reflected in the f-score.

7.3   Baselines

Two baseline retrieval methods were implemented to compare with EDR. The first one is TFIDF we used in our previous work [18]. It is based on the BOW model where cases are represented as vectors with words as dimensions and term frequency - inverse document frequency (tf-idf) weights as values. The similarity is computed as the cosine between vectors representing textual problem descriptions for a new and previous problems. The most similar case is retrieved and it's CaseGraph is then used in adaptation.

The second baseline retrieval method is based on Latent Semantic Indexing (LSI) [6]. It is a well-known semantic indexing method which applies singular value decomposition to a term-document matrix with tf-idf weights. Term dimensions in this matrix are transformed to maximize the variance between documents. In the transformed coordinates, similar terms become closer while distant terms become further from each other. The dimensions are also ranked by a so-called singular value which indicates the discriminative power of the dimension. With LSI, it is common to take a certain number of dimensions with the highest singular values. We tried different number of dimensions (50, 150, 300, 600) to maximise the results for the LSI baseline. The best results shown in Table 1 were obtained for 300 dimension.

For both baselines we attempted to use the whole report instead of the problem description part from the previous case when computing similarity. Intuitively, it makes sense because other parts of the report contain information not

Table 1. Score means in % obtained when evaluated on the complete test set.

Dataset	Measure	TFIDF	LSI	EVIDENCE
Aviation	Adaptability	47.46	21.47	**100.00**
	Precision	8.73	3.77	**14.04**
	Recall	6.04	1.78	**17.45**
	F-score	5.93	2.03	**12.56**
Marine	Adaptability	40.79	19.74	**100.00**
	Precision	7.19	3.62	**16.50**
	Recall	5.58	1.29	**12.91**
	F-score	5.58	1.73	**12.49**
Rail	Adaptability	37.50	25.00	**100.00**
	Precision	6.71	4.34	**11.62**
	Recall	4.81	2.93	**19.98**
	F-score	4.84	2.97	**13.41**

available in the problem description which might result in a more precise similarity assessment. In addition, EDR makes use of other parts of previous reports captured by a CaseGraph so it is fair that the baseline methods can utilize these parts as well. However, experimental results showed that including other parts of a report in the retrieval decreases the performance of the baselines. Therefore, the results presented in Sect. 7.4 were obtained when baselines use only the problem description part.

7.4 Results and Analysis

Table 1 shows the results obtained by our system on three datasets. Three different retrieval components were used: TFIDF, LSI and EVIDENCE. The first two are the baselines described in Sect. 7.3 and EVIDENCE is the implementation of EDR described in Sect. 6. These components are evaluated as part of the complete system following the evaluation procedure described in Sect. 7.2.

The results show significantly higher performance of the evidence-based retrieval. It has 100 % adaptability scores indicating that cases retrieved with EVIDENCE can always be adapted to a new problem. It is the consequence of the idea that the retrieval mechanism in the EDR is designed to facilitate the following adaptation process. In contrast, our baseline retrieval methods, TFIDF and LSI, are not aware of the adaptation process and as a result most of the cases they retrieve can not be adapted to the target problem. While adaptability is a binary score that indicates whether a retrieved case can be adapted, precision, recall and f-scores reflect the quality of the adapted solution. As with adaptability, these scores are significantly higher for EVIDENCE compared to TFIDF and LSI.

For all datasets, LSI demonstrated poor results, worse than a more primitive TFIDF baseline. Possible explanation for this is that the analysis task is driven by specific details rather than conceptual topics captured by LSI. Incident reports in the same domain cover very similar topics and mostly use the same vocabulary, which makes semantic indexing less useful. In addition, incident report collections we use for the experiments are relatively small in size, which might not be enough to learn a robust semantic representation.

8   Related Work

The idea of retrieving adaptable cases was proposed and thoroughly investigated by Smyth and Keane [19] in their work on adaptation-guided retrieval (AGR). They showed that similarity alone might not be enough to retrieve the most appropriate case to guide the problem solving process. AGR uses adaptation knowledge that provides the link between problem and solution features and allows to asses the importance of matches between features based on their influence on adaptation. AGR and similar approaches have been successfully used in several CBR systems for different tasks including plant-control software, example-based machine translation, and property-valuation [4,7,19]. EDR is based on the same general idea as AGR with reasoning chains providing explicit mapping between problem and solution features. However, unlike AGR, adaptation knowledge in EDR is not explicitly represented.

EDR is also inspired by explanation-based learning (EBL), where the relevance of features is determined by explaining their contribution to an example's solution [3]. Like in EBL, EDR uses solutions from previous cases to judge the relevance of features in the new problem description. These features become parts of the reasoning chains generated during adaptation, which can be viewed as the explanations. The major difference, however, is that EBL relies on high-quality domain knowledge to generate explanations while reasoning chains in our approach are generated from causal and entailment relations extracted from text.

Most of the systems using AGR and EBR approaches use cases and knowledge represented in structured form with a well-defined meaning. In contrast, EDR is designed for TCBR where knowledge is in textual form. Compared to structured representation, natural language has a much more complex and ambiguous semantics. It makes adaptation a very challenging task. Many adaptation methods are limited to substitution of textual units in the solution text such as in [11] where email responses are retrieved and adapted to new request. Although adaptability of cases is not considered in the retrieval process, there is an explicit link between a problem and solution spaces in a form of word associations. Problem-solution associations between larger textual units such as phrases and sentences have been also investigated in [17]. To enable more sophisticated forms of adaptation it is often necessary to convert textual cases to a structured form. A recent example of this approach is the work on extraction and adaptation of cooking workflows [5]. The possibility for AGR is briefly discussed but not implemented in their work.

9   Conclusion

In this paper we presented evidence-based retrieval (EDR), a case retrieval method for TCBR that aims to retrieve adaptable cases. EDR is based on the text reasoning graph representation which automatically captures reasoning knowledge contained in textual cases. EDR identifies evidence phrases in the textual problem description that serve as the starting points for adaptation. The cases are then ranked by the informativeness of these evidences.

We evaluated EDR on the incident analysis task in three domains (aviation, marine and railway) using incident reports from the Transportation Safety Board of Canada. Experimental results show significantly higher performance of EDR compared to commonly used IR methods. 100 % of cases retrieved with EDR are adaptable, while IR methods had adaptability below 50 %.

EDR is the latest addition to our TCBR approach for automated analysis. Previously we developed the representation and the adaptation technique specifically designed for this task. With EDR all the components in our approach are task-specific and well integrated with each other, which results in better performance. Still, there are many directions for future work. For instance, we plan to investigate the possibility for capturing textual context in the TRG representation. Currently, phrases contained in TRG nodes loose their context when extracted from text. A phrase without context can be ambiguous which reduces the accuracy of the textual similarity component used for generation of the representation, retrieval and adaptation. Another promising direction for future research is the development of a natural language generation component that could produce textual analysis from ReuseGraphs.

References

1. Aamodt, A., Plaza, E.: Case-based reasoning: foundational issues, methodological variations, and system approaches. AI Commun. **7**(1), 39–59 (1994)
2. Abedin, M.A.U., Ng, V., Khan, L.: Cause identification from aviation safety incident reports via weakly supervised semantic lexicon construction. J. Artif. Intell. Res. **38**(1), 569–631 (2010)
3. Cain, T., Pazzani, M.J., Silverstein, G.: Using domain knowledge to influence similarity judgements. In: Proceedings of the Case-Based Reasoning Workshop, pp. 191–198 (1991)
4. Collins, B., Cunningham, P.: Adaptation-guided retrieval in EBMT: a case-based approach to machine translation. In: Smith, I., Faltings, Boi V. (eds.) EWCBR 1996. LNCS, vol. 1168, pp. 91–104. Springer, Heidelberg (1996)
5. Dufour-Lussier, V.: Reasoning with qualitative spatial and temporal textual cases. Ph.D. thesis, Université de Lorraine (2014)
6. Dumais, S., Furnas, G., Landauer, T., Deerwester, S., Deerwester, S., et al.: Latent semantic indexing. In: Proceedings of the Text Retrieval Conference (1995)
7. Hanney, K., Keane, M.T.: Learning adaptation rules from a case-base. In: Smith, I., Faltings, Boi V. (eds.) EWCBR 1996. LNCS, vol. 1168, pp. 179–192. Springer, Heidelberg (1996)

 8. Khoo, C.S.G.: Automatic identification of causal relations in text and their use for improving precision in information retrieval. Ph.D. thesis, The University of Arizona (1995)
 9. Kireyev, K.: Semantic-based estimation of term informativeness. In: Proceedings of Human Language Technologies: The 2009 Annual Conference of the North American Chapter of the Association for Computational Linguistics, pp. 530–538. Association for Computational Linguistics (2009)
10. Kuhn, H.W.: The Hungarian method for the assignment problem. Nav. Res. Logist. Q. **2**(1–2), 83–97 (1955)
11. Lamontagne, L., Lee, H.-H.: Textual reuse for email response. In: Funk, P., González Calero, P.A. (eds.) ECCBR 2004. LNCS (LNAI), vol. 3155, pp. 242–256. Springer, Heidelberg (2004)
12. Leacock, C., Miller, G.A., Chodorow, M.: Using corpus statistics and WordNet relations for sense identification. Comput. Linguist. **24**(1), 147–165 (1998)
13. Manning, C.D., Surdeanu, M., Bauer, J., Finkel, J., Bethard, S.J., McClosky, D.: The Stanford CoreNLP natural language processing toolkit. In: Proceedings of 52nd Annual Meeting of the Association for Computational Linguistics: System Demonstrations, pp. 55–60 (2014)
14. Posse, C., Matzke, B., Anderson, C., Brothers, A., Matzke, M., Ferryman, T.: Extracting information from narratives: an application to aviation safety reports. In: 2005 IEEE Aerospace Conference, pp. 3678–3690. IEEE (2005)
15. Rooney, J.J., Heuvel, L.N.V.: Root cause analysis for beginners. Qual. Prog. **37**(7), 45–56 (2004)
16. Shokouhi, S.V., Aamodt, A., Skalle, P., Sørmo, F.: Determining root causes of drilling problems by combining cases and general knowledge. In: McGinty, L., Wilson, D.C. (eds.) ICCBR 2009. LNCS, vol. 5650, pp. 509–523. Springer, Heidelberg (2009)
17. Sizov, G., Öztürk, P.: Query-focused association rule mining for information retrieval. In: SCAI, pp. 245–254 (2013)
18. Sizov, G., Öztürk, P., Štyrák, J.: Acquisition and reuse of reasoning knowledge from textual cases for automated analysis. In: Lamontagne, L., Plaza, E. (eds.) ICCBR 2014. LNCS, vol. 8765, pp. 465–479. Springer, Heidelberg (2014)
19. Smyth, B., Keane, M.T.: Adaptation-guided retrieval: questioning the similarity assumption in reasoning. Artif. Intell. **102**(2), 249–293 (1998)
20. Weber, R.O., Ashley, K.D., Brüninghaus, S.: Textual case-based reasoning. Knowl. Eng. Rev. **20**(3), 255–260 (2005)
21. Wilson, D.C., Carthy, J., Abbey, K., Sheppard, J., Dunnion, J., Drummond, A., Wang, R.: Textual CBR for incident report retrieval. In: Kumar, Vipin, Gavrilova, Marina L., Tan, CJKenneth, L'Ecuyer, Pierre (eds.) ICCSA 2003, Part I. LNCS, vol. 2667, pp. 358–367. Springer, Heidelberg (2003)

Maintaining and Analyzing Production Process Definitions Using a Tree-Based Similarity Measure

Reinhard Stumptner[1](✉), Christian Lettner[1], Bernhard Freudenthaler[1],
Josef Pichler[1], Wilhelm Kirchmayr[2], and Ewald Draxler[2]

[1] Software Competence Center Hagenberg GmbH, Hagenberg, Austria
{reinhard.stumptner,christian.lettner,
bernhard.freudenthaler,josef.pichler}@scch.at
[2] Voestalpine Stahl GmbH, Linz, Austria
{wilhelm.kirchmayr,ewald.draxler}@voestalpine.com

Abstract. In this work a Case-Based reasoning system for managing production processes, declarative production process definitions in particular, with main focus on analysis and maintenance is introduced whereby each process task is represented by a case. A single process task definition includes among other elements, formulas, represented by fragmental program code. To get a meaningful similarity function among such cases, a new fuzzy tree edit distance metric on the formulas' abstract syntax tree has been developed. The fuzzy tree edit distance addresses two aspects of similarity – similarity in terms of similar structure and similarity in terms of similar wording. As such, the proposed method represents a multidisciplinary approach to production process maintenance that includes methods from Case-Based reasoning and code clone detection.

Keywords: Case base maintenance · Similarity measure · Tree edit distance · Code clone detection · Abstract syntax tree · Hierarchical clustering

1 Introduction

Production processes define the sequence of tasks that must be executed to produce a product in a required quality. In general, the quality of a final product is ensured by continuously monitoring particular quality parameters. If a quality divergence is detected, the production process will have to be adjusted, e.g. adding additional tasks that perform corrective actions.

If many product types with varying quality requirements are produced, then this leads to a large amount of different production processes. Often these production processes are similar in large part, especially if the final products distinguish themselves only in some special quality requirements.

Modern workflow management systems [13, 14] face this diversity by introducing inheritance or customizable process model definitions. Nevertheless, in industry there are many production process management systems in use that provide no such concepts.

© Springer International Publishing Switzerland 2015
E. Hüllermeier and M. Minor (Eds.): ICCBR 2015, LNAI 9343, pp. 366–380, 2015.
DOI: 10.1007/978-3-319-24586-7_25

Investigating the use case presented in this work showed that especially maintaining such diverse production process definitions is rather challenging. Some maintenance related questions regularly emerge in practice are:

- I have to define (a part of) a production process. Does a similar and possibly matching one already exist?
- Is a change I have to apply also relevant for other, similar process task definitions?
- Where can I find duplicate or similar parts in the production process definitions that could be merged?
- For quality control, find all groups of similar process task definitions. Maybe the definitions differ by mistake?

The production process management system we analyzed in the context of this work allows the definition of task sequences, whereby every task contains a set of parameters that describe target values and actual values. Moreover, the calculation of these values is specified within the parameter definition using formulas, which are implemented using PL/SQL source code fragments. If these fragments are implemented as so-called modules, they can be re-used within other process definitions. Unlike the term "module" in general has a slightly different meaning, in the scope of the underlying process management tool and consequently in the scope of this contribution it just marks a task as being re-usable (as a "task" within the process management tool cannot be re-used in different processes).

The decision to implement a process task specifically for a certain process or as a module is not always certain and easy. Individual implementations typically are clearer and easier to read, while modules that consider differences in processes tend to produce deeply nested if-then-else rules in the parameter value calculation formula. But, if there are only minor differences between formulas, it is advisable to create a module. Using a module, commonalties and differences are documented at a central point, which is a main advantage. But, if differences increase and exception handling consequently gets complicated, an individual implementation may be more feasible. But of course this can change over time, as a big number of individual implementations would lead to infeasible change management and high maintenance costs. Continuous changes and optimizations of the production process from the maintenance point of view require both, integrating individual tasks into modules as well as extracting individual tasks from modules.

To answer the questions from above a Case-Based reasoning (CBR) system was applied. The CBR system consists of two modules as shown in Fig. 1. The first module represents a classical CBR application to retrieve answers for questions of the analyst, like where to find similar process definitions. The second module implements a case base maintenance methodology to generate a cluster map of similar process task definitions. Obviously, there are many parallels between the above described modularization and case base maintenance concepts. The cluster map can be used by the analyst to get a general overview of the process definitions and to get support in deciding where modules should be built. As already mentioned, this map is used to create groups of similar cases as a basis for case base maintenance, thus reducing the number of cases stored in the case base, with the goal of improving the quality and performance of the CBR application. The focus of this work is on measuring the similarity between formulas and on case base maintenance.

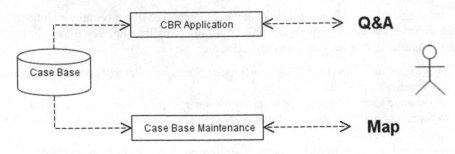

Fig. 1. Overview of the CBR system

Case grouping is based on hierarchical clustering which is to find duplicate and similar process tasks. Each process task along with its parameters is represented by a case. Case base maintenance [18], more precisely prototyping, is used to group similar cases into one representative case, to ensure a uniform distribution of cases across the feature space, which is very important for Case-Based reasoning systems [17].

In this work, case base maintenance is performed in the context of a certain similarity function. The characteristic of process tasks heavily depends on the formulas that calculate process parameter values for the task. To be able to calculate similarities between formulas, a special distance metric was developed, which is based on a tree edit distance between abstract syntax tree representations of formulas.

This paper is structured as follows. After giving a short overview of related re-search, background information to the investigated production process management system is provided, followed by an introduction into the approach applied for process definition maintenance and analysis. Section 4 outlines the developed fuzzy tree edit distance to compare process definitions. Finally, Sect. 5 gives an outline of results and experience of the approach as it is applied in industry.

2   Related Work

Whenever a case contains any kind of graphs, an effective is needed in order to compute case similarities in a suitable way and in reasonable time. This topic is investigated in CBR for a long period of time, like in [4] for instance.

Generally, the success of CBR systems highly depends on the quality of the case base, respectively, the quality of the cases stored in the case base. The retrieval process can become very time-consuming if the case base gets very large. Therefore, the case base necessarily has to be maintained.

Smyth et al. [18] presents in his paper how modeling the performance characteristics of a case base can provide a basis for automatic maintenance in CBR. He shows how performance models can be used for deletion of redundant cases from a case base to optimize the system performance. Furthermore, the presented models can be used for the detection of potential inconsistencies within a case base, the guidance of case authors during case acquisition and as an organizational framework for the construction of distributed case bases.

Smiti et al. [17] gives an overview of maintenance strategies for CBR systems. They classify case base maintenance algorithms in three classes. The first class follows a partitioning policy that builds an elaborate case base structure and maintains it continuously. The second class follows selection based data reduction methods that start with an empty set, select a subset of instances from the original set and add it into the new one. And finally the third class which follows a deletion policy based on cases' competence to optimize the case base. They remark that the most recently explicit algorithmic model of competence for CBR systems was suggested by Smyth et al. [19] by defining the two key fundamental concepts of coverage and reachability. Wilson et al. [21] categorizes the maintenance policies in terms of data collections which explain how to gather data relevant to maintenance, how to trigger maintenance, available types of maintenance operations and how selected maintenance operations are executed. Moreover, Pan et al. [11] classified Case-Based maintenance policies in search direction, order sensibility and evaluation criteria. All other research mainly relied on the deletion and the revision of irrelevant and redundant cases.

A more recent approach is presented by Jalali et al. [5]. The authors suggest an adaptation-guided case base maintenance approach which exploits the ability to dynamically generate new adaptation knowledge from cases. Case retention decisions are based both on cases' value as base cases for solving problems and on their value for generating new adaptation rules.

myCBR is an open-source CBR software tool which comes along with a workbench and a Software Development Kit (SDK). Implementing CBR applications from scratch remains a time consuming software engineering process and requires a lot of specific experience beyond pure programming skills. myCBR supports this process and is one of only few CBR software tools for supporting the development process. The workbench is used to design a knowledge model, which consists of concepts, cases and similarity functions, while the SDK can be used to incorporate the knowledge model into custom applications (see e.g. [1, 15, 20]).

Code clones are the origin of maintenance problems, to be solved by the approach presented in this paper. Code clones are well investigated by research community resulting in well understanding why source code contains code clones, types of code clones, and different techniques for detection of code clones. Researchers (e.g. Bellon et al. [3]) distinguish the following types of clones: Type 1 is an exact copy without modifications (except for white space and comments), Type 2 is a syntactically identical copy; only variable, type, or function identifiers were changed; Type 3 is a copy with further modifications, statements were changed, added, or removed. The system presented in this work is able to detect type 3 clones, as changed, added, or removed statements cause differences on subtree-level and can be handled accordingly. Techniques for code clone detection are based on text, tokens, metrics, abstract syntax trees (AST), or program dependency graphs (PDG).

Our CBR-based technique is based on the abstract syntax tree and, therefore, related to AST-based code clone detection. One of the first AST-based detection was proposed by Baxter et al. [2]. The proposed technique requires full parsing of the program source code and creating an abstract syntax tree. Subtrees of the abstract syntax tree are then partitioned based on a hash function; subtrees in the same partition are then compared through tree matching. Near-miss clones fails when the hash function includes every node of the AST. In order to mitigate this problem, leave nodes of the trees (e.g. identifiers) are ignored by the hash functions. Rather than comparing trees for exact equality, Baxter et al. compare instead for similarity using a few parameters such as number of nodes in subtrees. The similarity threshold parameter allows the user to specify how similar two subtrees should be. Small pieces of code (e.g. identifiers or small expressions) can be ignored in that way. Jiang et al. [6] present an efficient algorithm for identifying similar subtrees and apply it to AST representation of source code. The algorithm calculates numerical vectors in the Euclidean space for every subtree of the AST and clusters these vectors with respect to the Euclidean distance metric. Subtrees with vectors in one cluster are considered similar.

The majority of techniques detect code clones in source code without taking the structure of the software system into account. Clone pairs may be detected in source code of single function block or in source code of different modules of a software system. The problem setting of our work is a different one, because we are interested in duplicate or similar code formulas, however not in code pairs within a single formula. Method-level clone detection (e.g. [7, 10, 16]) fix the granularity to function or method level. Function/method clones are simply clones that are restricted to refer to entire function or method. Kodhai et al. [7] combines textual analysis and metrics for the detection of syntactic and semantic clones. CLAN [10] uses metric-based clone detection techniques for method-level granularity. CLAN gathers different metrics for code fragments and compares these metric vectors instead of comparing the code directly. An allowable distance (e.g. Euclidean distance) for these metric vectors can be used as a hint for similar code. NICAD [16] is multi-pass approach which is parser-based and language-specific but reasonably lightweight, using simple text line rather than subtree comparison to achieve good time and space complexity. Experiments indicate that the method is capable of finding near-miss clones with high precision and recall, and with reasonable performance.

3   Production Process Analysis and Maintenance

3.1   Background

As briefly introduced in Sect. 1, the investigated production process management system makes use of process task definitions that consist of multiple parameters. Figure 2 illustrates the underlying data structure. A task consists of several attributes like the name of the task ("taskName"), the product it produces ("product-Name"), on which type of machine it may be executed ("machineType"), a flag which specifies if the task is defined as a module ("isModule"), the version of the task ("version"), as well as the list of parameters the task contains ("parameterList"). Moreover, a parameter is defined by a name ("parameterName"), a sequence number which determines the order of evaluation of the formulas ("seqNumber"), a flag which defines weather the parameter is active ("isActive"), and finally the formula which calculates the value of the parameter ("formula").

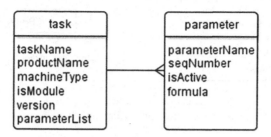

Fig. 2. Structure of a process task

The main part of every task definition is the formula for the parameter definitions. These formulas are responsible to calculate the actual value of the parameter and are implemented as PL/SQL code fragments. Predefined access paths are provided to read and write parameter values within formulas from other parameters formulas. The following example shows a formula that checks whether a value is within a predefined boundary:

```
IF cx.getParam('parameterName1') < 10.0 THEN
    cx.value := 'Value OK';
ELSE
    cx.value := 'Value not OK';
END IF;
```

[Example PL/SQL formula that evaluates the boundaries of a value]

The access path "cx.getParam(parameterName)" returns the value of the provided parameter name, while "cx.value" references the value of the actual parameter which the formula is defined for.

3.2 Preprocessing and Analysis Approach for Production Process Maintenance

This section describes the case base maintenance methodology we applied to support production process maintenance. The main steps are described in Fig. 3. For all formulas defined in the process task definitions the abstract syntax tree (AST) is generated using a PL/SQL parser. A part of the AST from the formula in Sect. 3.1 is shown in Fig. 4. The open source software myCBR is used to define the similarity function between process tasks. The GUI based environment allows a flexible definition of similarity functions, where analysts can design tailor-fit similarity functions.

The similarity function is defined in the myCBR Workbench. For the global similarity function (similarity function on case-level) of tasks we used a weighted Euclidian distance on all attributes. For the attribute in the parameter class, which contains the AST of the parameter formula, a fuzzy tree edit distance (in particular a normalized fuzzy tree edit similarity measure is used, which means sim = 1 for identical objects and sim = 0 for totally different objects) is used. This measure is described in detail in Sect. 4.

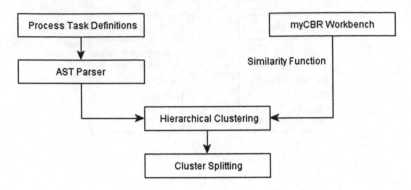

Fig. 3. Case base maintenance workflow

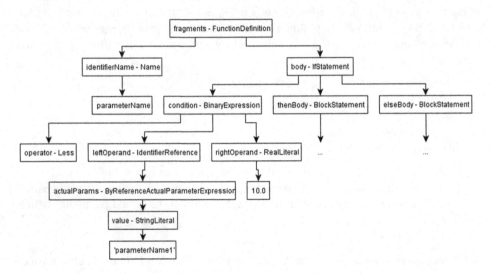

Fig. 4. AST from example code fragment

After the AST of the formulas are generated, hierarchical clustering is performed on all process tasks. We used a hierarchical clustering algorithm to get a topological map of all cases stored in the case base which is expressive and at the same time easily to understand by analysts and technicians.

Figure 5 shows an example hierarchical clustering result for five tasks, P1-T to P5-T, and eight references to modules, P6-M to P13-M. The distance between the clusters is shown in circles. The tasks P2-T and P3-T show a distance of 0, meaning P2-T and P3-T represent duplicate task definitions. The same applies to the tasks P4-T and P5-T, while P1-T apparently represents a unique task. Accordingly the example contains two different module definitions, P6-M to P9-M and P10-M to P13-M.

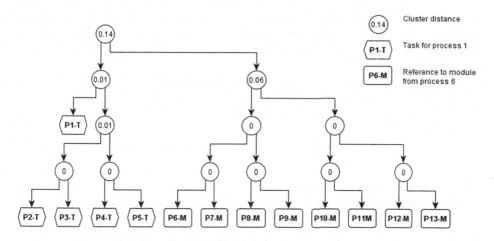

Fig. 5. Hierarchical clustering on tasks and modules

The resulting hierarchical cluster heavily depends on the chosen similarity function. Changing and adopting similarity functions gives analyst the possibility to switch the analysis focus to different aspects in production process maintenance. Last, the resulting hierarchical cluster is split and visualized based on a user specified distance threshold. If the distance between two sub-clusters is above this threshold, the two sub-clusters are split up to two separate clusters. In the example shown in Fig. 5, a threshold of 0.05 would lead to segmentation into three clusters. Finally, as a result of the case base maintenance process, for every sub-cluster a representative case is nominated (if the distance between cases within the cluster is reasonably low), which is called the prototype case. It seems reasonable to assume that the best prototype case is the case with minimal distance to all other cases. In graph theory this is also called the absolute median of a graph [9]. It is given by the vertex y_0 in an n-vertex graph if the sum of all distances between y_0 and every point y in the graph is minimal:

$$\sum_{i=1}^{n} d\left(v_i, y_0\right) <= \sum_{i=1}^{n} d\left(v_i, y\right) \tag{1}$$

The prototype case, i.e. the best case representing the sub-cluster, therefore is y_0.

4 A New Fuzzy Tree Edit Distance – A Similarity Measure for Process Definitions Represented by Abstract Syntax Trees

The Fuzzy Tree Edit Distance (FTED) is based on the "Robust Tree Edit Distance" (RTED) algorithm published by Pawlik et al. in 2011 [12]. RTED was chosen as a basis for FTED, because "RTED is robust to different tree shapes and always performs well" [12].

The distance function d(F, G) in terms of RTED is defined as the minimum cost to turn F into G, while cost correspond to the according sequence of node edit operations that transforms F into G, whereby forests F and G are graphs in which each connected component is a tree (in difference to a forest a tree must only have a single root). The elements w and v are arbitrary but not identical nodes of a tree or forest. F_v is the subtree of F with the root v.

According to [12], the distance function, is defined as

$$d\left(\emptyset,\emptyset\right) = 0 \tag{2}$$

$$d\left(F,\emptyset\right) = d\left(F - v,\emptyset\right) + c_d\left(v\right) \tag{3}$$

$$d\left(\emptyset,G\right) = d\left(\emptyset,G - w\right) + c_i\left(w\right) \tag{4}$$

if F is not a tree or G is not a tree:

$$d\left(F,G\right) = min\begin{cases} d\left(F - v,G\right) + c_d\left(v\right) \;[delete] \\ d\left(F,G - w\right) + c_i\left(w\right) \;[insert] \\ d\left(F_v,G_w\right) + d\left(F - F_v,G - G_w\right) \;[split] \end{cases} \tag{5}$$

if F is a tree and G is a tree:

$$d\left(F,G\right) = min\begin{cases} d\left(F - v,G\right) + c_d\left(v\right) \;[delete] \\ d\left(F,G - w\right) + c_i\left(w\right) \;[insert] \\ d\left(F - v,G - w\right) + c_r\left(v,w\right) \;[replace] \end{cases} \tag{6}$$

For the FTED let us take a closer look at the replacement cost c_r. Normally these costs are (statically) predefined (e.g. $c_d = c_i = c_r = 1$), but in our cast we want to calculate c_r dynamically corresponding to the string similarity of the labels of the according tree nodes.

Thus, in case of substitution the FTED algorithm takes into account the similarities between the according nodes. Subsequently, in terms of FTED it makes a difference if two nodes which have to be replaced have similar labels (similar in terms of string similarity, semantic and so on) or not. This shall be expressed by substitution costs. So, if the labels of the nodes are similar then the substitution or replacement costs are low, otherwise these costs are high. For expressing the (generalized) distance ([0, 1]) between the labels of nodes in this use case the Levenshtein Distance [8] was used. But this is not necessarily required. Meaning that also other algorithms could be used to express the substitution-costs between two nodes (numerical distances or even semantic measures by means of ontologies for instance). This can be varied depending on the use case. But in the scope of this contribution, let us stick to the Levenshtein Distance.

The Levenshtein Distance [8] between two strings a and b is defined by the following matrix D.

$$D_{i,j} = min \begin{cases} D_{i-1,j-1} \ [a_i = b_j] \\ D_{i-1,j} + 1 \ [delete] \\ D_{i,j-1} + 1 \ [insert] \\ D_{i-1,j-1} + 1 \ [replace] \end{cases} \qquad (7)$$

After the matrix is fully calculated the Levenshtein Distance is in the last cell ($D_{|a|, |b|}$). To make the Levenshtein Distance usable for FTED it has to be normalized to [0, 1], which can be done easily by $c_r = d(a, b) = \frac{D_{|a|,|b|}}{\max(|a|,|b|)}$ as the distance between two strings cannot be larger than their maximum length.

The following example shall demonstrate the FTED approach. The first example uses the original tree edit distance algorithm; the second one applies the FTED.

Figure 6 shows the two trees for which the distance should be calculated. The calculation algorithm and the final result are shown in Fig. 7. First, node "X" is deleted from the left tree. Second, node "Y" in the left tree is replaced by node "C". And the left tree has already been transformed into the right tree. Consequently their edit distance d = 2.

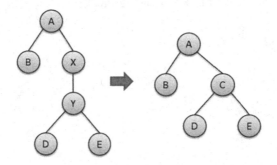

Fig. 6. Example 1: Problem

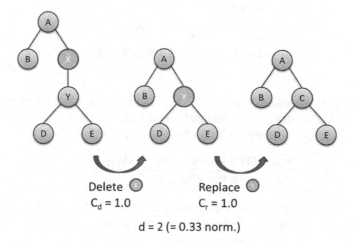

Delete ⓧ   Replace ⓨ
$C_d = 1.0$ $C_r = 1.0$

d = 2 (= 0.33 norm.)

Fig. 7. Example 1: RTED Result

To get a normalized measure for the distance between the trees, d is divided by the maximum distance between the trees which corresponds to the maximum number of nodes (tree1, tree2), which is 6 in our case.

The second example illustrates the calculation of the FTED. In Fig. 8 there are two trees again (tree1, tree2). Figure 9 shows the results of the FTED.

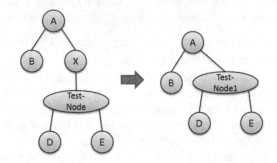

Fig. 8. Example 2: Problem

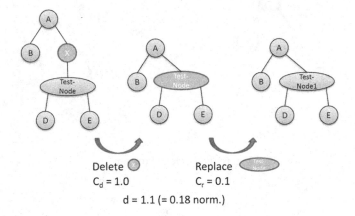

Fig. 9. Example 2: FTED Result

First, node "X" gets deleted like in the first example. But in case of replacing "Test-Node", its label is compared with the labels of the nodes in tree2, with "Test-Node1" in particular, which results in a value of 0.1 as replacement costs.

$$c_r = \frac{LevDist(\text{Test - Node, Test - Node1})}{len(\text{Test - Node1})} = 0.1 \tag{8}$$

Taking this into account, the normalized edit distance d between tree1 and tree2 is d = 0.18. As one can see, the measure addresses both aspects very well – similarity in the sense of similar structure and similarity in the sense of similar wording. Of course, regarding runtime this extension of the RTED is comparably expensive – depending on the costs of the local similarity measure. So it is advisable to do this extended calculation of replacement costs only if necessary. In the use case presented in this contribution the nodes could be divided into different classes and only for certain classes the extended calculation of replacement costs made sense (e.g. made sense for variable names, procedure names; not for any key words).

5 Results

Applying the approach presented in this paper to all available process task definitions of the investigated system, gives a cluster map as shown in Fig. 10. Every sub-cluster (represented by the respective "bubble" in the below figure) contains similar process tasks, whose maximal allowed distance is determined by the cluster splitting threshold (as described in Sect. 3.2).

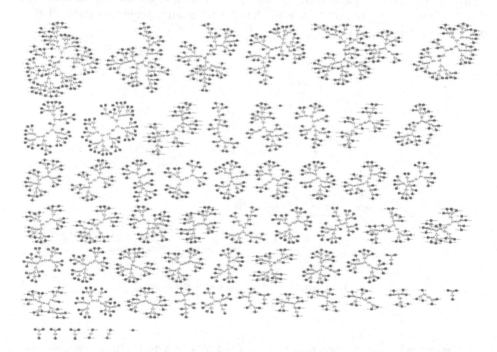

Fig. 10. Cluster map of process tasks

This map serves as a point of reference for the current situation of process definitions in the maintenance point of view and for the current situation of case distribution from a case base maintenance point of view (e.g. combination of cases to reduce the size of the case base without losing much information). The main benefits of this approach when deployed in practice are:

- Gives an awareness and overview to the current production process definitions, i.e. where are commonalities vs. where are differences and special treatments, compared to other processes
- Enables a continuously performed consolidation of process definitions in a controlled fashion. Naturally grown process definitions, especially duplicate definitions can be found and modularized easily
- Provides support for change management, especially if changes lead to tasks moving away from previously very similar tasks and approaching other task definitions
- A well maintained case base increases the diversity of answers returned by the retrieve module. Otherwise, almost duplicate results are omitted, leading to a more manifold and diverse result list

The following, two examples shall illustrate how the approach is applied and used in practice. Figure 11 shows an example where the task P9-T requires an additional "Quality Information" parameter. Having this information, the user has to decide, if this for instance is because task P9-T has higher quality requirements or if that parameter is missing in tasks P5-T to P8-T.

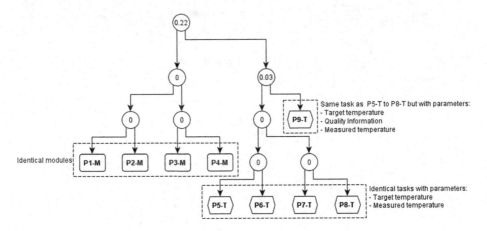

Fig. 11. Task with additional parameter example

The second example shown in Fig. 12 illustrates how production processes use different module versions. While processes 1 and 2 use the "Measure Temperature" module in version 7, processes 3 and 4 make use of the newer version 9. Applying the case base maintenance results on the case base, every of the 58 sub-clusters found in Fig. 10 will be substituted by a single representative case that best stands for the sub-cluster. Queries to the retrieve module now respond with a more versatile result list.

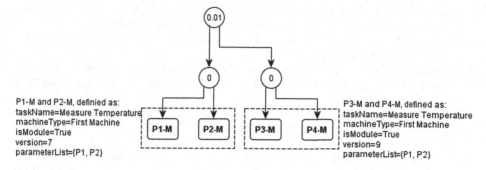

Fig. 12. Processes using different module versions

6   Conclusion

In this work we applied Case-Based reasoning to a production process maintenance use case. It represents a multidisciplinary approach including methods from Case-Based reasoning and code clone detection.

The presented work represents a bottom-up approach that provides an overview on difficulties in maintaining production process definitions and suggests a CBR based solution. It allows continuous consolidations and optimizations and to keep track on changes performed on production process definitions. AST-based techniques (as also text-based techniques), as used in conjunction with the fuzzy tree edit distance, have turned out to be effective in detecting near-miss clones [3]. The fuzzy tree edit distance addresses two aspects of similarity – similarity in terms of similar structure and similarity in terms of similar wording. For the presented use case it delivered good results and future investigations for instance will focus on exploring the combination of structural and semantical similarity in terms of FTED.

Nevertheless, the approach analyses each process task in isolation, neglecting predecessor and successor tasks. But especially the position of tasks in the overall process reveals valuable information for process definition maintenance. For this reason, future work will focus on extending the presented approach in the direction of analyzing sequences of process tasks, thus finding similarities between task chains.

Acknowledgements. This work has been supported by the COMET-Program of the Austrian Research Promotion Agency (FFG)

References

1. Bach, K. et al.: Knowledge modeling with the open source tool myCBR. In: Proceedings of 10th Workshop on Knowledge Engineering and Software Engineering (KESE10) Co-located with 21st European Conference on Artificial Intelligence (ECAI 2014), Prague, Czech Republic, 19 August 2014

2. Baxter, I.D., et al.: Clone detection using abstract syntax trees. In: Proceedings., International Conference on Software Maintenance, 1998, pp. 368–377. IEEE (1998)
3. Bellon, S., et al.: Comparison and evaluation of clone detection tools. IEEE Trans. Softw. Eng. 33(9), 577–591 (2007)
4. Bunke, H., Messmer, B.T.: Similarity measures for structured representations. In: Wess, S., Richter, M., Althoff, K.-D. (eds.) EWCBR 1993. LNCS, vol. 837. Springer, Heidelberg (1994)
5. Jalali, V., Leake, D.: Adaptation-Guided case base maintenance. In: Proceedings of the Twenty-Eighth AAAI Conference on Artificial Intelligence, July 27–31, 2014, pp. 1875–1881. Québec City (2014)
6. Jiang, L., et al.: Deckard: scalable and accurate tree-based detection of code clones. In: Proceedings of the 29th International Conference on Software Engineering, pp. 96–105. IEEE Computer Society (2007)
7. Kodhai, E., Kanmani, S.: Method-level code clone detection through LWH (Light Weight Hybrid) approach. J. Softw. Eng. Res. Dev. 2(1), 1–29 (2014)
8. Levenshtein, V.I.: Binary codes capable of correcting deletions, insertions and reversals. Sov. Phys. Dokl. 10, 707 (1966)
9. Marianov, V., Serra, D.: Median problems in networks. Available at SSRN: 1428362 (2009)
10. Mayrand, J., et al.: Experiment on the automatic detection of function clones in a software system using metrics. In: Proceedings of International Conference on Software Maintenance 1996, pp. 244–253. IEEE (1996)
11. Pan, R., et al.: Mining competent case bases for case-based reasoning. Artif. Intell. 171 (16–17), 1039–1068 (2007)
12. Pawlik, M., Augsten, N.: RTED: a robust algorithm for the tree edit distance. Proc. VLDB Endow. 5(4), 334–345 (2011)
13. Reichert, M., Weber, B.: Enabling Flexibility in Process-Aware Information Systems: Challenges, Methods. Technologies. Springer, Heidelberg (2012)
14. Rosa, M.L., et al.: Business process variability modeling: a survey. ACM Comput. Surv. (2013)
15. Roth-Berghofer, T., et al.: Building case-based reasoning applications with myCBR and COLIBRI studio. In: Proceedings of the UKCBR 2012 Workshop on Case-Based Reasoning, pp. 71–82 (2012)
16. Roy, C.K., Cordy, J.R.: NICAD: accurate detection of near-miss intentional clones using flexible pretty-printing and code normalization. In: The 16th IEEE International Conference on Program Comprehension, 2008 (ICPC 2008), pp. 172–181. IEEE (2008)
17. Smiti, A., Elouedi, Z.: Article: overview of maintenance for case based reasoning systems. Int. J. Comput. Appl. 32(2), 49–56 (2011)
18. Smyth, B.: Case-base maintenance. In: Mira, J., Moonis, A., de Pobil, A.P. (eds.) IEA/AIE 1998. LNCS, vol. 1416. Springer, Heidelberg (1998)
19. Smyth, B., McKenna, E.: Modelling the competence of case-bases. In: Smyth, B., Cunningham, P. (eds.) EWCBR 1998. LNCS (LNAI), vol. 1488, pp. 208–220. Springer, Heidelberg (1998)
20. Stahl, A., Roth-Berghofer, T.R.: Rapid prototyping of CBR applications with the open source tool myCBR. In: Althoff, K.-D., Bergmann, R., Minor, M., Hanft, A. (eds.) ECCBR 2008. LNCS (LNAI), vol. 5239, pp. 615–629. Springer, Heidelberg (2008)
21. Wilson, D.C., Leake, D.B.: Maintaining case-based reasoners: dimensions and directions. Comput. Intell. 17(2), 196–213 (2001)

Case-Based Plan Recognition Under Imperfect Observability

Swaroop S. Vattam[1](✉) and David W. Aha[2]

[1] NRC Postdoctoral Fellow, Naval Research Laboratory (Code 5514), Washington, DC, USA
swaroop.vattam.ctr.in@nrl.navy.mil
[2] Navy Center for Applied Research in Artificial Intelligence,
Naval Research Laboratory (Code 5514), Washington, DC, USA
david.aha@nrl.navy.mil

Abstract. SET-PR is a novel case-based recognizer that is robust to three kinds of input errors arising from imperfect observability, namely missing, mislabeled and extraneous actions. We extend our previous work on SET-PR by empirically studying its efficacy on three plan recognition datasets. We found that in the presence of higher input error rates, SET-PR significantly outperforms alternative approaches, which perform similarly to or outperform SET-PR in the presence of no input errors.

Keywords: Case-based reasoning · Plan recognition · Imperfect observability · Graph representation · Plan matching

1 Introduction

A plan recognizer observes the actions executed by an actor and attempts to infer the actor's plan. A plan recognizer typically receives its input observations (of actions) from a lower-level action recognition system that can be noisy. A sophisticated plan recognizer therefore needs to relax the assumption of perfect observability and expect at least three kinds of input errors: a *mislabeled* action occurs in an input when an actor's true action is recognized as some other action; a *missing* action occurs when a true action is unrecognized (i.e., classified as a non-action); and an *extraneous* action occurs when a non-action is classified as some valid action.

Single-Agent Error-Tolerant Plan Recognizer (SET-PR) is a novel case-based plan recognizer that has shown promise in tolerating these kinds of input errors. We previously introduced SET-PR and highlighted its representation and reasoning techniques (Vattam, Aha, & Floyd 2014). This paper extends our preliminary empirical study of SET-PR, which was limited to just one dataset (the Blocks World domain) (Vattam, Aha, & Floyd 2015). Here we conduct a more comprehensive empirical investigation of SET-PR by (1) expanding the scope of the investigation to three datasets (Blocks World, Linux, and Monroe), (2) adopting a wider range of plan-recognition performance metrics, and (3) comparing the performance of SET-PR to baseline algorithms.

© Springer International Publishing Switzerland 2015
E. Hüllermeier and M. Minor (Eds.): ICCBR 2015, LNAI 9343, pp. 381–395, 2015.
DOI: 10.1007/978-3-319-24586-7_26

382   S.S. Vattam and D.W. Aha

This paper is organized as follows. Section 2 describes related work on plan recognition. Section 3 gives an overview of SET-PR including its novel plan representation and retrieval mechanism. Section 4 presents our more comprehensive study of SET-PR, including the hypotheses we address, data used, evaluation method, empirical results, and their analysis. In this investigation, we found that SET-PR significantly outperformed the baseline algorithms in the presence of higher levels of input error, although the baselines performed similar to or outperformed SET-PR in the presence of no input errors. We conclude and discuss future research plans in Sect. 5.

2   Related Research

Early work on plan recognition (e.g., Kautz & Allen, 1986) assumed that the observed actor's actions follow a hierarchical plan structure, requiring the plan recognizer to infer plans and sub-plans at multiple abstraction levels. However, it assumed perfect observability, which is unrealistic. Since then, a number of important probabilistic (e.g., Charniak & Goldman, 1993; Bauer, 1994; van Beek, 1996) and statistical parsing approaches (e.g., Pynadath & Wellman, 1995; Geib & Goldman, 2009) have been proposed that address issues of uncertainty. They frame plan recognition as a problem of probabilistic inference in a stochastic process that models the actor's action execution. While this offers a general and coherent framework for modeling different sources of uncertainty, they have not focused on problems due to imperfect observability. In contrast, activity recognition (Duong et al., 2005) algorithms, which apply signal processing techniques to discretize sensor information into coherent actions, have addressed imperfect observability issues. Bridging the gap between low-level, often noisy activity models and higher-level plans remains a research challenge.

Recently Ramirez and Geffner (2010) proposed a novel approach to plan recognition by formulating it in terms of plan synthesis and solving it using off-the-shelf planners. They extended their approach to perform plan recognition in POMDP settings (Ramirez & Geffner 2011), which they claim can tolertae different kinds of input errors. They demonstrated that it tolerates one kind of input error, namely missing actions (i.e., incomplete observations). However, like most plan recognition approaches theirs is "model-heavy"; they require accurate models of an actor's possible actions and how those actions interact to accomplish different goals. Engineering these models is difficult and time consuming. Furthermore, these plan recognizers perform poorly when confronted with novel situations and are brittle when the operating conditions deviate from model parameters.

SET-PR exemplifies case-based plan recognition (CBPR), a model-lite, lesser studied approach to plan recognition. Existing CBPR approaches (e.g., Cox & Kerkez, 2006; Tecuci & Porter, 2009) eschew generalized models and instead use plan libraries that contain plan instances that can be gathered from experience. CBPR algorithms can respond to novel inputs outside the scope of their plan library using plan adaptation techniques. However, to our knowledge they have not been designed for imperfect observability, which is the unique focus of SET-PR.

Cox and Kerkez (2006) proposed a novel representation for storing and organizing plans in a plan library, modeled as action-state pairs and abstract states, which counts the number of instances of each type of generalized state predicate. SET-PR uses a similar representation, but stores and processes plans in an action-sequence graph. Our encoding was inspired by planning encoding graphs (Serina, 2010). These are syntactically similar to our graphs but encode a planning problem while ours instead encode a solution (i.e., a grounded plan).

Plan retrieval is an important step in CBPR algorithms and presents an efficiency bottleneck. Our previous contribution presented an algorithm for speeding plan retrieval in SET-PR that uses plan projection and clustering (Maynord, Vattam, & Aha 2015). Sánchez-Ruiz and Ontañón (2014) use Least Common Subsumer Trees for the same purpose, but they are not applicable to our representation.

3   SET-PR

3.1   Representation

SET-PR learns to recognize plans from a given plan library C (i.e., a set of cases). Each case is a tuple $c = (\pi, g)$, where π is a known plan (the problem part), and g is its corresponding goal (the solution part).

3.1.1   Action state sequences

Each case's plan $c.\pi$ is modeled as an *action state sequence* $\mathbb{S} = \langle (a_0, s_0), \dots, (a_n, s_n) \rangle$, where each action a_i is a ground operator in the planning domain, and s_i is a ground state obtained by executing a_i in s_{i-1}, with the caveat that s_0 is an initial state, a_0 is null, and s_n is a goal state. An action a in $(a, s) \in \mathbb{S}$ is a ground literal $p = p(o_1 : t_1, \dots, o_n : t_n)$, where $p \in P$ (a finite set of predicate symbols), $o_i \in O$ (a finite set of objects), and t_i is an instance of o_i (e.g., stack(block:A, block:B)). A state s in $(a, s) \in \mathbb{S}$ is a set of ground literals (e.g., {on(block:A,block:B), on(block:B,substrate:TABLE)}).

Inputs to SET-PR consist of sequences (observed parts of a plan). An input to SET-PR $\mathbb{S}^{query}$ is also modeled as an action state sequence. However, unlike a plan, s_0 and s_n in $\mathbb{S}^{query}$ need not be initial and goal states, and a_0 need not be null.

Each case's goal $c.g$ is modeled as a task to be achieved (using the HTN vocabulary) or as a state to be achieved depending on the domain. This reduces a goal to an instance of a task ($c.g$ is an a) or a state ($c.g$ is a s) respectively. The representation of a goal can be flexible because it is the solution part of a case and does not participate in matching during retrieval.

3.1.2   Action Sequence Graphs

An action sequence graph is a graphical representation of an action state sequence, which is propositional. This graph preserves the topology of the sequence it encodes (including the order of the propositions and their arguments). We mentioned that plans are modeled as action state sequences. SET-PR does not store the propositional representation of an action state sequence $\mathbb{S}$. Instead, $\mathbb{S}$ is encoded as an action sequence graph $\varepsilon^{\mathbb{S}}$ and stored in $c.\pi$. Similarly an input sequence $\mathbb{S}^{query}$ is also encoded as an action sequence graph $\varepsilon^{\mathbb{S}^{query}}$ and used in retrieval.

A labeled directed graph G is a 3-tuple $\ominus = (V, E, \lambda)$, where V is a set of vertices, $E \subseteq V \times V$ is a set of edges, and $\lambda : V \cup E \to 2^L$ assigns labels to vertices and edges. Here, an edge $e = [v, u] \in E$ is directed from v to u, where v is the edge's source node and u is the target node. Also, L is a finite set of symbolic labels and 2^L is a set of all the multisets on L; this permits multiple non-unique labels for a node or edge.

The union $G_1 \cup G_2$ of two graphs $G_1 = (V_1, E_1, \lambda_1)$ and $G_2 = (V_2, E_2, \lambda_2)$ is the graph $G = (V, E, \lambda)$, where $V = V_1 \cup V_2$, $E = E_1 \cup E_2$, and

$$\lambda(x) = \begin{cases} \lambda_1(x), & \text{if} x \in (V_1 \backslash V_2) \bigvee x \in (E_1 \backslash E_2) \\ \lambda_2(x), & \text{if} x \in (V_2 \backslash V_1) \bigvee x \in (E_2 \backslash E_1) \\ \lambda_1(x) \cup \lambda_2(x), & \text{otherwise} \end{cases}$$

Definition: Given ground atom p representing an action a or a fact of state s in the kth action-state pair $(a, s)_k \in \mathbb{S}$, a *predicate encoding graph* is a labeled directed graph $\varepsilon^p(p) = (V_p, E_p, \lambda_p)$ where:

$$V_p = \begin{cases} \left\{ A_{k_p}, o_1, \dots, o_n \right\}, & \text{if } p \text{ is an action} \\ \left\{ S_{k_p}, o_1, \dots, o_n \right\}, & \text{if } p \text{ is a state fact} \end{cases}$$

$$E_p = \begin{cases} \left[A_{k_p}, o_1 \right] \cup \bigcup_{i=1,n-1;j=i+1,n} [o_i, o_j] & \text{if } p \text{ is an action} \\ \left[S_{k_p}, o_1 \right] \cup \bigcup_{i=1,n-1;j=i+1,n} [o_i, o_j] & \text{if } p \text{ is a state fact} \end{cases}$$

$$\lambda_p \left(A_{k_p} \right) = \left\{ A_{k_p} \right\} ; \lambda_p \left(S_{k_p} \right) = \left\{ S_{k_p} \right\} ; \lambda_p(o_i) = \{ t_i \} \text{ for } i = 1, \dots, n$$

$$\lambda_p \left(\left[A_{k_p}, o_1 \right] \right) = \left\{ A_{k_p}^{0,1} \right\} ; \lambda_p \left(\left[S_{k_p}, o_1 \right] \right) = \left\{ S_{k_p}^{0,1} \right\} ;$$

$$\forall [o_i, o_j] \in E_p, \ \lambda_p([o_i, o_j]) = \begin{cases} \left\{ A_{k_p}^{i,j} \right\}, & \text{if } p \text{ is an action} \\ \left\{ S_{k_p}^{i,j} \right\}, & \text{if } p \text{ is a state fact} \end{cases}$$

Interpretation: Suppose we have a ground literal $p = p(o_1 : t_1, \dots, o_n : t_n)$. Depending on whether p represents an action or a state fact, the first node of the predicate encoding graph $\varepsilon^p(p)$ is either A_{k_p} or S_{k_p} (labeled $\left\{ A_{k_p} \right\}$ or $\left\{ S_{k_p} \right\}$). Suppose it is an action predicate. A_{k_p} is then connected to the second node of this graph, the object node o_1 (labeled $\{ t_1 \}$), through the edge $\left[A_{k_p}, o_1 \right]$ (labeled $\left\{ A_{k_p}^{0,1} \right\}$). Next, o_1 is connected to the third node

o_2 (labeled $\{t_2\}$) through the edge $[o_1, o_2]$ (labeled $\left\{A_{k_p}^{1,2}\right\}$), then to the fourth node o_3 (labeled $\{t_3\}$) through the edge $[o_1, o_3]$ (labeled $\left\{A_{k_p}^{1,3}\right\}$), and so on. Suppose also the third node o_2 is connected to o_3 through $A_{k_p}^{2,3}$, to o_4 through $A_{k_p}^{2,4}$, with appropriate labels, and so on.

Example: Suppose predicate $p = put\,(block{:}a, block{:}b, table{:}t)$ appears in the fifth $(k = 5)$ action-state pair of an observed sequence of actions. The nodes of this predicate are $\left\{A_{5_{put}}\right\}, \{a\}, \{b\},$ and $\{t\}$. The edges are $\left[A_{5_{put}}, a\right], [a, b], [a, t],$ and $[b, t]$, with respective labels $\left\{A_{5_{put}}^{0,1}\right\}, \left\{A_{5_{put}}^{1,2}\right\}, \left\{A_{5_{put}}^{1,3}\right\},$ and $\left\{A_{5_{put}}^{2,3}\right\}$. The predicate encoding graph for p is shown in Fig. 1.

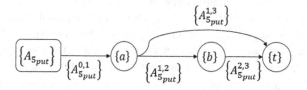

Fig. 1. A predicate encoding graph corresponding to $p = put\,(block{:}\,a,\ block{:}b,\ table{:}t)$

Definition: An action sequence graph of an action state sequence $\mathbb{S}$ is a labeled directed graph $\varepsilon^{\mathbb{S}} = \cup_{(a,s)\in\mathbb{S}}\left(\varepsilon\,(a)\,\cup_{p\in s}\,\varepsilon\,(p)\right)$, a union of the predicate encoding graphs of the action and state facts in $\mathbb{S}$.

Figure 2 shows an example of a sample action state sequence and its corresponding action sequence graph.

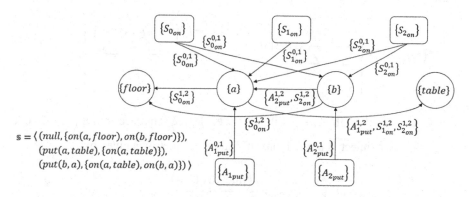

$$s = \langle\,(null, \{on(a, floor), on(b, floor)\}),$$
$$(put(a, table), \{on(a, table)\}),$$
$$(put(b, a), \{on(a, table), on(b, a)\})\,\rangle$$

Fig. 2. An example action-state sequence $\mathbb{S}$ and corresponding action sequence graph $\varepsilon^{\mathbb{S}}$

3.2   Retrieval

Case retrieval requires a similarity metric. Because we represent the input and the stored plans as graphs, our metric uses graph matching; this can be formulated as the task of computing the maximum common subgraph (MCS) of two graphs. Computing the MCS between two or more graphs is an NP-Complete problem, restricting applicability to only small plan recognition problems. Alternatively, a plethora of approximate graph similarity measures exist. For example, similarity metrics that compute graph degree sequences have been used successfully to match chemical structures (Raymond & Willett, 2002).

In SET-PR, we use one of the degree sequence similarity metrics called Johnson's similarity metric (Johnson, 1985). This metric, denoted as sim_{str}, computes the similarity between plans based on the approximate structural similarity of their graph representations. Previously, we tried alternative degree sequence metrics and found that Johnson's metric performed the best (Vattam, Aha & Maynord, 2015).

Let G_1 and G_2 be the two action-sequence graphs. To compute their similarity, we first separate the set of vertices in each graph into l partitions by label type, and then sort them in a non-increasing total order by degree (of a vertex v is the number of edges that touch v). Let L_1^i and L_2^i denote the sorted degree sequences of a partition i in the action-sequence graphs G_1 and G_2, respectively. An upper bound on the number of vertices $V\left(G_1, G_2\right)$ and edges $E\left(G_1, G_2\right)$ of the MCS of these two graphs can then be computed as:

$$\left|mcs(G_1, G_2)\right| = V\left(G_1, G_2\right) + E\left(G_1, G_2\right),$$

where

$$V\left(G_1, G_2\right) = \sum_{i=1}^{l} min\left(\left|L_1^i\right|, \left|L_2^i\right|\right),$$

and

$$E\left(G_1, G_2\right) = \left|\sum_{i=1}^{l} \sum_{j=1}^{min(|L_1^i|, |L_2^i|)} \frac{min\left(\left|E\left(v_1^{i,j}\right)\right|, \left|E\left(v_2^{i,j}\right)\right|\right)}{2}\right|,$$

where $v_1^{i,j}$ denotes the j th vertex of the L_1^i sorted degree sequence, and $E\left(v_1^{i,j}\right)$ denotes the set of edges connected to $v_1^{i,j}$. Johnson's similarity metric is given by:

$$sim_{str}\left(G_1, G_2\right) = \frac{\left(\left|mcs(G_1, G_2)\right|\right)^2}{\left|G_1\right| \cdot \left|G_2\right|}$$

Two plans that are similar in structure can differ drastically in semantics. For instance, a plan to travel to a grocery store to buy milk might coincidentally be structurally similar to a plan to travel to the airport to receive a visitor. To mitigate this issue, we use a weighted combination of structural and semantic similarity, denoted as sim_{obj}, as our final similarity metric:

$$\text{sim}\left(G_1, G_2\right) = \alpha\, \text{sim}_{str}\left(G_1, G_2\right) + (1 - \alpha)\, \text{sim}_{obj}\left(G_1, G_2\right),$$

where $\text{sim}_{obj}\left(G_1, G_2\right) = \frac{O_s \cap O_{\pi_i}}{O_s \cup O_{\pi_i}}$ is the Jaccard coefficient of the set of (grounded) objects in G_1 and G_2, and $\alpha(0 \leq \alpha \leq 1)$ governs the weights for sim_{str} and sim_{obj}.

SET-PR matches an input action-sequence graph $\mathbb{S}^{query}$ with each case $c = (\pi, g)$ in C using $\text{sim}(\mathbb{S}^{query}, c.\pi)$, and retrieves the top-ranked matching case. This case's plan is output as the recognized plan and its goal is output as the recognized goal.

SET-PR keeps track of its most recent previous prediction and uses it to resolve ambiguity if multiple cases are retrieved with nearly similar scores. In other words, selection preference favors a case that maintains continuity in plan prediction. If none of the cases in that set match the previous prediction, then one of them is selected randomly.

Table 1. Datasets used in this empirical study

Dataset	#Plans	Average plan length	#Plan classes	State information
Blocks	125	12.48 actions	5	Yes
Minroe	5000	9.6 actions	10	No
Linux	457	6.1 actions	20	No

3.3 Error Tolerance

The ability of SET-PR to tolerate input errors is a direct benefit of its representation and retrieval mechanism. By adding state information to plan representation, SET-PR reduces the overreliance on action information (which causes poor performance when they are error-prone) and increases the total amount of information that is used for recognition. SET-PR's graph representation of plans permits inexact matching, trading off higher recall for lower precision. We claim that this tradeoff allows SET-PR to generalize better in the presence of input errors compared to other approaches that favor propositional representations and symbol matching, and test this in Sect. 4.

4 Empirical Evaluation

We empirically test the following claim: for the task of plan recognition, SET-PR's approach, which employs a graph-based representation and similarity metric, offers more robustness to input errors compared to alternative CBPR approaches that use propositional representation and symbol matching. We test this claim by subjecting the approaches to increasing levels of input errors. At each level, we measure and compare their plan recognition performance. We perform this experiment across three different plan recognition datasets and note if similar performance trends emerge.

In the following sections we describe the approaches tested, performance metrics, datasets used, methodology, the results and their analysis.

4.1 Compared CBPR Approaches

1. **SET-PR**: This approach uses action sequence graph representation and Johnson's similarity metric for performing plan recognition as described above.
2. **EDIT**: This approach uses propositional representation and an ordered symbolic similarity metric for performing plan recognition. It treats inputs and plans as symbol sequences and computes their Edit distance (Levenshtein 1966).
3. **JACC**: This approach uses propositional representation and an unordered symbolic similarity metric for performing plan recognition. It treats inputs and plans as a set of action propositions (A_1 and A_2) and computes their Jaccard distance $(1 - (A_1 \cap A_2/A_1 \cup A_2))$.

Table 2. A sample convergence matrix

	a.1	a.2	a.3	a.4	a.5	#acts	# correct	r-Acc.	Converged?	Conv. point
p.1	c1✓	c1✓	c2✗	c2✗	c1✓	5	3	3/5 (0.6)	TRUE	5/5 (1.0)
p.2	c1✓	c3✗	c2✗	c1✓		4	2	2/4 (0.5)	TRUE	4/4 (1.0)
p.3	c3✗	c3✗	c1✗	c2✓	c3✗	5	1	1/5 (0.2)	FALSE	No value
p.4	c2✗	c3✓	c3✓	c3✓	c3✓	5	4	4/5 (0.8)	TRUE	2/5 (0.4)
c-Acc.	2/4 (0.5)	2/4 (0.5)	1/4 (0.25)	3/4 (0.75)	2/3 (0.67)					

4. **RAND**: This is the baseline condition. It performs plan recognition by randomly selecting a plan from the plan library in response to its inputs.

4.2 Datasets Used

In this study we used three datasets (Table 1). We repeated our evaluation method described below in all three datasets. Blocks World is a synthetic dataset that we generated using the HTN planner SHOP2 (Nau et al., 2003), which we modified to capture state information in the generated plans. Monroe (Blaylock & Allen, 2005) and Linux (Blaylock & Allen, 2004) are two datasets that are commonly used to assess plan recognizers. Monroe is also a synthetic dataset generated using SHOP2, while Linux is a corpus of plans collected from human users performing assigned tasks. Because the plans in these latter two datasets contain no state information, SET-PR's plans also contain only action information. This reduces the size of the encoding of the plans in these two datasets.

4.3   Evaluation Method

For each dataset, we developed an error simulator that takes as input a plan (π), an error-type (t), and an error-percentage (p). It outputs π^{err}, which contains $p\%$ errors of type t. The values for t include mislabeled (MLAB), missing (MSNG), extraneous (EXTR), and mixed (MXD). For MLAB, a specified percentage of actions was randomly chosen, and each was replaced with another action randomly chosen from the domain. For MSNG, a percentage of actions was randomly chosen, and each was replaced with an unidentified marker '*'. For EXTR, a percentage of randomly chosen actions from the domain were introduced at random locations in the plan. For MXD, a uniform distribution of all three types of errors was introduced.

For each dataset D, we obtained a set of datasets $D_{t,p}$ that combine $t = \{$MLAB, MSNG, EXTR, MXD$\}$ and $p = \{0, 0.15, 0.3, 0.45, 0.6\}$. For example, $Monroe_{MXD,0.6}$ is a Monroe version containing plans with 60 % mixed error.

For each $D_{t,p}$, we tested our compared conditions (SET-PR, EDIT, JACC, and RAND) using five-fold cross-validation (with shuffle). That is, for each plan in the test set, we *incrementally queried* the training set to predict a plan. For example, if a test plan had four actions $\{a1, a2, a3, a4\}$, the evaluator performed 4 queries $\{a1\}, \{a1, a2\}$, $\{a1, a2, a3\}$, and $\{a1, a2, a3, a4\}$ to obtain a predicted plan after observing each action in succession. For a prediction to be correct, the plan class of the predicted plan must match the plan class of the test plan.

Table 3. Confusion matrix

Pred. Actl.	c1	c2	c3	Recall
c1	5	3	1	5/9
c2	1	1	3	1/5
c3	0	1	4	4/5
Precision	5/6	1/5	4/8	

For each compared condition for each $D_{t,p}$, the results of the cross validation was tabulated in a *convergence matrix* (example in Table 2). The rows in matrix are plan indices and columns are action indices. After observing the jth action of the ith plan in the test set, $cell_{ij}$ registers (1) the predicted value (the goal class), and (2) a Boolean value indicating a correct or incorrect prediction. We maintain two additional columns, total number of actions (#acts) per row and correct predictions (#correct) per row.

From the convergence matrix, we derive a *confusion matrix* (Table 3) by counting the instances where the predicted plan class agrees or disagrees with the actual class.

4.4   Performance Metrics

We defined the following four plan recognition performance metrics from the convergence matrix depicted in Table 2:

Percent convergence: Convergence indicates whether the final prediction in each row was correct. For each condition, the percentage of True values is computed.

Convergence point: If a prediction converged, the convergence point (CP) is the point in the input that the recognizer starts to output only the correct prediction. A smaller value for this metric indicates a better performance. For each condition, we also compute the average convergence point.

r-Accuracy: We can compute row-wise prediction accuracy as the ratio of the total number of correct predictions in a row versus the total number of actions observed in that row (#correct/#acts). We compute r-Accuracy as the average of this value for each test plan. It has often been referred to as "precision" (e.g., Blaylock & Allen 2006) in plan recognition literature, which differs from the traditional meaning of precision in the general classification literature. Table 3 displays the traditionally-defined precision and recall values from the confusion matrix.

c-Accuracy: We calculate column-wise prediction accuracy as the ratio of the total number of or correct predictions in a column versus its total number of plans (#correct in col/#plans in col). c-Accuracy is the average of these values.

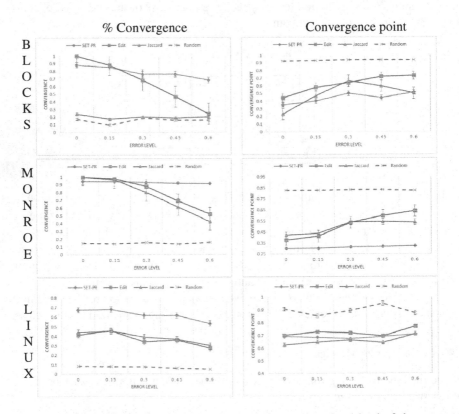

Fig. 3. Percent Convergence and Convergence-point vs. Error level for the 3 datasets

From the confusion matrix depicted in Table 3, we define a final performance metric, *F1-Score*, which is the harmonic mean of average precision and recall.

4.5 Results

Figure 3 shows the plots for percent convergence and convergence point of SET-PR, EDIT, JACC, and RAND at varying levels of input errors of type MXD (mixed error) for the three datasets. These are mean values obtained using five-fold cross validation. Similarly, Fig. 4 shows the plots for mean *r-Accuracy* and *c-Accuracy*, and Fig. 5 shows plots for the mean *F1-Score*. Due to space restrictions, we do not show the plots and other significance test results for other error types. However, we note that the trends observed in MXD hold for other error conditions as well. We highlight MXD because it contains a uniform distribution of the three kinds of errors (we chose uniform distribution because we currently lack domain-specific error models).

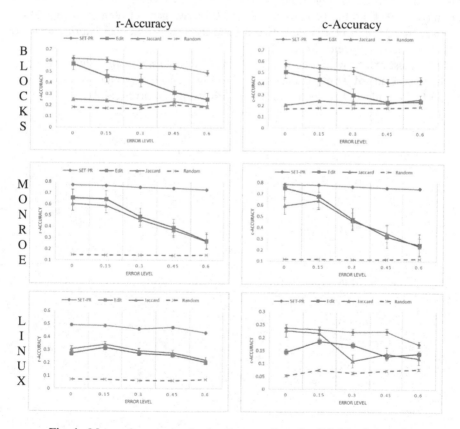

Fig. 4. Mean r-Accuracy and c-Accuracy vs. Error level for the 3 datasets

4.6   Analysis

At 0 % error level, the plots in Figs. 3, 4 and 5 indicate the following. (1) For Blocks: with the exception of one metric, EDIT and SET-PR perform comparably, but JACC performs poorly, closer to RAND. (2) For Monroe: with the exception of one metric, all three perform comparably. (3) For Linux: SET-PR shows performance advantage in 3 out of 5 metrics, while EDIT and JACC perform comparably with each other. Overall, for 0 % error, in majority of the experiments, SET-PR's performance is comparable to EDIT, JACC or both.

At 15 % error level, we see small to negligible performance declines for SET-PR, but more declines for EDIT and JACC. Finally, at higher levels of error, we see moderate declines in performance for SET-PR, but steep declines for EDIT and JACC. This trend can be observed across datasets and across different error types in a majority of experiments.

To assess the impact of the two independent factors (CBPR approach and error level) on the value of a performance metric, we compared the means of the performance metric values across these two factors. For each dataset and for each error type, we subjected this two factor data to a two-way ANOVA test to measure the statistical significance of the outcomes of the comparison, amounting to a total of 60 tests. In all 60 tests, there was a statistically significant effect observed for both factors as well as for their interaction ($p < 0.05$ for all tests; for *error level*, $F(4,59)$ ranged between 650 and 2229; for *CBPR approach*, $F(3,59)$ ranged between 1042 and 17654; and their *interaction* factor, $F(12,59)$ ranged between 100 and 409).

From these results we can conclude that SET-PR has a *significantly* higher tolerance for the three kinds of input errors compared to EDIT and JACC although the latter two can perform similarly to or outperform SET-PR in the 0 % error condition.

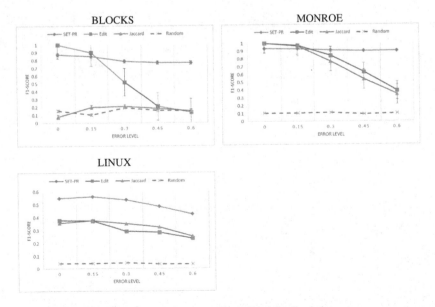

Fig. 5.   Mean F1-Score vs. Error level for the 3 datasets

5   Discussion

In Sect. 3.3 we argued that the superior performance of SET-PR under imperfect observability can be attributed to two factors: (1) the content of the plans, which includes action and state information, and (2) the graph representation of the plans, which permits inexact matching. Our evaluation lends support to (2) because only SET-PR uses graph representations. Regarding (1), in our earlier pilot studies with Blocks world (Vattam, Aha, & Floyd 2015), we compared SET-PR with and without state information, keeping all else constant. There, we found preliminary evidence to support (1), but our current investigation does not focus on (1) because no state information is included in SET-PR for the Monroe and Linux plan libraries.

One of the limitations of our study is that we do not compare SET-PR with other state-of-the-art plan recognizers. In the future, we plan to obtain and run these experiments with other well-known plan recognizers.

Given that graph matching is generally considered a hard problem, what can we say about the computational efficiency of SET-PR's matching process? Using its degree sequence metric, others showed that the similarity between two graphs can be computed in $O(n \cdot logn)$ time, where $n = max_i \left(\left| L_1^i \right|, \left| L_2^i \right| \right)$ (Raymond & Willett, 2002).

Without efficient indexing techniques, plan retrieval time scales linearly with the size of SET-PR's library C. This can be prohibitively expensive for online plan recognition. Thus, we use Plan Projection Clustering (PPC), a method to increase the plan retrieval speed of SET-PR (Maynord, Vattam, & Aha 2015). PPC is a domain-general approach for organizing SET-PR's plans in a hierarchy. It employs a metric d (e.g., Johnson's similarity metric) that measures distances among plans. PPC computes $d(p_1, p_2)$ for each pair of plans $p_1, p_2 \in C$ to produce a distance matrix M. PPC then projects M into N-dimensional Euclidean space by applying multi-dimensional scaling (Kruskal, 1964). All cases are placed into a single group constituting the top level of a hierarchy. We then recursively apply a clustering algorithm m to these cases until the desired depth of the hierarchy, k, is reached. Hyper-parameters d, N, m, and k can be tuned for optimal performance.

PPC processes a query q by recursively matching it down the hierarchy. At each level, d is used to determine the distance between q and the case c closest to each candidate cluster center. At each step, q is matched to a cluster for which this distance is smallest. Once a leaf is reached, q's nearest neighbor is retrieved.

Our pilot study (Maynord, Vattam, & Aha 2015) indicated that PPC can reduce retrieval time by up to 72 % while sacrificing only a small amount in retrieval accuracy (approximately 4 %), because queries are partial (rather than complete) plans.

6   Conclusions and Future Work

We described SET-PR, a case-based plan recognition algorithm that represents plans as action sequence graphs. Unlike most prior algorithms, we designed SET-PR to be tolerant of input errors in the observed actions (i.e., missing, mislabeled, or extra action labels). We use Johnson's (1985) similarity metric for plan retrieval in SET-PR because it is an approximation of the maximal common subgraph function for matching graphs. In our empirical studies on plan recognition tasks involving three data sets, which we modified by adding input errors, we compared the performance of SET-PR with alternative approaches that use propositional representation and similarity functions for plan retrieval. We found that SET-PR's use of a graph representation for plans contributed to its superior performance when error rates are high. This complements our earlier work (Vattam, Aha, & Floyd 2015), in which we showed the incorporation of state information in its plan representation is another positive contributing factor.

In our future work, we will compare the performance of SET-PR versus other state-of-the-art plan recognition algorithms. We also plan to investigate more sophisticated graph similarity functions (e.g., graph kernels) and compare them versus SET-PR's current similarity function. Current plan recognizers, including SET-PR, assume that the observed actor's plans remains static during plan recognition. We will relax this assumption and extend SET-PR to tolerate dynamic changes to an actor's plans. Finally, we plan to integrate SET-PR with sensory perceptual systems on simulated and real robotic platforms so that we can study its performance on the ground in real time.

Acknowledgements. Thanks to OSD ASD (R&E) for sponsoring this research. Swaroop Vattam performed this work while an NRC post-doctoral research associate at NRL. Thanks also to the reviewers for their comments. The views and opinions contained in this paper are those of the authors and should not be interpreted as representing the official views or policies of NRL or OSD.

References

Bauer, M.: Integrating probabilistic reasoning into plan recognition. Proceedings of the Eleventh European Conference on Artificial Intelligence, pp. 620–624. Wiley & Sons, Amsterdam, The Netherlands (1994)

Blaylock, N., Allen, J.: Statistical goal parameter recognition. In: Proceedings of the Fourteenth International Conference on Automated Planning and Scheduling, pp. 297–304. Whistler, BC, Canada (2004)

Blaylock, N., Allen, J.: Generating Artificial Corpora for Plan Recognition. In: Ardissono, L., Brna, P., Mitrović, A. (eds.) UM 2005. LNCS (LNAI), vol. 3538, pp. 179–188. Springer, Heidelberg (2005)

Blaylock, N., Allen, J.: Hierarchical instantiated goal recognition. In: Kaminka, G., Pynadath, D., Geib, C. (eds.) Modeling Others from Observations: Papers from the AAAI Workshop (Technical Report WS-06-13). AAAI Press, Boston, MA (2006)

Charniak, E., Goldman, R.P.: A bayesian model of plan recognition. Artif. Intell. **64**(1), 53–79 (1993)

Cox, M.T., Kerkez, B.: Case-based plan recognition with novel input. Control Intell. Syst. **34**(2), 96–104 (2006)

Duong, T.V., Bui, H.H., Phung, D.Q., Venkatesh, S.: Activity recognition and abnormality detection with the switching hidden semi-Markov model. In: Proceedings of the IEEE Computer Society Conference on Computer Vision and Pattern Recognition, pp. 838–845. IEEE Press, San Diego, CA (2005)

Geib, C.W., Goldman, R.P.: A probabilistic plan recognition algorithm based on plan tree grammars. Artif. Intell. **173**(11), 1101–1132 (2009)

Johnson, M.: Relating metrics, lines and variables defined on graphs to problems in medicinal chemistry. Wiley, New York (1985)

Kautz, H., Allen, J.F.: Generalized plan recognition. In: Proceedings of the Fifth National Conference on Artificial Intelligence, pp. 32–37. AAAI Press, Philadelphia, PA (1986)

Levenshtein, V.I.: Binary codes capable of correcting deletions, insertions, and reversals. Sov. Phys. Dokl. **10**(8), 707–710 (1966)

Maynord, M., Vattam, S., Aha, D.W.: Increasing the runtime speed of case-based plan recognition. In: Proceedings of the Twenty-Eighth Florida Artificial Intelligence Research Society Conference. AAAI Press, Hollywood, FL (2015, to appear)

Nau, D.S., Au, T.C., Ilghami, O., Kuter, U., Murdock, J.W., Wu, D., Yaman, F.: SHOP2: an HTN planning system. J. Artif. Intell. Res. **20**, 379–404 (2003)

Pynadath, D.V., Wellman, M.P.: Accounting for context in plan recognition with application to traffic monitoring. In: Proceedings of Uncertainty in Artificial Intelligence, pp. 472–481. Morgan Kaufmann, Montreal, Quebec (1995)

Ramirez, M., Geffner, H.: Probabilistic plan recognition using off-the-shelf classical planners. In: Proceedings of the Conference of the Association for the Advancement of Artificial Intelligence. AAAI Press, Atlanta, GA (2010)

Ramirez, M., Geffner, H.: Goal recognition over POMDPs: inferring the intention of a POMDP agent. In: Proceedings of the Twenty-Second International Joint Conference on Artificial Intelligence, pp. 2009–2014. AAAI Press, Barcelona, Spain (2011)

Raymond, J.W., Willett, P.: Maximum common subgraph isomorphism algorithms for the matching of chemical structures. J. Comput. Aided Mol. Des. **16**, 521–533 (2002)

Sánchez-Ruiz, A.A., Ontañón, S.: Least Common Subsumer Trees for Plan Retrieval. In: Lamontagne, L., Plaza, E. (eds.) ICCBR 2014. LNCS, vol. 8765, pp. 405–419. Springer, Heidelberg (2014)

Serina, I.: Kernel functions for case-based planning. Artif. Intell. **174**(16), 1369–1406 (2010)

Tecuci, D., Porter, B.W.: Memory based goal schema recognition. In: Proceedings of the Twenty-Second International Florida Artificial Intelligence Research Society Conference. AAAI Press, Sanibel Island, FL (2009)

van Beek, P.: An investigation of probabilistic interpretations of heuristics in plan recognition. In: Proceedings of the Fifth International Conference on User Modeling, pp. 113–120 (1996)

Vattam, S.S., Aha, D.W., Floyd, M.: Case-Based Plan Recognition Using Action Sequence Graphs. In: Lamontagne, L., Plaza, E. (eds.) ICCBR 2014. LNCS, vol. 8765, pp. 495–510. Springer, Heidelberg (2014)

Vattam, S., Aha, D.W., Floyd, M.: Error tolerant plan recognition: an empirical investigation. In: Proceedings of the Twenty-Eighth Florida Artificial Intelligence Research Society Conference. AAAI Press, Hollywood, FL (2015, to appear)

Author Index

Printed in the United States
By Bookmasters